# Optimizing Methods in Statistics

# Optimizing Methods in Statistics

*Edited by*
**Jagdish S. Rustagi**
*Division of Statistics*
*The Ohio State University*
*Columbus, Ohio*

*Proceedings of a Symposium*
*Held at The Center for Tomorrow*
*The Ohio State University*
*June 14–16, 1971*

**Academic Press New York and London 1971**

ACADEMIC PRESS, INC.
111 Fifth Avenue, New York, New York 10003

*United Kingdom Edition published by*
ACADEMIC PRESS, INC. (LONDON) LTD.
24/28 Oval Road, London NW1 7DD

LIBRARY OF CONGRESS CATALOG CARD NUMBER: 73-182612

PRINTED IN THE UNITED STATES OF AMERICA

# CONTENTS

CONTRIBUTORS . . . . . . . . . . . . . . . . . . . . . . . ix
PREFACE . . . . . . . . . . . . . . . . . . . . . . . . . xiii

The Efficient Estimation of a Parameter Measurable by Two Instruments of Unknown Precisions . . . . . . . . . . . . 1
*Herman Chernoff*

Optimization Problems in Simulation . . . . . . . . . . . . . 29
*Herman Rubin*

Some Optimization Problems in Parameter Estimation . . . . . . 33
*H. Leon Harter*

Optimal Designs and Spline Regressions . . . . . . . . . . . 63
*W. J. Studden*

Isotonic Approximation . . . . . . . . . . . . . . . . . . . 77
*R. E. Barlow* and *V. Ubhaya*

Asymptotically Efficient Estimation of Nonparametric Regression Coefficients (Abstract) . . . . . . . . . . . . . . . . . . 87
*L. Weiss* and *J. Wolfowitz*

Comparisons of Order Statistics and of Spacings from Heterogeneous Distributions . . . . . . . . . . . . . . . . . . . . . . 89
*Gordon Pledger* and *Frank Proschan*

Moment Problems with Convexity Conditions I . . . . . . . . . 115
*J. H. B. Kemperman*

Variational Methods in Adaptive Filtering . . . . . . . . . . 179
*Andrew P. Sage*

CONTENTS

Non Linear Filtering . . . . . . . . . . . . . . . . . . . . 211
*G. Kallianpur*

A Convergence Theorem for Non Negative Almost Supermartingales and Some Applications . . . . . . . . . . . . . . . . . . 233
*H. Robbins and D. Siegmund*

On Relationships between the Neyman-Pearson Problem and Linear Programming . . . . . . . . . . . . . . . . . . . . . . 259
*Richard L. Francis*

Statistical Control of Optimization . . . . . . . . . . . . . . 281
*H. O. Hartley* and *R. Pfaffenberger*

Current Capabilities in Mathematical Programming (Abstract) . . . 301
*Philip Wolfe*

Patterns and Search Statistics . . . . . . . . . . . . . . . . 303
*Allen Klinger*

Necessary Conditions for a Local Optimum without Prior Constraint Qualification . . . . . . . . . . . . . . . . . . . . . . 339
*G. V. Reklaitis* and *D. J. Wilde*

Mathematical Models for Statistical Decision Theory . . . . . . . 369
*Bernard Harris*

Chance-Constrained Programming: An Extension of Statistical Method . . . . . . . . . . . . . . . . . . . . . . . . . 391
*A. Charnes, W. W. Cooper,* and *M. J. L. Kirby*

Stochastic Allocation of Spare Components . . . . . . . . . . . 403
*Bennet P. Lientz*

Outlier Proneness of Phenomena and of Related Distributions . . . 413
*Jerzy Neyman* and *Elizabeth L. Scott*

Problem Areas Requiring Optimizing Methods . . . . . . . . . . 431
*John V. Armitage*

Stochastic Approximation . . . . . . . . . . . . . . . . . . 439
*Václav Fabian*

CONTENTS

Allocation of Observations in Ranking and Selection with Unequal Variances (Abstract) . . . . . . . . . . . . . . . . . . . . . 471
*Edward J. Dudewicz* and *Siddhartha R. Dalal*

Sequences of Minimal Fractions of $2^n$ Designs of Resolution V (Abstract) . . . . . . . . . . . . . . . . . . . . . . . . 475
*Peter W. M. John*

Optimum Interval Estimation for the Largest Scale Parameter . . 477
*K. M. Lal Saxena* and *Yung Liang Tong*

c-Sample Tests of Homogeneity against Ordered Alternatives (Abstract) . . . . . . . . . . . . . . . . . . . . . . . . 479
*Z. Govindarajulu* and *H. Smith Haller, Jr.*

PARTICIPANTS . . . . . . . . . . . . . . . . . . . . . . 481

# CONTRIBUTORS

*Armitage, John V.*, U. S. Air Force Headquarters, Washington, D. C.

*Barlow, R. E.*, Department of Statistics and Operations Research Center, University of California, Berkeley, California

*Charnes, A.*, Center of Cybernetic Studies, Graduate School of Business, University of Texas, Austin, Texas

*Chernoff, Herman*, Department of Statistics, Stanford University, Stanford, California

*Cooper, W. W.*, Carnegie-Mellon University, Pittsburgh, Pennsylvania

*Dalal, Siddhartha R.*, Department of Statistics, University of Rochester, Rochester, New York

*Dudewicz, Edward J.*, Department of Statistics, University of Rochester, Rochester, New York

*Fabian, Vaclav*, Department of Statistics, Michigan State University, East Lansing, Michigan

*Francis, Richard L.*, Department of Industrial and Systems Engineering, University of Florida, Gainsville, Florida

*Govindarajulu, Z.*, Department of Statistics, University of Kentucky, Lexington, Kentucky

*Haller, H. Smith, Jr.*, B. F. Goodrich Company, Akron, Ohio

*Harris, Bernard*, Mathematics Research Center, University of Wisconsin, Madison, Wisconsin

*Harter, H. Leon*, Applied Mathematics Laboratories, Wright-Patterson Air Force Base, Dayton, Ohio

*Hartley, H. O.,* Institute of Statistics, Texas A & M University, College Station, Texas

*John, Peter W. M.,* Department of Statistics, University of Kentucky, Lexington, Kentucky

*Kallianpur, G.,* Department of Mathematics, University of Minnesota, Minneapolis, Minnesota

*Kemperman, J. H. B.,* Department of Mathematics, University of Rochester, Rochester, New York

*Kirby, M. J. L.,* Dalhousie University, Halifax, Nova Scotia

*Klinger, Allen,* Department of Computer Science, University of California, Los Angeles, California

*Lientz, Bennet P.,* Department of Industrial Engineering, University of Southern California, Los Angeles, California

*Neyman, Jerzy,* Department of Statistics, University of California, Berkeley, California

*Pfaffenberger, R.,* Institute of Statistics, Texas A & M University, College Station, Texas

*Pledger, Gordon,* Department of Statistics, Florida State University, Tallahassee, Florida

*Proschan, Frank,* Department of Statistics, Florida State University, Tallahassee, Florida

*Reklaitis, G. V.,* School of Chemical Engineering, Purdue University, Lafayette, Indiana

*Robbins, H.,* Department of Mathematical Statistics, Columbia University, New York, New York

*Rubin, Herman,* Department of Statistics, Purdue University, Lafayette, Indiana

*Sage, Andrew P.,* Information and Control Sciences Center, Southern Methodist University, Dallas, Texas

*Saxena, K. M. Lal,* Department of Mathematics, University of Nebraska, Lincoln, Nebraska

*Scott, Elizabeth L.,* Department of Statistics, University of California, Berkeley, California

*Siegmund, D.,* Department of Mathematical Statistics, Columbia University, New York, New York

*Studden, W. J.,* Department of Statistics, Purdue University, Lafayette, Indiana

*Tong, Yung Liang,* Department of Mathematics, University of Nebraska, Lincoln, Nebraska

*Ubhaya, V.,* Department of Industrial Engineering, University of California, Berkeley, California

*Weiss, L.,* Department of Industrial Engineering and Operations Research, Cornell University, Ithaca, New York

*Wilde, D. J.,* Department of Chemical Engineering, Stanford University, Stanford, California

*Wolfe, Philip,* Thomas J. Watson Research Center, IBM, Yorktown Heights, New York

*Wolfowitz, J.,* Department of Mathematics, University of Illinois, Urbana, Illinois

# PREFACE

Problems of optimization in statistics arise in many different contexts. The methods of estimation require maximization or minimization of certain functions. The classical methods of least squares, maximum likelihood, minimum variance, and minimum chi square are associated by their very name with optimization. The statistical decision theory deals with problems of finding optimum decision rules. In the theory of tests of hypotheses, methods of obtaining most powerful tests are directly associated with the process of optimization. Techniques of mathematical programming also have been used in statistics in many contexts, such as in estimation of parameters of a Markov chain and in regression analysis. Statisticians have also contributed in the development of many optimizing methods such as the variational techniques provided by J. Neyman and E. S. Pearson. In solving practical problems, statisticians have contributed significantly to the development of many numerical methods of optimization.

In order to bring active research workers together to exchange ideas and to stimulate young researchers in this important area, a symposium on optimizing methods in statistics was held at The Ohio State University during June 14-16, 1971. The program of the symposium was organized in sessions on: Variational Methods, Regression Analysis, Optimum Seeking Methods, Mathematical Programming, Stochastic Control, Optimum Design of Experiments, Optimum Spacings and Order Statistics, Problems and Applications. Speakers were invited from the government, industry, and universities. This book contains the proceedings of the symposium. The abstracts of the contributed papers are also included.

The organizing committee consisted of Professor Herman Chernoff, Stanford University; Professor Bernard Harris, University of Wisconsin, Mathematics Research Center; Lt. Col. W. R. Trott, U. S. Air Force Office of Scientific Research, and myself. I am very grateful for the excellent advice and cooperation I received from all the members of the organizing committee. I am highly obliged to the following persons who acted as chairmen of various sessions: W. R. Trott, S. S. Gupta, B. Harris, P. R. Krishnaiah, R. L. Anderson, I. Olkin, B. Mond, J. Gastwirth, D. R. Whitney, L. Wolaver, W. A. Thompson, W. J. Hall, and G. Kulldorff. The banquet speaker was Professor J. Wolfowitz. All sessions were held at The Center for Tomorrow, The

Ohio State University. For excellent arrangements at The Center for Tomorrow, I wish to thank the members of the local arrangements committee which consisted of Professors J. Singh, R. C. Strivastava, T. Willke, J. Sullivan, P. Anderson, A. Barron, and D. R. Whitney (chairman). The advice and encouragement of Professor D. R. Whitney is highly appreciated.

The symposium was partially sponsored by the U. S. Air Force Office of Scientific Research under Grant No. AFOSR-71-2052. I am specially grateful to Professor Herman Chernoff who gave constant encouragement and advice during the development of a graduate course on Optimizing Methods in Statistics at The Ohio State University. His help in the formulation of the program of the symposium is also highly appreciated. I am grateful for the editorial assistance of Miss Diane Marting who also typed the manuscripts for the publishers, and to the publishers for their help and cooperation.

Jagdish S. Rustagi

# Optimizing Methods in Statistics

# THE EFFICIENT ESTIMATION OF A PARAMETER MEASURABLE BY TWO INSTRUMENTS OF UNKNOWN PRECISIONS[1]

Herman Chernoff

Stanford University

## 1. Introduction and Summary

Let $\mu$ be the unknown value of a parameter for which an estimate is required. Suppose that this parameter may be measured by either of two instruments whose precisions are not known in advance. It is desired to obtain an estimate of satisfactorily low variability with as little waste in sampling cost as possible. To do so, the investigator must decide after each observation whether to select another observation and, if so, which instrument to use. When he estimates that enough information has been accumulated, he will stop. While continuing, he will tend to use the instrument with greater precision, except insofar as the estimate of the precision of the apparently poorer instrument is itself based on too few observations. In other words, there is a conflict between the need to use what seems to be the more precise instrument and the need to be reasonably sure that the other instrument is indeed less precise.

A rule for procedure is presented and evaluated in the context of normally distributed data. Reference is made to a related problem treated by Mallik [4], where one instrument has known precision. This problem was originally presented to me by Harold Gumbel. The rule used combines an approach used by Robbins [5] for stopping sampling in a one-sample problem and the solutions of a one-armed and two-armed bandit problem to be described in more detail later.

---

[1]This work was carried out with the partial support of the National Science Foundation under NSF Grant GP-277550.

Monte Carlo simulations have been carried out to measure the efficiency of the rule. The efficiency was measured by comparing the loss with the case where the precisions are known, in which case one can sample with the better instrument the appropriate number of times. Roughly speaking, this difference of loss measuring the cost of ignorance is shown to be equivalent to the cost of 6, 10, and 17 observations in the cases where the appropriate sample sizes are

$$10^{3/2},\ 100,\ 10^{5/2}$$

respectively.

The indications are that as the sample size increases, the loss due to ignorance goes up more slowly than the typical order of magnitude, which is the square root of the sample size. To make this point clear, cases with much larger sample sizes would have to be studied.

In Section 2 the estimation problem is stated. In Section 3 the related stopping problem analogous to that of Robbins and the bandit problems are briefly described with some indication of their relevance to the estimation problem. In Section 4 the method developed in this paper is presented. In Section 5 the method is motivated in part by illustrating how an essential component of the estimation problem resembles the two-armed bandit problem. In Section 6 some theoretical background on the bandit problems is presented to justify the particular technique with which the solution of the two-armed bandit paroblem is applied in our estimation problem. In Section 7 the Monte Carlo simulation results are presented, and finally Section 8 has some summarizing comments.

## 2. Statement of Problem

Let $X_i$ be normally distributed with common unknown mean $\mu$ and unknown variance $\sigma_i^2$, $i = 1,2$. If $n_i$ independent observations are taken on $X_i$, the mean is estimated by

$$\hat{\mu} = (n_1\hat{\sigma}_1^{-2}\bar{X}_1 + n_2\hat{\sigma}_2^{-2}\bar{X}_2)/(n_1\hat{\sigma}_1^{-2} + n_2\hat{\sigma}_2^{-2}) \qquad (2.1)$$

where $\bar{X}_i$ and $\hat{\sigma}_i^2$ are the sample mean and estimated variance respectively of $X_i$. (We shall use a slight variation of the usual estimate of $\sigma_i^2$.) The cost of error in estimation is $L(\hat{\mu},\mu) = k(\hat{\mu} - \mu)^2$ and the cost per observation is c.

After each observation, one must choose to stop sampling or to take another observation, and in the latter case one must decide between $X_1$ and $X_2$. It is desired to select a good sequential sampling rule.

Given a proper Bayesian framework, the problem of finding an optimal procedure is well defined. The approach used here is to settle for a procedure whose plausibility depends on the solution of several other related optimality problems which are of substantial depth in themselves.

3. Related Problems

One of these is the stopping problem for the case of one instrument. More precisely, if observations are available on $X_1$ only, when should the investigator stop sampling and use $\bar{X}_1$ to estimate $\mu$. The relation between this problem and ours is evident and hardly needs the elaboration of Section 5. Suffice it to say that the approach, proposed by Robbins for a related problem where the loss was $|\hat{\mu} - \mu|$, though not claimed to be optimal in a Bayesian sense, is highly satisfactory. This approach leads to stopping as soon as the sample size n satisfies

$$n \geq \max\,[n_o, \hat{\sigma}_1 (k/c)^{\frac{1}{2}}] \qquad (3.1)$$

where $n_o$ is a specified minimum sample size and $\hat{\sigma}_1$ is the sample standard deviation. This method is related in an obvious way to the case where $\sigma_1$ is known and one selects n to minimize the risk

$$cn + \frac{k\sigma^2}{n} .$$

The difference in risks for the case where the Robbins approach is applied and the case where $\sigma$ is known can be called the cost of ignorance. As a consequence of results of Starr and Woodroofe [6], it follows that the cost of sampling is a bounded multiple of c independent of k and $\sigma$.

A second problem is a <u>one-armed bandit</u> problem. Let $X$ be normally distributed with mean $\mu$, and variance $\sigma^2$ where $\mu$ is unknown but $\sigma^2$ is known. If a player selects $n$ observations $X_i$, $i = 1, 2, \ldots, n$ he recieves $X_1 + X_2 + \ldots + X_n$ as payoff. The player is restricted to select $n$ sequentially, i.e., after each observation he may decide on the basis of the past whether to proceed or stop. He is restricted to $n \leq N$ where $N$ is a specified <u>horizon</u>. Given a prior normal distribution on $\mu$, this problem has a Bayes solution which can be computed by backward induction. A continuous time version of this problem yields a limiting solution described in [3]. Since $E(X_1 + \ldots + X_n) = E(n\mu)$, the problem can be phrased in terms of finding a stopping rule subject to $n \leq N$ which maximizes $E(n\mu)$. One easily formulates an expression for the loss due to ignorance of $\mu$ which has been shown to be of the order of $\sigma(\log N)^2$.

A two-armed bandit problem involves two random variables $X_i$, $i = 1,2$ which are normally distributed with unknown means $\mu_i$ and known variance $\sigma^2$. A specified total of $N$ observations must be selected. The player keeps

$$X_{11} + \ldots + X_{1n_1} + X_{21} + \ldots + X_{2n_2}, \quad n_1 + n_2 = N \tag{3.2}$$

where

$$X_{i1}, X_{i2}, \ldots, X_{in_i}$$

are the observations on $X_i$, $i = 1,2$. Two independent normal prior distributions are given for $\mu_1$ and $\mu_2$. Having taken $m_i$ observations on $X_i$, $i = 1,2$, $m_1 + m_2 < N$, how should one decide between $X_1$ and $X_2$ for the next observation? Note that

$$E(X_{11} + \ldots + X_{1n_1} + X_{21} + \ldots + X_{2n_2}) = \tag{3.3}$$

$$E(n_1 \mu_1 + n_2 \mu_2) = N\mu_1 + E[n_2(\mu_2 - \mu_1)]$$

and hence one is concerned with minimizing $E[n_2(\mu_1 - \mu_2)]$.

The two-armed bandit problem where the prior distribution on $\mu_1$ has 0 variance reduces to the one-armed bandit problem. One expects a strong relationship between the behavior of the solution of the two-armed bandit problem where N is large and many observations have been taken on $X_1$ and relatively few on $X_2$, and the solution of the one-armed bandit problem for large N.

The relationship between the two-armed bandit problem and our estimation problem will be elaborated on later. For the moment, we merely indicate that if

$$\sigma_1^2 < \sigma_2^2$$

in the estimation problem then $N = \sigma_1 (k/c)^{\frac{1}{2}}$ observations on $X_1$ are called for if $\sigma_1$ and $\sigma_2$ are known. Furthermore, it can be shown that if $n_1$ observations are taken on $X_1$ and $n_2$ on $X_2$, the extra loss (due to ignorance of $\sigma_1$ and $\sigma_2$) is approximately

$$n_2(\sigma_2^2 - \sigma_1^2)/\sigma_2^2$$

If we associate the unknown $\sigma_1^2$ and $\sigma_2^2$ of the estimation problem with the unknown parameters $\mu_1$ and $\mu_2$ of the two-armed bandit problem, the similarity in the expressions minimized suggests that a reasonable procedure in the estimation problem would treat the choice of next observation as it is treated in the two-armed bandit problem after marking suitable adjustments for the differences in parameters involved.

In an analogous fashion the one-armed bandit problem relates to that version of the estimation problem treated by Mallik [4] where $\sigma_1$ is known.

## 4. The Method

Having barely hinted at some of the considerations that are involved in determining a sequential sampling procedure for the estimation problem, we outline a rule and leave a more detailed motivation to Sections 5 and 6. Without loss of generality we assume that k = 1 hereafter.

If $n_i$ observations have been taken on $X_i$, let

$$Z_{i,n_i} = \sum_{j=1}^{n_i} (X_{ij} - \bar{X}_i)^2 \tag{4.1}$$

$$z_i = Z_{i,n_i} + z_{io} \tag{4.2}$$

$$m_i = (n_i - 1) + m_{io} \tag{4.3}$$

$$\hat{\sigma}_i^2 = z_i/m_i \tag{4.4}$$

$$Y_i = \log \hat{\sigma}_i^{\,2} + \log \frac{m_i}{2} - \Psi\left(\frac{m_i}{2}\right) \tag{4.5}$$

and

$$s_i = \Psi'(m_i/2), \tag{4.6}$$

where

$$\Psi(x) = d[\log \Gamma(x)]/dx. \tag{4.7}$$

Here $z_{io}$ and $m_{io}$ are specified to correspond to prior knowledge, $\Psi$ is sometimes call the digamma function, $Y_i$ corresponds to a Bayesian estimate of $\log \sigma_i^{\,2}$, and $s_i$ corresponds to the variance of $Y_i$. These will be explained in more detail in Section 6.

Suppose $\hat{\sigma}_1 < \hat{\sigma}_2$. Then let

$$\hat{N} = \hat{\sigma}_1 c^{-\frac{1}{2}} \tag{4.8}$$

and

$$N^* = \tfrac{1}{2}\{m_{10} + m_{20} + (\hat{N}-2) + n_2(1-\hat{\sigma}_1^{\,2}/\hat{\sigma}_2^{\,2})\} \quad . \tag{4.9}$$

As long as $n_1 + n_2 < n_0(c)$, a specified minimum sampling size, sampling is continued. Otherwise sampling is stopped if

$$n_1\hat{\sigma}_1^{-2} + n_2\hat{\sigma}_2^{-2} \geq c^{-\frac{1}{2}}\,\hat{\sigma}_1^{-1} \tag{4.10}$$

or if

$$(s_1^{-1} + s_2^{-1}) \geq N^* \tag{4.11}$$

If sampling is to be continued, select $X_1$ or $X_2$ according to whether or not

$$Y_2 - Y_1 \geq (N^*)^{-\frac{1}{2}} g(N^*s_1, N^*s_2) \tag{4.12}$$

where g is a function representing an approximation to the

solution of a normalized version of the two-armed bandit problem described in Section 6.

## 5. Motivation for the Method

### a) First Stopping Rule

In the spirit of Robbins' method for his estimation problem, suppose $n_1$ observations are taken on $X_1$ and $n_2$ observations on $X_2$ and that the variances $\sigma_1^2$ and $\sigma_2^2$ are known. Then an appropriate estimate of $\mu$ would be

$$(n_1\sigma_1^{-2}\bar{X}_1 + n_2\sigma_2^{-2}\bar{X}_2)/(n_1\sigma_1^{-2} + n_2\sigma_2^{-2})$$

which would yield a risk of

$$(n_1\sigma_1^{-2} + n_2\sigma_2^{-2})^{-1} + c(n_1 + n_2).$$

If $\sigma_1 < \sigma_2$, we may ask how large should $n_1$ be made (keeping $n_2$ fixed) to minimize the risk. Minimizing with respect to $n_1$ gives us

$$n_1\sigma_1^{-2} + n_2\sigma_2^{-2} = c^{-\frac{1}{2}}\sigma_1^{-\frac{1}{2}} \tag{5.1}$$

where the left-hand side of the above equation represents the amount of information cumulated for estimating $\mu$. There is no point to taking more observations if the left-hand side exceeds the right-hand side. The first stopping rule (4.10) applies this principle with $\sigma_i$ replaced by $\hat{\sigma}_i$, $i = 1,2$. The second stopping rule (4.11) is asymptotically approximately equivalent to the first. As we shall see, its supplementary use is dictated by the inapplicability of the two-armed bandit problem when that inequality obtains. With it the first rule could be discarded but has been retained because of its evident interpretation.

### b) Relevance of the Two-Armed Bandit

Having stopped sampling after $n_i$ observations on $X_i$, $i = 1, 2$, we use

$$\hat{\mu} = (n_1\hat{\sigma}_1^{-2}\bar{X}_1 + n_2\hat{\sigma}_2^{-2}\bar{X}_2)/(n_1\hat{\sigma}_1^{-2} + n_2\hat{\sigma}_2^{-2}) \tag{5.2}$$

as an estimate of $\mu$. Given $\hat{\sigma}_1$, $\hat{\sigma}_2$, the conditional risk is

$$R = \frac{n_1\hat{\sigma}_1^{-4}\sigma_1^2 + n_2\hat{\sigma}_2^{-4}\sigma_2^2}{(n_1\hat{\sigma}_1^{-2} + n_2\hat{\sigma}_2^{-2})^2} + c(n_1 + n_2) \tag{5.3}$$

which we shall compare with the target $2c^{\frac{1}{2}}\sigma_1^{\frac{1}{2}}$ which corresponds to the case where $\sigma_1 < \sigma_2$ are known and the appropriate numbers of observations have been taken, i.e., $N_{10} = \sigma_1 c^{-\frac{1}{2}}$ on $X_1$ and 0 on $X_2$. Assuming that stopping takes place when

$$n_1\hat{\sigma}_1^{-2} + n_2\hat{\sigma}_2^{-2} = \hat{\sigma}_1^{-1}c^{-\frac{1}{2}} \tag{5.4}$$

with $\hat{\sigma}_1 \leq \hat{\sigma}_2$ and letting

$$\hat{N} = \hat{\sigma}_1 c^{-\frac{1}{2}} \tag{5.5}$$

we have

$$R = \frac{[\hat{N} + (n_1 - \hat{N})]\hat{\sigma}_1^{-2}\left(\frac{\sigma_1^2}{\hat{\sigma}_1^2} - 1\right) + n_2\hat{\sigma}_2^{-2}\left(\frac{\sigma_2^2}{\hat{\sigma}_2^2} - 1\right)}{(n_1\hat{\sigma}_1^{-2} + n_2\hat{\sigma}_2^{-2})^2}$$

$$+ (n_1\hat{\sigma}_1^{-2} + n_2\hat{\sigma}_2^{-2})^{-1} + c[\hat{N} + (n_1 - \hat{N}) + n_2]$$

$$R = c\hat{\sigma}_1^2\left[(n_1-\hat{N})\hat{\sigma}_1^{-2}\left(\frac{\sigma_1^2}{\hat{\sigma}_1^2} - 1\right) + n_2\hat{\sigma}_2^{-2}\left(\frac{\sigma_2^2}{\hat{\sigma}_2^2} - 1\right)\right]$$

$$+ c[(n_1 - \hat{N}) + n_2] + c^{\frac{1}{2}}\hat{\sigma}_1\left[\frac{\sigma_1^2}{\hat{\sigma}_1^2} - 1 + 2\right]$$

Replacing $\hat{\sigma}_i^2$ by

$$\sigma_i^2[1 + (\hat{\sigma}_i^2 - \sigma_i^2)/\sigma_i^2] = \sigma_i^2[1+\eta_i]$$

the coefficient of $c^{\frac{1}{2}}$ in R is

$$\hat{\sigma}_1^{-1}[\sigma_1^{\ 2} + \hat{\sigma}_1^{\ 2}] = \sigma_1^{-1}(1 + \eta_1)^{-\frac{1}{2}}\, \sigma_1^{\ 2}(2 + \eta_1)$$

$$= \sigma_1[\ 2 + \tfrac{1}{4}\,\eta_1^{\ 2} + \dots\ ]$$

Since the assumed stopping rule equality implies

$$(\hat{N} - n_1)\hat{\sigma}_1^{-2} = n_2\hat{\sigma}_2^{-2} ,$$

the coefficient of c in R is

$$\hat{\sigma}_1^{\ 2}\left[(n_1 - \hat{N})\hat{\sigma}_1^{-2}\left(\frac{\sigma_1^{\ 2}}{\hat{\sigma}_1^{\ 2}} - 1\right) + n_2\hat{\sigma}_2^{-2}\left(\frac{\sigma_2^{\ 2}}{\hat{\sigma}^2} - 1\right)\right]$$

$$+ [(n_1 - \hat{N}) + n_2] = n_2\left[\frac{-\hat{\sigma}_1^{\ 2}}{\hat{\sigma}_2^{\ 2}}\left(\frac{\sigma_1^{\ 2}}{\hat{\sigma}_1^{\ 2}} - 1\right) + \frac{\hat{\sigma}_1^{\ 2}}{\hat{\sigma}_2^{\ 2}}\left(\frac{\sigma_2^{\ 2}}{\hat{\sigma}_2^{\ 2}} - 1\right)\right.$$

$$\left. - \frac{\hat{\sigma}_1^{\ 2}}{\hat{\sigma}_2^{\ 2}} + 1\right] = n_2\left[1 - \frac{\sigma_1^{\ 2}}{\sigma_2^{\ 2}} - \frac{\sigma_1^{\ 2}}{\sigma_2^{\ 2}}\,\eta_2(\eta_1 - \eta_2) + \dots\right]$$

Let $r = R - 2c^{\frac{1}{2}}\sigma_1$ be the <u>conditional cost of ignorance</u>. Then

$$r = cn_2\left[1 - \frac{\sigma_1^{\ 2}}{\sigma_2^{\ 2}}\right] + c^{\frac{1}{2}}[\ \tfrac{1}{4}\eta_1^{\ 2} + \dots\ ] \qquad (5.6)$$

$$- cn_2\,\frac{\sigma_1^{\ 2}}{\sigma_2^{\ 2}}\,[\eta_2(\eta_1 - \eta_2) + \dots\ ]$$

When the number of observations on $X_1$ is of the order of $c^{-\frac{1}{2}}$, we would certainly expect $\eta_1$ to be of the order of magnitude of $c^{\frac{1}{4}}$ and the second term in (5.6) would be of the order of $c$ and the main term in $r$ is essentially given by

$$cn_2[1 - \sigma_1^{\ 2}/\sigma_2^{\ 2}] \quad \text{when} \quad \sigma_1^{\ 2} < \sigma_2^{\ 2}. \qquad (5.7)$$

The case when $\sigma_1^2 < \sigma_2^2$ but $\hat{\sigma}_1^2 > \hat{\sigma}_2^2$ involves somewhat more careful analysis. This possibility is of concern when $\sigma_1^2$ and $\sigma_2^2$ are relatively close together $(0(c^{\frac{1}{4}}))$ and then one expects both $n_1$ and $n_2$ to be $0_p(c^{\frac{1}{2}})$. One may still derive the same major contribution to risk. (The interested reader may find it more convenient to apply symmetry after treating the case $\sigma_1^2 > \sigma_2^2$ and $\hat{\sigma}_1^2 < \hat{\sigma}_2^2$ for which most of the above algebra remains unchanged though r must be replaced by $R - 2c^{\frac{1}{2}}\sigma_2$.)

Since the case where $\sigma_1^2$ and $\sigma_2^2$ are relatively close is the one requiring sophisticated treatment and $\sigma_2^2/\sigma_1^2$ appears as a basic parameter in this problem, let

$$\nu_i = -\log \sigma_i^2 .$$

Then $1 - \sigma_1^2/\sigma_2^2 \approx \nu_1 - \nu_2$ and the conditional loss due to ignorance is the product of c with

$$n_2(1 - \sigma_1^2/\sigma_2^2) \approx n_2(\nu_1 - \nu_2). \qquad (5.8)$$

If we think of the observations as providing information on $\nu_1$ and $\nu_2$, the loss due to ignorance relates to that of the two - armed bandit except that the data are not quite independent normally distributed observations with mean $\nu_i$ and constant known variance.

Insofar as the estimate of $\nu_i$ based on successive observations on $X_i$ resembles a process of successive means of i.i.d. normal random variables, it seems reasonable to apply the solution of the two-armed bandit problem here.

The additional discussion of the bandit problems of Section 6 will help explain how the solution of the two-armed bandit problem is applied in the method of Section 4.

## 6. Bandit Problems and Their Applicability

In Section 3 the bandit problems were introduced. Consider the one-armed bandit problem. Suppose $\mu$ has a normal prior distribution with mean $\mu_o$ and variance $\sigma_o^2$. Then the posterior distribution of $\mu$, given n observations $X_1, X_2, \ldots, X_n$, is normal with mean $Y_n$ and variance $s_n$ where

$$Y_n = \frac{\mu_o\sigma_o^{-2} + (X_1 + \ldots + X_n)\sigma^{-2}}{\sigma_o^{-2} + n\sigma^{-2}}$$

and

$$s_n^{-1} = \sigma_o^{-2} + n\sigma^{-2}$$

It has been shown [2] that when $\mu$ is treated as random, the Y process is a Gaussian process of independent increments with

$$E(Y_{n+1} - Y_n) = 0$$

$$\text{Var}(Y_{n+1} - Y_n) = s_n - s_{n+1} \quad .$$

A continuous time version of this problem involves the observation of a process X(t) which for fixed $\mu$ is a Wiener process with drift $\mu$ and variance $\sigma^2$ per unit time. At the same time let

$$Y(s) = (\mu_o\sigma_o^{-2} + X(t)\sigma^{-2})/(\sigma_o^{-2} + t\sigma^{-2}) \tag{6.1}$$

and

$$s^{-1} = \sigma_o^{-2} + t\sigma^{-2} \quad . \tag{6.2}$$

Regarding $\mu$ as random, Y is a Wiener process in the -s scale starting from $Y = \mu_o$ at $s = \sigma_o^2$. That is, Y has mean drift 0 and variance 1 per unit decrease in s. The payoff in the discrete time case is $X_1 + X_2 + \ldots + X_n$ and in the continuous time case, it is

$$X(t) = (Y(s)/s - \mu_o\sigma_o^{-2})\sigma^2$$

Thus it is desired to select a stopping variable which maximizes E(X(t)) subject to t less than or equal to a specified value N. This is equivalent to maximizing E(Y(s)/s) subject to

$$s \geq (\sigma_0^{-2} + N\sigma^{-2})^{-1}.$$

Let

$$T = \sigma_o^{-2} + N\sigma^{-2} \tag{6.3}$$

which represents the total amount of information (data plus prior information) potentially available for estimating $\mu$. Let

$$Y^* = \sqrt{T}\, Y \tag{6.4}$$

and

$$s^* = Ts. \tag{6.5}$$

It has been shown that $Y^*$ is a Wiener process in the $-s^*$ scale with mean drift 0 and variance 1 per unit decrease in $s^*$. Since the payoff is proportional to $Y/s = \sqrt{T}\,(Y^*/s^*)$ and $s^* = 1$ when $s = T^{-1}$, it follows that the optimal procedure can be expressed in terms of the solution of the normalized version of the problem where $s^* \geq 1$.

The solution of this problem has been characterized in some detail [3]. It consists of stopping as soon as $Y^* \leq -f(s^*)$ or equivalently as soon as

$$Y \leq -T^{-\frac{1}{2}} f(Ts) \tag{6.6}$$

where f is a special function. One way of approximating f is to apply backward induction to solve the following related discrete time stopping problem.

Observe a process $Y(s)$ which starts at an integer $y_o$ for $s = s_o$ where $s_o$ is also an integer. Each time one decides to continue, s decreases by one and Y either increases or decreases by one with probability ½ each. One is permitted to continue as long as $s \geq T$, a specified integer. The payoff upon stopping $Y(s)/s$. This problem which involves only integer values of Y and s is subject to a relatively easy backward induction which is computationally feasible with special attention to allow truncating the range of Y. The relation between this problem and the one-armed bandit is clear, and the solution consists of stopping whenever $Y \leq -T^{-\frac{1}{2}} f_T(Ts)$ where $f_T$ converges to f as $T \to \infty$. Table 1 presents an estimate of f.

In the one-armed bandit problem it is convenient to think of $Y/\sqrt{s} = Y^*/\sqrt{s^*}$ as the number of standard deviations away from 0 of the current estimate of $\mu$, and of

$$\beta = \Phi(Y/\sqrt{s}) \tag{6.7}$$

as the significance[1] of the data for rejecting the hypothesis $\mu = 0$. Let

$$t^* = (\sigma_o^{-2} + t\sigma^{-2})/T = s^{*-1} \qquad (6.7)$$

represent the proportion of the potentially available information achieved. Then the optimal procedure, determined by f, can be regarded as stopping as soon as the hypothesis $\mu = 0$ is rejected in favor of $\mu < 0$ at a <u>nominal significance</u> level

$$\beta(t^*) = \Phi[-f(s^*)/\sqrt{s^*}] \qquad (6.9)$$

which varies with $t^*$. It makes sense to use the nominal significance level in variations of the one-armed bandit problem where the data are not normally distributed and it is desired to maximize $E(n\mu)$. Mallik has applied this function in studying the estimation problem with one known variance. Chernoff [1] applied it to a problem of clinical trials involving two treatments, one of which has known probability of success.

In the continuous time version of the two-armed bandit problem the natural extension of the preceding calculus leads us to

$$Y_i = \frac{\mu_{io}\sigma_{io}^{-2} + X_i(t_i)\sigma^{-2}}{\sigma_{io}^{-2} + t_i\sigma^{-2}} \qquad (6.10)$$

$$s_i^{-1} = \sigma_{io}^{-2} + t_i\sigma^{-2} \qquad (6.11)$$

where the prior distribution of $\mu_i$ are independent $N(\mu_{io}, \sigma_i^2)$ and the posterior distribution given $X_i(t_i)$ are independent $N(Y_i, s_i)$, $i = 1,2$. Both $Y_1$ and $Y_2$ are Wiener processes in the $-s_1$ and $-s_2$ scales respectively.

Since the payoff is $X_1(t_1) + X_2(t_2)$ when $t_1 + t_2$ reaches a apecified value N, the object is to select a sesequential procedure for observing $X_1$ and $X_2$ so as to maximize the expectation of

[1] The symbol $\Phi$ represents the standard normal cdf.

$$\frac{Y_1}{s_1}+\frac{Y_2}{s_2}=\left(\frac{Y_1+Y_2}{2}\right)\left(\frac{1}{s_1}+\frac{1}{s_2}\right)+\left(\frac{Y_1-Y_2}{2}\right)\left(\frac{1}{s_1}-\frac{1}{s_2}\right) \tag{6.12}$$

when

$$s_1^{-1}+s_2^{-1}=T=\sigma_{10}^{-2}+\sigma_{20}^{-2}+N\sigma^{-2} \quad . \tag{6.13}$$

Since the expectation of the first term on the right-hand side of (6.12) is unaffected by the procedure, our problem reduces to maximizing

$$E\left\{Y\left[\frac{1}{s_1}-\frac{1}{s_2}\right]\right\} \qquad \text{when } s_1^{-1}+s_2^{-1}=T \tag{6.14}$$

where $Y = Y_1 - Y_2$ is a Gaussian process with mean drift 0 and variance of 1 per unit decrease is $s_1$ or $s_2$, whichever is changing. Finally the transformation

$$Y^*=\sqrt{T}\,Y\ , \quad s_1^*=Ts_1, \quad s^*=Ts_2 \tag{6.15}$$

normalizes the problem to that of maximizing

$$E\{Y^*(s_1^{*-1}-s_2^{*-1})\} \quad \text{when } s_1^{*-1}+s_2^{*-1}=1 \tag{6.16}$$

where $Y^*$ is a Gaussian process with mean drift 0 and variance 1 per unit decrease in $s_1^*$ or $s_2^*$ and the process begins at

$$Y^*=\sqrt{T}\,(\mu_{10}-\mu_{20}),\ s_1^*=T\sigma_{10}^2\ ,$$

$$s_2^*=T\sigma_{20}^2 \quad .$$

The solution of the two-armed bandit problem can be represented by a function g such that the appropriate variable to observe is $X_1$ or $X_2$ depending on whether or not

$$Y \geq T^{-\frac{1}{2}}\, g(Ts_1,Ts_2). \tag{6.17}$$

The two-armed bandit problem can also be simulated by a discrete time, discrete valued process problem. Here one starts at $Y = y_o$, $s_1 = s_{10}$ and $s_2 = s_{20}$ where $y_0$, $s_{10}$, $s_{20}$ are integers. At each stage one decreases either $s_1$ or $s_2$ by one and Y changes by $\pm 1$ with probability $\frac{1}{2}$. This procedure is carried on till a value $(s_1, s_2)$ is reached where decreasing one of these by one will lead to a violation of

$$s_1^{-1} + s_2^{-1} \leq 2n^{-1}, \text{ n a specified integer.} \tag{6.18}$$

If decreasing $s_1$ by one leads to the violation, one effectively permits fractional decrease by letting the payoff corresponding to decreasing $s_1$ be

$$2Y \left\{ \frac{1}{n} - \frac{1}{s_2} \right\} .$$

A two-dimensional backward induction is feasible. With normalization, one may use the solution of this discrete problem to approximate g as well as the expected payoff using the optimal strategy in the two-armed bandit problem. The larger the value of n, the finer is the approximation to the solution of the normalized problem and the greater is the computational effort involved to apply the backward induction to a specified value of $(Y^*, s_1^*, s_2^*)$.

In the one-armed bandit problem $\beta = \Phi(Y/\sqrt{s}) = \Phi(Y^*/\sqrt{s^*})$ and $t^* = s^{*-1}$ were intuitively convenient variables. Here $\beta = \Phi(Y/\sqrt{s_1 + s_2}) = \Phi(Y^*/\sqrt{s_1^* + s_2^*})$ and $t_i^* = s_i^{*-1}$ are corresponding variables. Asymptotic expansions suggest that for $s_1^*$ and $s_2^*$ large, the ratio of $s_1^*/s_2^*$ plays an important role, while for $s_1^*$ large but $s_2^*$ relatively small, the relative amount $t_1^*$ has to increase before

$$s_1^{*-1} + s_2^{*-1} = t_1^* + t_2^* = 1$$

is an important normalizing constant, i.e., $(1 - s_2^{*-1})s_1^*$ is an important parameter. A study of the results of computations carried out for $s_1^* \leq 20$, $s_2^* \leq 20$ with $n = 8$ and for $s_1^* \leq 12$, $s_2^* \leq 12$ with $n = 32$ led to the following approximation for g which leaves much to be desired theoretically but which seems to fit the computational results well. Let $s_1^{*-1} + s_2^{*-1} \leq 1$. Then

$$g(s_1^*, s_2^*) = -g(s_2^*, s_1^*) \qquad \text{for } s_1^* \leq s_2^* \tag{6.19}$$

$$g(s_1^*, s_2^*) = \sqrt{s_1^*}\,\alpha_1(t') \qquad \text{for } s_2^* \leq 2,\ s_1^* \geq s_2^* \tag{6.20}$$

$$g(s_1^*, s_2^*) = (t')^{-\frac{1}{2}}\alpha_2(t'') \qquad \text{for } s_2^* > 2,\ s_1^* \geq s_2^* \tag{6.21}$$

where

$$t' = s_2^*/(s_2^* - 1)s_1^* \tag{6.22}$$

$$t'' = s_2^*/s_1^* \tag{6.23}$$

$$\alpha_1(t) = t^{\frac{1}{2}} f(t^{-1}) = f(s)/\sqrt{s} \qquad (s = t^{-1}) \tag{6.24}$$

and $\alpha_2(t)$ is tabulated in Table 1.

Note that in the two-armed bandit problem $\alpha_1(t^*) = f(s^*)/\sqrt{s^*}$ represents the number of standard deviations which leads to stopping.

We terminate this section by indicating how to apply the approximate solution of the two-armed bandit problem to the estimation problem. The bandit problem may be regarded as that of minimizing $En_2(\mu_1 - \mu_2)$ while we have indicated that the experimental choice aspect of the estimation problem involves minimizing

$$E[n_2(-\log \sigma_1^2 + \log \sigma_2^2)]. \tag{6.25}$$

The essential variables in the bandit problem are Y, the estimated value of $\mu_1 - \mu_2$, $s_1^{-1} = \sigma_{10}^{-2} + t\sigma^{-2}$ and $s_2^{-1} = \sigma_{20}^{-2} + t\sigma^{-2}$, the amounts of information accumulated for estimating $\mu_1$ and $\mu_2$, and finally

$$T = N + \sigma_{10}^{-2} + \sigma_{20}^{-2},$$

the total potential information available. The variables $\mu_{10}$, $\mu_{20}$, $\sigma_{10}^2$, and $\sigma_{20}^2$ which enter in Y relate to the prior distributions.

In studying the parameters $-\log \sigma_i^2$, we observe that if $V_i$ are i.i.d. normal random variable with mean $\mu$ and variance $\sigma^2$, then

$$Z_n = \sum_{i=1}^{n} (V_i - \bar{V})^2$$

has the $\chi^2$ distribution with n-1 degrees of freedom. A convenient form of prior distribution is the gamma distribution for $\sigma^{-2}$. The gamma distribution with positive parameters a and b is given by

$$g(x) = e^{-ax} a^b x^{b-1}/\Gamma(b) \quad \text{for } x > 0.$$

If $\sigma^{-2}$ has a prior gamma distribution with parameters $a = z_0/2$ and $b = m_o/2$ the posterior distribution of $\sigma^{-2}$ given $Z_n$ is also a gamma distribution with parameters $a' = z/2 = (z_o^n + Z_n)/2$ and $b' = m/2 = (m_o + n - 1)/2$. Thus $z_o$ and $m_o$ can be interpreted as providing a prior distribution corresponding to the equivalent of $m_o+1$ observations with a sum of squared deviations of $z_o$.

In this Bayesian framework, $\hat{\sigma}^2 = z/m$ is a reasonable estimate of $\sigma^2$. For an estimate of $-\log \sigma^2$ we note that the mgf of the posterior distribution of $\log \sigma^2$ is

$$(z/2)^t \Gamma(m/2 - t)/\Gamma(m/2)$$

from which the mean and variance are calculated to be $\log \hat{\sigma}^2 + \log m/2 - \Psi(m/2)$ and $\Psi'(m/2)$ respectively. In the large sample situation at least it is natural to apply the solution of the two-armed bandit to our estimation problem.

Representing prior knowledge of $\sigma_i^2$ by the non-negative parameters $z_{i0}$ and $m_{i0}$, the role of $Y^1$ is naturally played by $Y_2$-$Y_1$ of Section 4, i.e.,

$$Y \sim [\log \hat{\sigma}_2^2 + \log \frac{m_2}{2} - \Psi\left(\frac{m_2}{2}\right)] - [\log \hat{\sigma}_1^2 + \log \frac{m_1}{2} - \Psi\left(\frac{m_1}{2}\right)]$$

while $\Psi'(m_i/2) \approx 2m_i^{-1}$ corresponds to $s_i$. The potential amount of information anticipated, T, can be estimated. If $\hat{\sigma}_1^2 < \hat{\sigma}_2^2$ where these are based on $n_1$ and $n_2$ observations respectively, we anticipate taking extra observations on $X_1$, i.e., increasing $n_1$ till

$$c(n_1 + n_2) + (n_1\hat{\sigma}_1^{-2} + n_2\hat{\sigma}_2^{-2})^{-1}$$

is minimized. This occurs when

$$n_1\hat{\sigma}_1^{-2} + n_2\hat{\sigma}_2^{-2} = c^{-\frac{1}{2}}\,\hat{\sigma}_1^{-1}.$$

Thus we anticipate the eventual value of $n_1$ to be

$$n_1{}^* = (c^{-\frac{1}{2}}\,\hat{\sigma}_1^{-1} - n_2\hat{\sigma}_2^{-2})\hat{\sigma}_1^{2}.$$

The corresponding values of the $s_i^{-1}$ are $\Psi'[n_1{}^* - 1 + m_{10})/2]$ and $\Psi'[(n_2 - 1 + m_{20})/2]$ which approximately sum to

$$\tfrac{1}{2}[n_1{}^* + n_2 - 2 + m_{10} + m_{20}] = N^*$$

as given in Section 4. This provides the rationale for the method described in Section 4.

## 7. Simulations

The method described in Section 4 was applied to a number of Monte Carlo simulations. Each case was defined by values of c, $\sigma_1^2$, $\sigma_2^2$, $z_{10}$, $m_{10}$, $z_{20}$, $m_{20}$. Sampling started with 2 observations from each $X_i$, and termination was prohibited before $n_o(c) = 8$ observations were taken. (If the stopping rule applied before 8 observations, the $X_i$ with smaller estimated variance was selected next.) Upon termination of sampling, the loss due to ignorance was computed to be

$$r = (n_1\hat{\sigma}_1^{-4}\sigma_1^{2} + n_2\hat{\sigma}_2^{-4}\sigma_2^{2})/(n_1\hat{\sigma}_1^{-2} + n_2\hat{\sigma}_2^{-2})^2$$

$$+ \; c(n_1 + n_2) - 2c^{\frac{1}{2}}\min(\sigma_1,\sigma_2)$$

To measure this loss in terms of the cost of an <u>equivalent number of observations</u>, we take

$$r^* = r/c \tag{7.1}$$

Scale invariance permits the choice of $\sigma_1^2 = 1 \leq \sigma_2^2$ without loss of generality. Three values of c were examined exhaustively. These were $c = 10^{-3}$, $10^{-4}$, and $10^{-5}$ for which the appropriate sample sizes would be $10^{-3/2} = 32$, 100, and $10^{5/2} = 316$. Most cases were selected with $z_{10} = m_{10} = z_{20} = m_{20} = 2$ though a few cases with different priors were examined to see if the prior had much effect. In each case the method was applied 100 times with 2 exceptions. For $\sigma_2^2 = 1$ when $c = 10^{-3}$ and $c = 10^{-4}$, 400 repetitions were carried out.

At the bottom of Figures 1, 2, and 3 are presented estimates of the expected cost of ignorance as a function of $\sigma_2^2$ for the 3 values of c. Two estimates are given. The first, consisting of raw averages $\overline{r^*}$, is sensitive to the occasionally very large values of r*. To reduce this effect, the following procedure was followed. A brief examination of the histograms indicated that the density of r* is a mixture of three components for values of $\sigma_2^2$ not very close to one. The first component has most of the probability and yields relatively low values of r*, and this component is remarkably insensitive to variations in $\sigma_2^2$. The second component corresponds to large values of r*. The probability going into this component and the distribution within this component is sensitive to $\sigma_2^2$ but behaves in an orderly fashion. Finally the third component corresponds to very rare and very large outliers.

To separate the first from the second components, critical values $r_c$ of r* were selected to be 10, 20, and 40 for $c = 10^{-3}$, $10^{-4}$, and $10^{-5}$ respectively. For each case, 3 quantities were estimated. These are

$$p_u = P\{r^* > r_c\},$$

$$\mu_u = E\{r^* | r^* > r_c\} \tag{7.2}$$

$$\mu_d = E\{r^* | r^* \leq r_c\}$$

estimated by $\hat{p}_u$, the proportion of observations exceeding $r_c$ and $\hat{\mu}_u$ and $\hat{\mu}_d$, the sample means of those $r^* > r_c$ and those $r^* \leq r_c$ respectively. The $\hat{p}_u$ and $\hat{\mu}_d$ were plotted as functions of $\sigma_2^2$ and a smooth curve drawn by eye through the plotted points. Before doing this for $\hat{\mu}_u$, a minor finagle was performed to modify the effect of the few very large outliers. The effect of each outlier was distributed among the 4 or 5 points with nearby values of $\sigma_2^2$. Since the outliers, which are listed below, are rare and erratic, it was felt that this smoothing would give a more accurate estimate of the long run effect of outliers on Er* as a function of $\sigma^2$.

Outliers

| | | | | | | | | | | |
|---|---|---|---|---|---|---|---|---|---|---|
| c | $10^{-3}$ | $10^{-3}$ | $10^{-3}$ | $10^{-3}$ | $10^{-3}$ | $10^{-3}$ | $10^{-4}$ | $10^{-4}$ | $10^{-5}$ | $10^{-5}$ |
| $\sigma_2^2$ | 1.8 | 2.1 | 2.2 | 2.5 | 2.6 | 3.5 | 1.8 | 2.5 | 1.5 | 1.7 |
| r* | 29.9 | 55.5 | 34.6 | 34.8 | 56.7 | 74.1 | 75.4 | 114.5 | 137.4 | 96.8 |

In Figure 1, the data for $c = 10^{-3}$ are presented. The top graph has $\hat{p}_u$ and the smooth curve $\tilde{p}_u$. The next graph has $\hat{\mu}_u$, the modified version $\mu_u{*}$, and the smooth curve $\tilde{\mu}_u$. The next graph has $\tilde{\mu}_d$. Finally the bottom graph has the raw averages $\overline{r*}$ and

$$\tilde{r}* = \tilde{p}_u\tilde{\mu}_u + (1 - \tilde{p}_u)\tilde{\mu}_d \tag{7.3}$$

With the graphs of $\hat{p}_u$, $\hat{\mu}_u$, and $\hat{\mu}_d$ are attached legends which indicate approximate variances for the raw estimates. The $\sigma_2^2$ scale is broken at the right to present large values of $\sigma_2^2$. The same presentations for $c = 10^{-4}$ and $c = 10^{-5}$ appear in Figures 2 and 3.

Some results of the graphs indicating minimal value of E(r*), peak value of E(r*), and the value of $\sigma_2^2$ at which the peak is attained are summarized in the following table.

| c | 0.001 | 0.0001 | 0.00001 |
|---|---|---|---|
| Minimal Value of E(r*) | 1.4 | 1.8 | 2.3 |
| Peak Value of E(r*) | 6.7 | 10.5 | 17.5 |
| Coresponding Value of $\sigma_2^2$ | 1.85 | 1.40 | 1.17 |

The minimal value of E(r*) attained at $\sigma_2^2 = \sigma_1^2$ is quite low. The main contributions to it are the fact that the stopping rule may lead to the wrong sample size and that the error in estimating the variance may lead to the use of a suboptimal wighted average for estimating $\mu$.

It is difficult to verify the conjecture that the losses due to ignorance are proportional to the square of the log of the appropriate sample size. If that were the case, they would be in the ratio of 9, 16, and 25 for $c = 10^{-3}$, $10^{-4}$, and $10^{-5}$. These peak losses are very roughly in this proportion, but they could just as easily fit some other rules. Actually, the proportion should be applied to an average over $\sigma^2$, and it would make sense to subtract out the stopping rule effect. For the range of values of c considered here, which seems to be a reasonable range, it does not seem useful to belabor the issue further.

The choice of $z_{10} = m_{10} = z_{20} = m_{20} = 2$ was made in a somewhat arbitrary fashion. A few trial simulations suggested that the use of the no-information prior, $z_{10} = m_{10} = z_{20} = m_{20} = 0$, might lead to occasionally large values of $r^*$ when the first few observations are highly misleading. To minimize this possibility, it seemed reasonable to "stabilize" the estimates of $\sigma_1^2$ and $\sigma_2^2$ by putting in somewhat arbitrary prior information. The one used leads to some loss when $\sigma_2^2$ is large because it may take longer than necessary to recognize the fact and the reduced estimate of $\sigma_2^2$ gives $\bar{X}_2$ a greater weight than it deserves. Perhaps it would have been wise to select a prior not so carefully attuned to $\sigma_1^2 = 1$. In any case a few simulations were carried out with other priors. The results do not seem to be affected very much. Perhaps the only striking result was the expected one that the use of $z_{10} = m_{10} = z_{20}/2 = m_{20} = a$ leads to improved performance for $\sigma_2^2 = 2$, $c = 10^{-3}$ as a increases from 0 to 4; the simulations gave $\overline{r^*}$ decreasing from 6.88 to 3.05 with estimated standard deviations of 0.3.

## 8. Comments

1. Inasmuch as the procedure applied in this paper applies properties of the chi-square distribution, it is evidently sensitive to the assumption of normality. To produce a more robust procedure, a reasonable approach would involve estimating the variance of the estimates of $\sigma_i^2$ or $\log \sigma_i^2$ by direct reference to the data without assuming normality. This can be done in several ways. One simple way is to group observations on X in sets of 5. The sample variances of these groups are variables whose average estimates the variance of X, and whose sample variance is a robust estimate of the variance of this estimate.

2. In non-sequential estimation problems, ignorance typically leads to losses which are large compared to the cost of an observation. For example, in the simple case of estimating the mean of a variable with known variance the loss

$$n^{-1}\sigma^2 + cn$$

is minimized for $n = c^{-\frac{1}{2}}\sigma$ and the minimum value of the loss is $2c^{\frac{1}{2}}\sigma$. If $\sigma$ is not known but is guessed to be $\sigma_1$, the case of $n_1 = c^{-\frac{1}{2}}\sigma_1$ leads to an increased cost which is

$$n_1^{-1}\sigma^2 + cn_1 - 2c^{\frac{1}{2}}\sigma = c^{\frac{1}{2}}(\sigma-\sigma_1)^2/\sigma_1$$

which is of the order of magnitude of $c^{\frac{1}{2}}$ and large compared to the cost of an observation. The point of the Robbins procedure is to reduce the cost of ignorance by using the information that comes in sequentially. This reduction is considerable. Starr and Woodroofe [6] have shown that for fixed c, the cost of ignorance is bounded uniformly in $\sigma$. A scale transformation implies that the cost is bounded by that of a finite number of observations independent of c and $\sigma$. The data on the values of Er* at $\sigma_2^2 = 1$, giving values of 1.4, 1.8, and 2.3 for $c = 10^{-3}$, $10^{-4}$, and $10^{-5}$ respectively, suggest that the minimum value of Er* is small but proportional to -log c. However, this value of Er* involves other factors besides the stopping rule. In any case the cost of ignorance is small compared to the risk of estimating $\mu$. Here the concept of optimality is highly refined and concerned with what may be regarded as higher order terms, and traditional concepts of asymptotic optimality which compare ratios of risks are inappropriate, easy to achieve and of dubious value.

In the bandit problem, the possibly costly decisions are made at each stage, and ignorance is costly as long as it lasts. One would expect a substantially larger risk than in the Robbins problem. In the bandit problems these risks are probably of the order of magnitude of $(\log c)^2$. Unfortunately, the range of values of c taken in our simulations are not adequate to distinguish this effect which should exist in our estimation problem. The study of much smaller values of c has been deferred because it would have dubious practical value and involve considerable computing expense.

3. Several natural extensions readily suggest themselves. These include the case of more than two measuring instruments and the case where the cost of sampling depends on the measuring instrument used.

4. The simulations carried out produced a considerable amount of data, some of which remains to be analyzed. The detailed history was recorded for 10 or 20 simulations for a few cases. These produced vast amounts of paper. While they were instructive in guiding the later large computer runs, they have not been studied further. Each simulation was accompanied by additional output. These included 1) the number of observations on each $X_i$, 2) the final estimates of each $\sigma_i^2$, 3) the first sample size at which the number of observations on $X_1$ exceeded an initial number n*. The values of n* taken were 7, 10, and 20 for c = 10 , $10^{-4}$, and $10^{-5}$. Hopefully a study of the correlations between these variables and r could reveal suggestions for improvement in the procedure.

5. I wish to thank Mrs. Elizabeth Hinkley for her efficient work in programming the Monte Carlo simulations on rather short notice.

References

[1] Chernoff, H. (1967), "Sequential Models for Clinical Trials", Proc. of Fifth Berkeley Symp. on Math. Statist. and Prob., Univ. of Calif. Press, Vol. 4, 805-812.

[2] Chernoff, H. (1968) "Optimal Stochastic Control", Sankhyā, Series A, 30, 221-252.

[3] Chernoff, H. and Ray, S. N. (1965), "A Bayes Sequential Sampling Inspection Plan", Ann. Math. Statist., 36, 1387-1407.

[4] Mallik, A., Unpublished dissertation to be completed.

[5] Robbins, H. (1959), "Sequential Estimation of the Mean of a Normal Population", Probability and Statistics, ed. by U. Grenander, Wiley, 235-245.

[6] Starr, N. and Woodroofe, M. B. (1969), "Remarks on Sequential Point Estimation", Nat. Acad. of Sci., 63, 285-288.

Table 1

Tabulations Relevant to Solutions of Bandit Problems

| $t^*$ | $f(t^{*-1})$ | $\beta(t^*)$ | $\alpha_1(t^*)$ | $\alpha_2(t^*)$ |
|---|---|---|---|---|
| 0.0001 | 354.70 | 1.95(-4) | 3.547 | 3.891 |
| 0.0002 | 237.87 | 3.84(-4) | 3.364 | 3.719 |
| 0.0005 | 138.77 | 9.58(-4) | 3.103 | 3.481 |
| 0.001 | 91.80 | 1.85(-3) | 2.903 | 3.291 |
| 0.002 | 59.94 | 3.68(-3) | 2.684 | 3.090 |
| 0.005 | 33.52 | 8.89(-3) | 2.370 | 2.807 |
| 0.01 | 22.45 | 0.0124 | 2.245 | 2.576 |
| 0.02 | 13.14 | 0.0316 | 1.852 | 2.326 |
| 0.04 | 7.72 | 0.0613 | 1.544 | 2.054 |
| 0.06 | 6.09 | 0.0680 | 1.491 | 1.880 |
| 0.08 | 4.91 | 0.082 | 1.389 | 1.751 |
| 0.10 | 4.07 | 0.099 | 1.287 | 1.645 |
| 0.15 | 2.82 | 0.137 | 1.094 | 1.439 |
| 0.20 | 2.18 | 0.165 | 0.975 | 1.282 |
| 0.25 | 1.76 | 0.189 | 0.882 | 1.150 |
| 0.30 | 1.47 | 0.210 | 0.805 | 1.036 |
| 0.40 | 1.07 | 0.248 | 0.680 | 0.842 |
| 0.50 | 0.818 | 0.281 | 0.579 | 0.675 |
| 0.60 | 0.627 | 0.314 | 0.486 | 0.524 |
| 0.70 | 0.477 | 0.345 | 0.399 | 0.385 |
| 0.80 | 0.347 | 0.378 | 0.311 | 0.253 |
| 0.85 | 0.283 | 0.397 | 0.260 | 0.189 |
| 0.90 | 0.222 | 0.417 | 0.210 | 0.126 |
| 0.92 | 0.194 | 0.426 | 0.186 | 0.100 |
| 0.94 | 0.157 | 0.440 | 0.152 | 0.753 |
| 0.96 | 0.130 | 0.449 | 0.128 | 0.502 |
| 0.98 | 0.087 | 0.466 | 0.086 | 0.251 |
| 0.99 | 0.063 | 0.475 | 0.062 | ... |
| 1.00 | 0.000 | 0.500 | 0.000 | 0.000 |
| 1.000 | 0.000 | 0.500 | 0.000 | 0.000 |

1. $f(s^*)$ designates solution of normalized one-armed bandit problem.
2. $\beta$ is nominal significance level for one-armed bandit problem.
3. $\alpha_1$ is corresponding number of standard deviations.
4. $\alpha_2$ is function used in the approximation to $g(s_1^*, s_2^*)$

which represents the solution of the two-armed bandit problem.

$$t^* = s^{*-1}$$

$$\alpha_1 = (t^*)^{\frac{1}{2}} f(t^{*-1})$$

$$\beta = \Phi(-\alpha_1)$$

For $0.99 < t^* < 1.000$ use $\alpha_1 = 0.639(1-t^*)^{\frac{1}{2}}$.

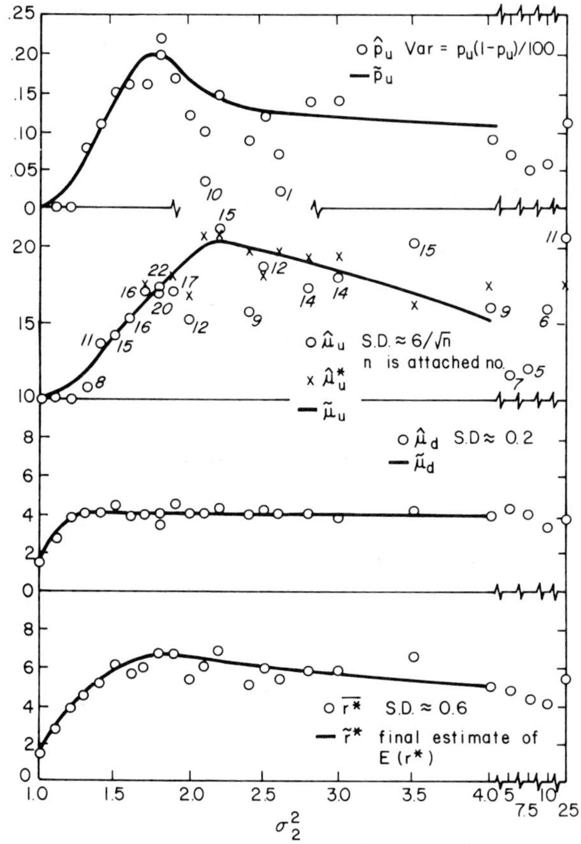


Figure 1. Estimates of $p_u$, $\mu_u$, $\mu_d$, $E(r^*)$; $c = 10^{-3}$

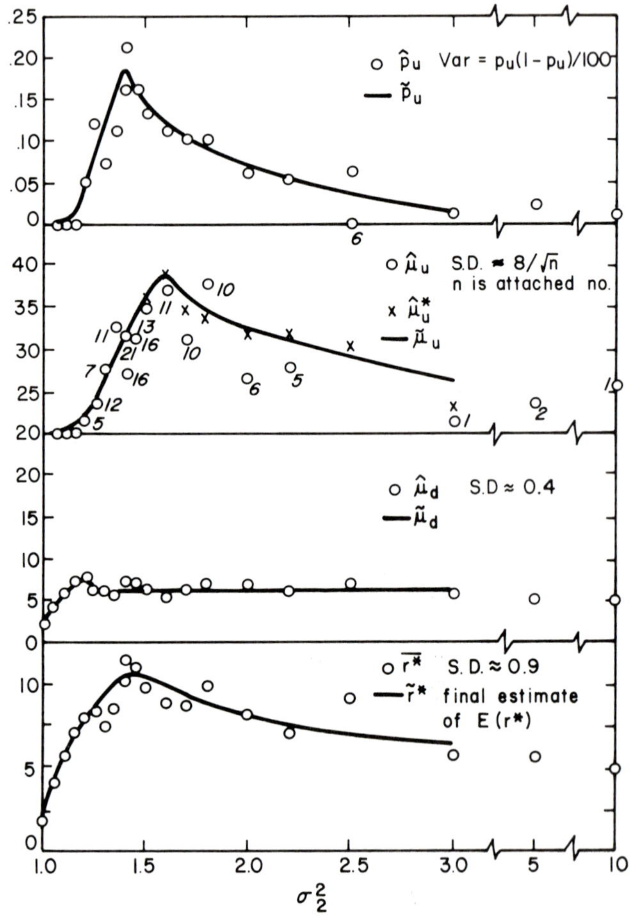


Figure 2. Estimates of $p_u$, $\mu_u$, $\mu_d$, $E(r^*)$; $c = 10^{-4}$

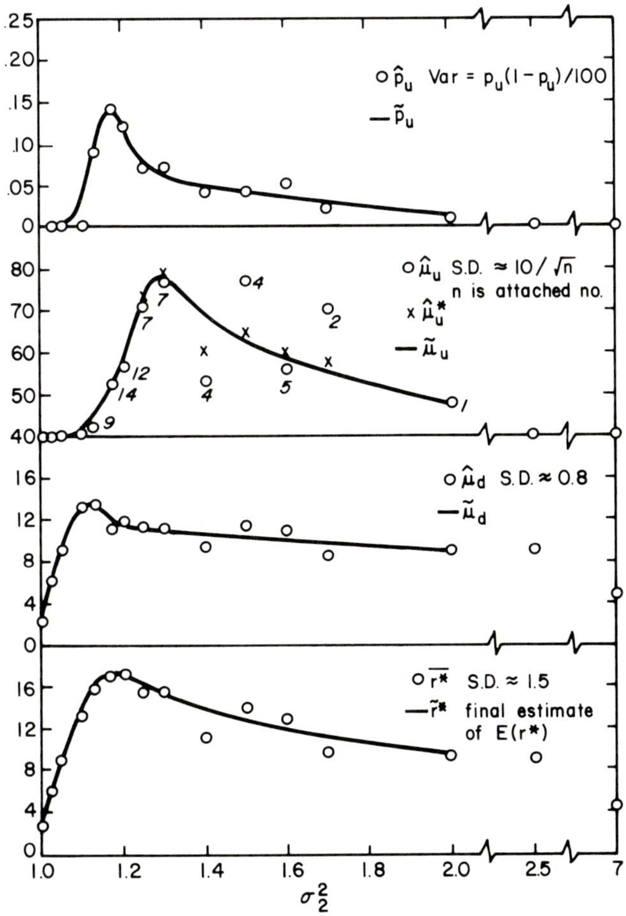


Figure 3. Estimates of $p_u$, $\mu_u$, $\mu_d$, $E(r^*)$; $c = 10^{-5}$

# OPTIMIZATION PROBLEMS IN SIMULATION[1]

Herman Rubin

Purdue University

When it is necessary to resort to simulation, large numbers of random variables are used, typically in rather complicated ways. Since this uses a great deal of computer time, optimization is clearly called for. However, the optimization methods depend on the computer and the programming methods available. Even the sources of random variables can make a sizable difference in the procedures used.

Now there has been a considerable amount of work done on methods, such as antithetical variables, importance sampling, etc., to increase the effectiveness of simulation procedures, but few of the papers have considered the effect of the computer on which methods to use. The author's acquaintance with the hardware structure of "third generation" computers is not great enough to give the best examples, but these should point out the problems.

For example, the 6500 and 6600 have identical arithmetic codes. A buffer procedure suggested by the author for the exponential distribution gives a cost of about 2.5 times the cost of uniform random variable on the 6500. The same procedure would have comparable relative cost on the 6600. A slightly less efficient (6500) procedure on the 6600 working on three values at once would decrease the cost to about 1.9 times the cost of a uniform because of the time-overlapping of the instructions.

Other hardware effects are even more obvious. If a square root is built into hardware, the maximum of two

[1]Research was supported in part by the Office of Naval Research Contract N 0014-67-A-226-008. project number NR 042-216. Reproduction in whole or in part is permitted for any purpose of the United States Government.

uniform random variables can be obtained much faster by it than by any other non-microprogrammed procedure. A very intricate and ingenious procedure can be derived for the exponential distribution which requires micro-micro-programming to be effective. None of the other much-used distributions are readily adapted to this procedure, which requires for input a random bit stream, but somewhat more complicated ingenious procedures of this type may be practical.

Problem 1. To obtain an ordered sample of size N from a uniform distribution.

Two methods immediately come to mind.

A. Take a sample of size N and order it.

B. Let

$$X_i = \sum_{j=1}^{i} E_j \Big/ \sum_{j=1}^{N+1} E_j$$

where the $E_j$ are independent exponential.

Clearly the cost of B is proportional to N, and the cost of ordering an array of size N is roughly proportional to $N \log_2 N$. If we take the cost of a uniform random variable to be 1, the cost of an exponential random variable will vary from 1.5 to 20 - good procedures will limit the upper bound to 5. The cost of computing the X's from the E's should make the total cost of B approximately kN, $2 < k < 7$. The cost of A is approximately $N(\qquad {}_2 N)$, where $0 \cdot 1 < w < 1$, $0 < \lambda < 1$, (some software has such bad sorting routines). Hence the break-even point is for $\log_2 N$ near $(k-1-\lambda)/w$. Now all the bounds are not independent, so this, say between 2 and 50. That means that the decision to change can be for N as low as 4 or 5, or may not occur until N is well beyond the capacity of all machines in existence put together.

Let us consider briefly an unusual, and probably impractical, variant of Problem 1, which will have significance later.

Problem 1'. To obtain an ordered sample of size M, M Poisson with mean $\mu$, from a uniform distribution.

Here we have the obvious analogue A' of A, but the procedure

B'. Let

$$X_i = \sum_{j=1}^{i} E_j/\mu \quad \text{as long as} \quad \sum_{j\ 1}^{i} E_j < \mu$$

is even better than B.

Problem 2. To obtain an ordered sample of size N from a preassigned distribution F.

Again we have two obvious choices.

C. Take a sample of size N and order it.

D. Take a sample of size N from a uniform distribution and apply $F^{-1}$.

Now the cost of a random variable with distribution F can be typically reduced to about 3 uniforms with an overhead, depending on F, of 100 - 10,000 uniforms, or about 4.5 uniforms with an overhead of 5 - 100 uniforms. The cost of $F^{-1}$ typically runs 5 - 50 uniforms, possibly with a sizeable overhead.

However, the solution to Problem 1' is even better. A modification of either of the procedures can be made to produce an ordered sample of size M, $E(M) = \mu$, M Poisson, from an ordered sample of uniform random variables [2] of size K, $E(K) = c\mu$, K Poisson, about as cheaply per unit as a single one can be produced with the same overhead. Typically, c is in the range (1.02,1.2). Now if the sample size M is too large, it can be reduced at a cost of about 1.5M - N uniform variables; if it is too small, an additional sample can be produced, and the cost of amalgamating two ordered samples is small and proportional to the total size.

Note also that if F is sufficiently important, the overhead may have already been done and available from the

[2] Ordered exponential variables may also be needed, but these are straightforward.

program library.[3]

Problem 3. To obtain a single order statistic from a distribution F.

If the sample size is large, there seems to be no alternate to computing $F^{-1}$ of a uniform order statistic if only one case is wanted for the particular order statistic-this case is likely to be unusual. If the sample is not too large, the breakpoint being in the range 50-1000 depending again on the problem, a modification of one of the author's procedures is quite effective. If the density is logarithmically concave (monotone likelihood ratio for a translation parameter), so is the distribution of all order statistics, and the methods for that type of distribution can be used.

References

[1] Halton, J. H. "On the relative merits of correlated and importance sampling for MonteCarlo integration," Proc. Camb. Phil. Soc. 61, (1965), pp. 497-498.

[2] Hammersley, J. M. and Morton, K. W., "A new Monte Carlo technique: antithetic variables," Proc. Camb. Phil. Soc. 52, (1956), pp. 449-475.

[3] Trotter, H. F. and Tukey, J. W., "Conditional Monte Carlo for normal samples," Symposium on Monte Carlo Methods, H. A. Mayer,ed., Wiley, New York; (1956), pp. 64-79.

[4] Rubin, H., "Some fast methods of generating random variables with preassigned distributions," to be issued as a Purdue technical report.

[5] Marsaglia, G., MacLaren, M. D., Bray, T. A., "A fast procedure for generating normal random variables," Comm. Assn. Comp. Mach., 7, (1964) pp. 4-10.

[6] Marsaglia, G., MacLaren, M. D., Bray, T. A., "A fast procedure for generating exponential random variables," Comm. Assn. Comp. Mach. 7, (1964) pp. 298-300.

[3] It will be necessary, in general, to extract the necessary tables and procedures from the programs in the library.

# SOME OPTIMIZATION PROBLEMS IN PARAMETER ESTIMATION

H. Leon Harter

Aerospace Research Laboratories

## 0. Summary

Various optimization problems arise in connection with the estimation of parameters of statistical populations. Without making any attempt to present an exhaustive treatment, the author offers some observations concerning three such problems with which he has had experience. These problems are: (1) iterative procedures for maximum likelihood estimation, based on complete or censored samples, of the parameters of various populations; (2) optimum spacings of quantiles for linear estimation; and (3) optimum choice of order statistics for linear estimation. Optimization procedures employed in problem (1) include the rule of false position, the Newton-Raphson method, and the gradient method, as well as combinations of these methods. Solutions of problem (2), which are based on asymptotic theory, are, at best, only approximations for small or even moderate samples. Some of the difficulties that arise in the use of optimum spacings are inherent in the problem, while others are due to the method used to determine the sample quantiles, on which the author offers some pointed comments. Solutions of problem (3) depend on computations which, in many cases, are feasible only for small samples; in such cases, one must settle for nearly optimum estimators based on solutions of problem (2) when the sample is large enough so that exact solution of problem (3) is not feasible.

## 1. Maximum Likelihood Estimation

Several years ago the author carried out an extensive program of research on maximum likelihood estimation, from complete or censored samples, of the parameters of various populations. Studied were three two - parameter populations (normal, logistic and Type I extreme - value), three three-parameter populations (lognormal, Weibull, and Gamma), and one four - parameter population (generalized Gamma). Albert H. Moore collaborated in all of these studies except that on the four - parameter generalized Gamma population. The results were reported in papers by Harter and Moore (1965, 1966a,b,1967a,b,1968) and in a technical report and a paper by Harter (1966, 1967); all are summarized in a book by the author [Harter (1970), Chapter VII].

For each of the populations mentioned above, the likelihood equations were found by equating to zero the first partial derivative, with respect to each of the parameters, of the natural logarithm of the likelihood function (the joint probability density function of the ordered sample values). [For complete samples, all sample values are treated alike, and it is not necessary to take account of order.] The likelihood equations have explicit solutions only in the case of complete samples from a normal population; in all other cases it is necessary to resort to iterative procedures, which the author programmed in FORTRAN. These procedures involve, with exceptions to be noted later, use of the rule of false position to estimate the parameters, one at a time, in cyclic order, omitting a parameter if it is assumed to be known. One starts by choosing initial estimate(s) for the unknown parameter(s). At each step the rule of false position is used to determine the value of the parameter then being estimated which satisfies the appropriate likelihood equation, in which the latest estimate(s) [or known value(s)] of the other parameter(s) has (have) been substituted. At the end of each cycle the latest estimate (or known value) of each parameter is printed out, along with the corresponding value of the natural logarithm of the likelihood function. Iteration continues for a specified number of cycles or until the estimates at the end of successive cycles all agree to within some assigned tolerance. A glance at the column of natural logarithms of the likelihood function is sufficient to verify that the likelihood is indeed increasing from cycle to cycle, so that the set of parameter values to which the procedure converges corresponds to maximum, not minimum, likelihood.

For the three two - parameter populations mentioned above, estimation is known to be regular, so that the likelihood function has no minimum and a single maximum, to which experience shows that the iterative procedure usually converges after a few cycles, especially if the initial estimates are reasonably good. Bad initial estimates cannot affect the final results, but can and do increase the number of iterations required. The maximum likelihood estimators are asymptotically unbiased and their joint distribution is asymptotically bivariate normal with covariance matrix equal to the inverse of the information matrix, whose elements are equal to the limits, as the sample size $n \to \infty$, of the negatives of the expected values of the second partial derivatives of the natural logarithm of the likelihood function with respect to the parameters. The maximum likelihood estimators, from complete or symmetrically censored samples, of the location parameters (means) of the normal and logistic populations are unbiased even for small samples, but those from asymmetrically censored samples are biased. The maximum likelihood estimators of the scale parameters of the normal and logistic populations and of both parameters of the Type I extreme - value population are biased even for complete or symmetrically censored samples. The bias of all the maximum likelihood estimators may be considerable if censoring is strongly asymmetric. Monte Carlo studies for samples of size 10 and 20 from all three populations have shown, however, that even when the censoring is strongly asymmetric, the mean square errors of the estimates are only slightly greater than the theoretical variances given by the asymptotic formulas. The mean square errors of the estimates of the scale parameter when the location parameter is known are often less than the theoretical variances given by the asymptotic formulas, and sometimes (when censoring is severe but symmetric or almost so) they may be substantially less. The mean square errors of the maximum likelihood estimates are approximately equal to those of the best linear invariant estimates [see Mann (1967a,b,1968)]. When the maximum likelihood estimates are substantially biased, their mean square errors are considerably less than those of the best linear unbiased estimates.

Each of the three - parameter populations mentioned above involves a location parameter which is the lower limit of the range of the population, and the same is true for the four - parameter generalized Gamma population. When the location parameter is one of those being estimated, estimation may be nonregular and problems may arise in connection with the iterative estimation procedure. Specifically, if the estimate of the location parameter, at any cycle, is equal to the smallest value in the sample (or the smallest one not already censored), the corresponding estimates of the other parameters become infinite or indeterminate, and the iterative procedure cannot be continued without modification. In order to circumvent this difficulty, the author modified the iterative procedure so that when the estimate of the location parameter, at any cycle, is equal to the smallest value(s) in the (perhaps already censored)sample, that value (or those values, in the case of two or more equal smallest values) is (are) censored, and plays (play) no further role in the procedure except that of an upper bound on the estimate of the location parameter. In this way, one ensures that the estimates of the location parameter will, from this point on, always be strictly less than the smallest remaining sample value, and the iterative procedure can continue without further interruption. From a purely mathematical viewpoint, the location parameter has no lower limit. Practically, however, in applications to life distributions, which were the motivation of the author's study, negative values of the location parameter (minimum life) make no sense. Therefore, when a negative estimate of the location parameter occurred at any cycle of the iterative procedure, it was replaced by zero.

For the three - parameter lognormal population, Hill (1963) has shown that, whatever the sample values and whatever the true values of the parameters, the likelihood function attains its global maximum when the estimate of the location parameter is equal to the smallest value in the sample and the estimates of $\mu$ and $\sigma$ (the mean and standard deviation of the parent normal population) are $-\infty$ and $+\infty$ respectively. Such estimates are, of course, of no practical interest. Iterative solution of the likelihood equations, however, using when necessary the modification of the iterative procedure mentioned above, usually leads to local - maximum - likelihood estimates which are reasonable and appear to possess most of the desirable properties usually associated with maximum-likelihood estimates. Exceptions may occur for small samples, for which the likelihood function may have no clearly defined local maximum. The number of cycles required for the iterative procedure to converge tends to be somewhat larger than for the two - parameter populations discussed above, but the required computer time is not excessive, again with occasional exceptions for very small samples. If the iterative procedure does not converge within a reasonable number of cycles (say 550) it may as well be abandoned, since, even if it does eventually converge, the estimates are unlikely to be reasonable ones. In other words, the value of the estimates tends to be inversely proportional to their cost. The results of a Monte Carlo study [see Harter (1970), Chapter VII, subsection 2.5 and Appendix G, Table G2], for samples of sizes 50, 100, and 200, indicate that the asymptotic variances and covariances are valid only for very large samples.

Next let us consider estimation of the parameters of the three - parameter Weibull and Gamma populations and the four - parameter generalized Gamma population, which is a generalization of the Weibull and other populations as well as the Gamma. Contrary to assertions by several earlier writers (including the present author) that estimation of the location parameter is nonregular for these populations when the shape parameter is less than or equal to two, but regular when it is greater than two, Blischke et al (1970) have pointed out that the regularity conditions given by Cramér (1946), Section 33.3 [and even the somewhat weaker ones proposed by Huber (1964)] are not satisfied for the location parameter even when the shape parameter is greater than two. Blischke (1971) has shown, however, that new regularity conditions due to LeCam (1970) are satisfied for the Gamma population with shape parameter greater than two when the location parameter is the only unknown parameter. It appears that the same holds also for the Weibull and generalized Gamma populations, but it is not clear whether it can be extended to the case of simultaneous estimation of the location parameter and one or more other parameters. The case in which both the location and the shape parameters are unknown, so that one does not know whether or not the latter is greater than two, is especially nebulous. Failure of the regularity conditions to be satisfied can affect both the iterative estimation procedure and the computation of asymptotic variances and covariances. The latter is impossible for complete samples (or for samples censored only from above) from Weibull, Gamma, and generalized Gamma populations with unknown location parameters and shape parameters less than or equal to two; it becomes possible, however, as soon as some proportion $q_1 > 0$ of the sample is censored from below [see Harter and Moore (1967) and Harter (1966); or see Harter (1970), Chapter VII]. For each of these three populations, the iterative estimation procedure can continue without modification, even though the location parameter is unknown and the true value of the shape parameter is less than or equal to two, so long as the known value of the shape parameter (or the estimate of the unknown shape parameter at each cycle) is greater than one. Whenever the location parameter is unknown and the known value of the shape parameter (or the estimate of the unknown shape parameter at any cycle) is less than or equal to one, the estimate of the location parameter at that cycle is equal to the smallest value in the (perhaps already censored) sample. At that point the unmodified iterative procedure cannot continue, so it necessary to go to the modified procedure.

As for the three - parameter lognormal population, the number of cycles required for convergence of the iterative procedure for estimating the parameters of the three - parameter Weibull population is somewhat larger than for the two - parameter populations studied. If, as rarely occurs, the procedure does not converge within 550 iterations, it should be abandoned. Except in rare instances of slow convergence, computer time is not excessive. The results of a small Monte Carlo study [see Harter and Moore (1967) or Harter (1970), Chapter VII, subsection 3.6 and Appendix G, Table G3], for samples of sizes 50 and 100, indicate that when all three parameters are unknown the variances and the absolute values of the covariances of the estimates exceed their asymptotic values, with the excess closely proportional to $n^{-2}$ (where n is the sample size), a phenomenon previously observed in the case of the local - maximum - likelihood estimates of the parameters of a three-parameter lognormal population. When at least one of the parameters is known, the sample variances and covariances agree quite well with their asymptotic values, even for sample size as small as 50.

Both the average number of cycles to convergence [abandonement is recommended only after 1100 cycles] and the time required per cycle are somewhat greater for the three-parameter Gamma iterative estimation procedure than for the three - parameter lognormal and Weibull procedures. The result is that computer time becomes something of a problem. Therefore, the author has incorporated into the procedure a speed - up feature, which is too complicated to describe in detail here, based on first and second differences of the results of successive iterations. This feature is applied on the twentieth cycle and every twentieth cycle thereafter until the procedure converges or is abandoned after 1100 cycles. Though some time can be saved in this way, the computer time for the iterative procedure is still somewhat excessive, especially when the number of observations (after censoring, if any) is small. For this reason, no Monte Carlo study was made for the Gamma estimation procedure; it is conjectured that if one had been made, the results would have been quite similar to those for the Weibull procedure.

As one would expect, both the average number of cycles to convergence and the time required per cycle are greater for the four - parameter generalized Gamma iterative estimation procedure than for the corresponding procedures for the three - parameter populations. Excessive computer time becomes so serious a problem that the Newton - Raphson method and the gradient method (steepest ascent) were tried as alternatives to the rule of false position. Somewhat surprisingly, the gradient method was found to be the most slowly converging of the three; it was therefore dropped. The Newton - Raphson method converges most rapidly of the three if the estimates are already quite good, but behaves erratically if they are not, as is likely to be the case at the outset. The best procedure found is a hybrid one which employs the rule of false position for the first 120 cycles but, at the 121st cycle, changes over to the Newton - Raphson method, except for the estimation of the location parameter, which, because it is restricted to the closed interval $[0, x_1]$, where $x_1$ is the smallest observation, does not lend itself to estimation by the Newton - Raphson method, which might yield a value outside this interval. Superimposed upon this hybrid procedure is the same speed - up feature used in the three - parameter Gamma procedure. Even so, computer time tends to be quite excessive and, when the number of observations (after censoring, if any) is small, there are numerous cases in which the procedure does not converge within 1100 cycles and has to be abandoned. For obvious reasons, no Monte Carlo study was made of this procedure.

Hager and Bain (1970) [see also Bain and Antle (1970)] have made a study of inferential procedures for the generalized Gamma distribution, and have concluded that the Weibull model is quite robust relative to the generalized Gamma distribution with regard to reliability estimation. Because of the computational complexity and other difficulties encountered with the generalized Gamma distribution, they recommend use of the Weibull model instead of the generalized Gamma model unless the sample size is quite large [Bain and Antle suggest 200 or more]. The present author concurs in this recommendation.

Annotated listings of the FORTRAN programs for the above - mentioned iterative estimation procedures are available from the author on request.

Barnett (1966a) has studied the case in which the likelihood equation has multiple roots. In this case, a method is required which can be applied systematically to locate all roots of the likelihood equation and choose the one which corresponds to the absolute maximum of the likelihood function. Barnett tried five methods, including the Newton-Raphson method and the rule of false position, and found that the rule of false position is the only one which is suitable in this situation. He devised a slight modification of the rule of false position which he applied successively to the estimation of the location parameter of the Cauchy distribution from small samples, a situation in which the likelihood equation does have multiple roots. This modification involves a systematic search for relative maxima in the interval between the smallest and largest observations. It is not clear how this modification would be adapted to the estimation of parameters which do not have clearly defined limits. The author has not used this modification in his iterative procedures, but he has, in some cases, varied the starting values, and found that the procedure always converged to the same value(s) as before. This tends to confirm the conjecture that the likelihood functions studied have only a single maximum for finite values of the parameters. Richards (1967) has proposed a method, somewhat different from the one used by the author [Harter (1967, 1970)], which uses the rule of false position to prevent divergence which may occur when the Newton-Raphson method is started with a bad estimate, but retains the advantages of the latter method.

## 2. Linear Estimation

In view of the fact that, until the present century, the probable error was more widely used than the standard deviation (root-mean-square error) as a measure of the dispersion of a population, it is not surprising that the corresponding sample statistic, the quartile deviation or semi-interquartile range, is mentioned in the literature as far back as the time of Quetelet (1846). Other nineteenth century writers, including Galton (1875, 1889), Edgeworth (1886), and Sheppard (1899), explored the use of other quantiles. Sheppard discussed the choice of the best pair of quantiles for linear estimation of the standard deviation (or the probable error) of a normal population, but the first detailed study of the question was made by Karl Pearson (1920), who determined the best one, two or three pairs of quantiles for this purpose. While these results and subsequent one on the linear estimation of population parameters from sample quantiles are optimum only for infinitely large samples, estimators of this type have been employed by many later authors for samples of moderate or even small size.

Mosteller (1946) proposed estimators, based on linear combinations of order statistics, which he called systematic statistics. Godwin (1949) tabulated the coefficients of the order statistics of samples of size $n = 2(1)10$ in the best linear unbiased estimators of the standard deviation of a normal population, and Yamanouchi (1949) showed that these estimators are considerably more efficient than those used by Mosteller. Ogawa (1951) pointed out that, by application of the Gauss-Markov theorem, the method of least squares can be used to obtain best linear unbiased estimators. Ogawa also determined the optimum spacings of k quantiles for linear estimation of the mean of a normal population when the standard deviation is known, of the standard deviation when the mean is known, and of both parameters simultaneously, for $k = 1(1)10$, $k = 1(1)6$, and $k = 2$, respectively. Lloyd (1952) used the method of least squares to determine the coefficients of the order statistics of complete samples in the best linear unbiased estimators of the location and scale parameters of symmetric populations, especially the rectangular and normal populations. Gupta (1952) tabulated the coefficients of the order statistics in good (not best) linear unbiased estimators of population mean and standard deviation from singly censored samples of size $n = 2(1)10$ from a normal population.

During a period of less than two decades, well over a hundred journal articles have been published, besides numerous technical reports and theses, dealing with the estimation of parameters of various populations from linear combinations of selected quantiles or of some or all available order statistics of complete or censored samples. No attempt will be made here at a complete listing of these publications. Mention should be made, however, of the work of Dalcher(1955), Sarhan & Greenberg(1956,1958a,b)[see also Sarhan & Greenberg(1962),Chapter 10], Kulldorff(1963c), Hoksch(1967), Simpson(1967), Govindarajulu & Rao(1968) and Chan & Cheng(1971) on the normal distribution; Dalcher(1955), Ogawa(1957), Sarhan & Greenberg(1957,1958c)[see also Sarhan & Greenberg(1962),Chapter 11], Harter(1961b), Kulldorff(1963a, b), Sarhan, Greenberg & Ogawa(1963), Siddiqui(1963) and Dubey (1965b) on the exponential distribution; Lieblein(1954), Lieblein & Zelen(1956), White(1964), Richardson(1966), Shelnutt(1966), Elwer(1967), Fratzke(1967), Mann(1967a,b,1969), Hassanein(1968,1969a) and Chan & Kabir(1969) on the Type I extreme-value distribution; Dubey(1960,1965a,1967a,b), Govindarajulu & Joshi(1962,1968), Quayle(1963), Clark(1964), Carlton(1965), Musson(1965), Salisberry(1966), Herman(1968) and Polfeldt(1969) on the Weibull distribution; Aitchison & Brown(1957) and Highfield(1964) on the lognormal distribution; Plackett(1958), Kjelsberg(1962), Gupta, Qureishi & Shah (1965,1967), Beyer(1966), Richardson(1966), Hoksch(1967), Simpson(1967), Govindarajulu & Rao(1968), Hassanein(1969b) and Chan and Cheng(1971) on the logistic distribution; Karns (1963), Bruce(1964), Särndal(1964), Hill(1965), Musson(1965), Salisberry(1966) and Polfeldt(1969) on the Gamma distribution; Tukey(1958), Govindarajulu(1966) and Govindarajulu & Rao(1968) on the double exponential distribution; Govindarajulu & Eisenstat(1965) on the chi (1 d.f.) distribution; and Dalcher(1955), Barnett(1966b), Bloch(1966), Govindarajulu & Rao(1968), Chan(1970) and Chan & Cheng(1971) on the Cauchy distribution.

A number of the results obtained by the above authors deal with conditional estimation of one parameter, assuming that the other parameter(s) is (are) known. Herman (1968) has shown that if one starts with reasonable estimates of two parameters and applies the conditional estimators alternately, each time assuming that the latest estimate of the other parameter is its known value, the procedure converges and the result is the same as that given by the simultaneous estimators.

Most of the above authors have obtained best linear unbiased estimators by the method of least squares; however, the authors of several of the AFIT Master's theses mentioned above, which were prepared under the guidance of Albert H. Moore, have used instead the method of Lagrange multipliers [see Taylor(1965)], which yields equivalent results. Mann (1967a,b,1968) has advocated the use of best linear invariant estimators (minimum mean-square-error estimators which are invariant under transformations of location and scale) instead of best linear unbiased estimators. In cases of strongly asymmetric censoring, the best linear invariant estimators, like the maximum likelihood estimators, may be substantially biased, but if so they also have substantially smaller mean square error than the best linear unbiased estimators, of which they are simple linear functions.

The best linear unbiased estimators of scale parameters of symmetric populations have coefficients for the (r + 1)st and (n - r)th order statistics which are equal in numerical value but opposite in sign, so that an estimator which is a linear combination of 2k order statistics can be expressed as a linear combination of k quasi-ranges. Mosteller (1946) proposed estimators based on quasi-ranges. Carlton (1946) showed that the range of a sample from a rectangular population is a sufficient statistic for the population range. Cadwell (1953) studied the optimum quasi-range for estimating the standard deviation of a normal population. Harter (1959)[see also Harter(1970), Chapters I andII] tabulated the coefficients in the best linear unbiased estimators of the standard deviations of a normal population, based on one or two quasi-ranges, and of a rectangular population, based on the sample range.

Problems of optimization abound in the derivations of best linear unbiased (or best linear invariant) estimators based on order statistics, quasi-ranges, or quantiles. Except when all available order statistics or quasi-ranges are used, there is the problem of optimum choice of order statistics, quasi-ranges, or quantiles. Even in this case, the problem of determining the optimum coefficients (weights) arises. When the estimator is based on a single order statistic or quasi-range, the expected values and the variances of the order statistics or quasi-ranges must be known; when two or more are used, the covariances must also be known. Closed-form expressions for expected values, variances, and covariances of order statistics are known only for a few populations, e.g. rectangular and exponential. For most populations, values must be obtained by numerical integration or some other time-consuming process on an electronic computer and, as sample size increases, computer time soon increases to the point where the computation of the necessary expected values, variances, and covariances is no longer feasible. Even when the expected values, variances and covariances are known, it is still time-consuming to choose the $1, 2, \ldots, \ell \leq m \leq n$ of the m available order statistics of a (possibly censored) sample of size n which will yield the best estimator. Since there is, in general, no analytical method for choosing the $\ell$ order statistics to be used, it is necessary first to determine the linear unbiased estimators for all $\binom{m}{\ell}$ possible choices (or at least a very large subset of them) and then pick the one which has the smallest variance. Still, it is the computation of the expected values, variances, and covariances of the order statistics which requires the most time, so it is appropriate to pause here and take inventory of available tables.

Teichroew (1956) tabulated the expected values of the order statistics and of their squares and products for samples up through size n = 20 from a standard normal population. Sarhan & Greenberg (1956) used Teichroew's results to tabulate the corresponding variances and covariances. These tables were reproduced as Tables 10B.1-10B.3 of Sarhan & Greenberg (1962). Tabulations of the expected values, variances, and covariances of extreme-value order statistics were made by Lieblein (1954), White (1964), and Mann (1965) up through sample sizes n = 6,20, and 20 (later extended through n = 25). Govindarajulu & Joshi (1962,1968) tabulated the expected values, variances, and covariances of Weibull order statistics up through sample size n = 10 for shape parameter K = 1.0(1.0)10.0, also for K = 2.5. The 1962 version of their tables contained gross errors; the most flagrant of these were corrected in the 1968 version, which still contains errors as large as five units in the fifth (and last) decimal place. In connection with work on their AFIT Master's theses, Clark (1964) computed, and Musson (1965) spot checked, the expected values, variances, and covariances of Weibull order statistics up through sample size n = 15 for shape parameter K = 0.5 0.5(0.25)2.0(0.5)4.0. Both carried 10 decimal places, and their results agree to within two units in the eighth place. Unfortunately, neither included these tables in his thesis. Shah (1965,1966) tabulated the variances and covariances of logistic order statistics up through samples of size n = 10, and Gupta, Qureishi & Shah (1965,1967) extended the tables up through sample size n = 25. The tables in the 1965 report are complete, but those in the 1967 journal article are abridged, with results for sample sizes above 10 included only when the sample size is a multiple of 5.

In connection with work on their AFIT Master's theses, Bruce (1964) and Hill (1965) independently computed, by different methods, the expected values, variances and covariances of Gamma order statistics for sample sizes up through n 15 and shape parameter 1(1)6. Their results agree to eight decimal places. Unfortunately, neither included these tables in his thesis. Harter (1964)[see also Harter (1970), Table C4] tabulated the expected values of Gamma order statistics for sample size up through n 40 and shape parameter 0.5(0.5)4.0; the table in the 1964 technical report contains a few errors of up to five units in the last place for the first order statistic when 0.5, which were corrected in the 1970 book, Gupta (1960) and Breiter & Krishnaiah (1967,1968) tabulated the first four moments about the origin, from which one can readily obtain the variances, of Gamma order statistics for various values of the shape parameter and sample sizes up through n 10 and n 16 respectively (abridged to n 9 in the 1968 journal article). The author is not aware of the existence of any published tables of covariances of Gamma order statistics. Govindarajulu (1966) tabulated the expected values, variances and covariances of the order statistics of samples of size n 2(1)20 from a double exponential population. Govindarajulu & Eisenstat (1965) did the same for the chi (1 d.f.) population; their table of expected values extends up through n 100. Barnett (1966b) tabulated the finite expected values, for samples of size n 3(1)20, and the finite variances and covariances, for samples of size n 5(1)16(2)20, of Cauchy order statistics. The first and last order statistics have infinite expected values and the first two and the last two have infinite variances and covariances; hence these order statistics are not useful in estimation.

The search for linear estimators which do not depend on knowledge of expected values, variances, and covariances of the order statistics, yet retain insofar as possible the properties of best linear unbiased estimators, dates back at least as far as 1952. The property of unbiasedness depends only on the expected values of the order statistics, but the property of minimum variance depends on their variances and covariances. Gupta (1952) used only the expected values of the order statistics, and hence his linear estimators of the mean and standard deviation of a normal population from censored samples are unbiased, but not of minimum variance. Other authors, notably Bennett and Jung, went even further, not even assuming knowledge of the expected values of the order statistics. Bennett (1952) considered the asymptotic properties of ideal linear estimators of location and scale parameters of continuous univariate populations and their use in obtaining estimators for comparatively small samples.

He succeeded in obtaining approximations, based on asymptotic theory, which are quite close to the exact ideal linear (best linear unbiased) estimators for sample sizes as small as ten. Jung (1955) proposed linear estimators in which the weight (coefficient) of the i-th order statistic of a sample of size n is 1/n times a continuous function, depending only on the population, of $i/(n + 1)$. He determined the asymptotically best weight function, subject to certain regularity conditions, and applied the results to the Student t and Pearson Type III (Gamma) distributions.

Epstein (1956) derived "almost best" linear unbiased estimators, from censored samples, of parameters of exponential distributions; the coefficients depend upon the expected values of the order statistics, but not upon their variances and covariances. Blom (1956,1958) proposed nearly best unbiased estimators, in which the coefficients depend upon the expected values of the order statistics, but not upon their variances and covariances. For cases in which the expected values of the order statistics have not been tabulated, he proposed nearly best, nearly unbiased estimators, in which the coefficients are determined by approximating the expected value of the i-th order statistic of a sample of size n by $F^{-1}[(i-\alpha)/(n-\alpha-\beta+1)]$, where $F^{-1}$ is the inverse of the cumulative distribution function of the population under consideration and $\alpha$ and $\beta$ are suitably chosen constants (for symmetric populations, $\alpha = \beta$). Plackett (1958) proposed linearized maximum likelihood estimators. He and Weiss (1963) independently showed that suitable linear combinations of the available order statistics of censored samples are asymptotically normally distributed. McCool (1965) suggested good linear unbiased estimators which can be constructed from the best linear unbiased estimators for a smaller sample size. Chernoff, Gastwirth & Johns (1967) studied the asymptotic distribution of linear combinations of order statistics and their application in estimating location and scale parameters from uncensored or censored samples from logistic, Cauchy, and normal populations. Prescott (1970) used the method of Gupta (1952) to obtain linear unbiased estimators of the standard deviation of a normal population from complete samples up through size $n = 50$ and from doubly censored samples up through $n = 25$. He made use of the expected values of normal order statistics, which have been tabulated by Harter (1961a) [see also Harter (1970), Table C1] for samples of size $n = 2(1)100$ and for selected values of n up through 400. The expected values of exponential, Weibull, and Gamma order statistics, for sample sizes up through n = 120,40, and 40 respectively, have also been tabulated by Harter (1964) [see also Harter (1970), Tables C2-C4].

Asymptotically optimum quantiles for linear estimation, and the corresponding weights (coefficients) have been tabulated by Ogawa (1951), Dalcher (1955), Kulldorff (1963c, 1964) and Chan & Cheng (1971) for the normal distribution; by Dalcher (1955), Ogawa (1957), Kulldorff (1962), Siddiqui (1963) and Saleh (1966) for the exponential distribution; by Hassanein (1968) and Chan & Kabir (1969) for the extreme-value distribution; by Chan (1969), Hassanein (1969b) and Chan & Cheng (1971) for the logistic distribution; by Särndal (1964) for the Gamma distribution; and by Dalcher (1955), Chan (1970) and Chan & Cheng (1971) for the Cauchy distribution. The asymptotically optimum quantiles for the normal and extreme-value (log Weibull) distributions can be used indirectly to estimate the parameters of the lognormal and Weibull distributions, respectively.

In using linear estimators based on asymptotically optimum quantiles, one is faced with the problem of using the order statistics of a sample to determine the sample quantiles or to estimate the quantiles of the population. Many different methods of doing this have been proposed. All of them are asymptotically equivalent, so for sufficiently large samples it makes no difference which method is chosen. For sufficiently small samples from most populations of interest, we have available the best linear unbiased estimators based on order statistics, which should certainly be used, whenever they are available, instead of quantile estimators. For samples of intermediate size, however, one is forced to make a choice among the methods of determining quantiles from order statistics, and it is important that the choice be made so as to avoid bias and imprecision. Let us therefore consider the various methods that have been proposed.

Ogawa (1951) defined the $\lambda$ quantile of a sample of size n as the i-th order statistic of the sample, where $i = n\lambda$ if $n\lambda$ is an integer and $i = [n\lambda] + 1$ otherwise. This method of determining quantiles from order statistics has the unfortunate property that one obtains the same value of i for quite different values of $\lambda$. For example, when n = 20, $\lambda = .001$ and $\lambda = .049$ both yield i = 1 and $\lambda = .951$ and $\lambda = .999$ both yield i = 20. Now the first order statistic may not be a bad value to take as the .049 quantile, but it is certainly too high for .001 quantile; likewise, the twentieth order statistic may not be a bad value to take as the .951 quantile, but it is certainly too low for the .999 quantile. Thus this method tends to yield values for the quantiles which lie too close to the median; in other words, values whose dispersion is too small. This does not cause any bias, though it does lead to some loss of precision, in estimating the location parameter of a symmetric population, since the $\lambda$ and $1 - \lambda$ quantiles have equal weights and their biases counteract each other. It does, however, lead to a negative bias in the scale parameter estimator for a symmetric population, since the $\lambda$ and $1 - \lambda$ quantiles have weights which are equal in magnitude but opposite in sign and their biases augment each other. Difficulty of another sort arises when $n\lambda$ is an integer, since in this case $n(1 - \lambda)$ is also an integer; thus, when one is estimating the parameters of a symmetric population, the values of i corresponding to $\lambda$ and $1 - \lambda$ sum to n, not to n + 1 as they should, and the estimator of the location parameter is negatively biased. If this method of determining quantiles from order statistics is used, one should never use an odd number of quantiles to estimate the parameters of a symmetric population from a sample of even size, since $\lambda = .5$ is one of the optimum quantiles and $n\lambda = n/2$ (an integer).

Now let us consider some other methods of determining quantiles from order statistics. This problem is related to that of plotting positions on probability paper, which has a long history, dating back at least as far as a paper by Fuller (1914), who proposed the the i-th order statistic of a sample of size n be plotted at cumulative probability $\lambda = (i - 1/2)/n$. Weibull (1939) suggested $\lambda = i/(n + 1)$. A good case can be made for the latter plotting position since, as Margolin & Mosteller (1969) put it, the expected coverage to the left of the i-th order statistic of a sample of size n is $i/(n + 1)$. Nevertheless, the i-th order statistic of a sample of size n, as an estimator of the $\lambda$ quantile of the population, where $\lambda = i/(n + 1)$, though unbiased in the sense of Uhlmann (1963), is biased according to the usual definition. If one sets $\lambda = (i-1/2)/n$, the bias is in the opposite direction. Thus, if one defines the $\lambda$ quantile of a sample of size n as the i-th order statistic of the sample, where $i = \lambda(n + 1)$, the result is an estimator of the scale parameter of a symmetric population which is biased in one direction (upward), but for $i = n\lambda + 1/2$ the bias is in the opposite direction (downward). Blom (1958) has proposed a compromise plotting position $\lambda = (i - \alpha)/(n - \alpha - \beta + 1)$ where $\alpha$ and $\beta$ are chosen so that $F^{-1}(\lambda)$ is an overall good approximation to $E(x_i)$, the expected value of the i-th order statistic ($\alpha = \beta$ for symmetric populations). This leads to a definition of the $\lambda$ quantile as the i-th order statistic of a sample of size n, where $i = (n - \alpha - \beta + 1)\lambda + \alpha$ [$i = (n - 2\alpha + 1)\lambda + \alpha$ for a symmetric population]. Blom (1958) found that $\alpha = 3/8$ is a good choice for a normal distribution; the results of Mann (1968) indicate that it is also a good choice for an extreme - value distribution. A good choice can be made for any distribution for which the expected values of the order statistics have been tabulated. Harter (1961a) [see also Harter (1970), Chapter III] has found, however, that for the normal distribution the optimum value of $\alpha$ is a slowly increasing function of n, the sample size. The reader should note that, for all the methods mentioned in this paragraph, the value of i corresponding to a given value of $\lambda$ is not, in general, an integer, so that interpolation [extrapolation, if $i < 1$ or $i > n$] between adjacent order statistics is required.

In conclusion, the author offers the following recommendations concerning linear estimation: (1) For small samples, use the best linear unbiased estimator based on all m available order statistics (or on a subset of size $\ell$, where $\ell$ is sufficiently large to yield the desired efficiency) whenever the expected values, variances and covariances of the order statistics have been tabulated, so that the appropriate weights (coefficients) are known or can be easily determined; when only the expected values of the order statistics are known, use the nearly best linear unbiased estimator; if even the expected values of the order statistics are not known, use the nearly best, nearly unbiased estimator. (2) For samples of intermediate size, use the linear estimator based on asymptotically optimum quantiles (spacings), determining the value of i corresponding to each $\lambda$ by Blom's method for a suitable choice of $\alpha$ and $\beta$ [$\alpha = \beta$ for symmetric populations] and interpolating [extrapolating for $i < 1$ or $i > n$] between adjacent order statistics. It is usually best to avoid using so many quantiles that extrapolation becomes necessary. (3) For very large samples, use the linear estimator based on asymptotically optimum quantiles, but determine i corresponding to each $\lambda$ by $i = [n\lambda] + 1$, which does not require interpolation between adjacent order statistics or the choice of $\alpha$ (symmetric populations) or of $\alpha$ and $\beta$ (asymmetric populations).

An alternative method for intermediate sample sizes has been proposed by Johns & Lieberman (1966), who have given asymptotically optimum linear estimators for the location and scale parameters of an extreme-value distribution, based on the first m order statistics of a sample of size n, and have tabulated the coefficients for n = 10,15,20,30,50, 100,m = [i(n/4)], i = 1,2,3,4. Extension to other distributions is possible.

References

[1] Aitchison, J.; Brown, J.A.C. (1957), The Lognormal Distribution. Cambridge University Press, Cambridge, England.

[2] Bain, Lee J.; Antle, Charles E. (1970). Inferential Procedures for the Weibull and Generalized Gamma Distributions. ARL 70-0266, Aerospace Research Laboratories, Wright-Patterson AFB. AD 718103.

[3] Barnett, V.D. (1966a). Evaluation of the maximum likelihood estimator where the likelihood equation has multiple roots. Biometrika 53, 151-166.

[4] Barnett, V. D. (1966b). Order statistics estimators of the location of the Cauchy distribution. J. Amer. Statist. Assoc. 61, 1205-1218.

[5] Bennett, Carl Allen (1952). Asymptotic Properties of Ideal Linear Estimators. Ph.D. dissertation, University of Michigan. University Microfilms, Ann Arbor, Michigan.

[6] Beyer, Jon N. (1966). Conditional Estimation of the Scale Parameter of the Logistic Distribution by the Use of Selected Order Statistics. Unpublished M. S. thesis (GRE/Math/66-1), Air Force Institute of Technology.

[7] Blischke, W. R. (1971). Further Results on Estimation of the Parameters of the Pearson Type III Distribution. ARL 71-0063, Aerospace Research Laboratories, Wright-Patterson AFB.

[8] Blischke, W. R.; Brady,E. J.; Mundle, P. B. (1970). Further Results on Estimation of the Parameters of the Pearson Type III Distribution in the Regular and Nonregular cases. ARL 70-0017, Aerospace Research Laboratories, Wright-Patterson AFB. AD 705207.

[9] Bloch, Daniel (1966). A note on the estimation of the location parameter of the Cauchy distribution. J. Amer. Statist. Assoc. 61, 852-855.

[10] Blom, Gunnar (1956). On linear estimates with nearly minimum variance. Arkiv för Matematik, 3 (31), 365-369.
[11] Breiter, M. C.;Krishnaiah, P. R. (1967). Tables for the Moments of Gamma Order Statistics. ARL 67-0166, Aerospace Research Laboratories, Wright-Patterson AFB. AD 639729.
[12] Breiter,M. C.; Krishnaiah, P. R. (1968). Tables for the moments of Gamma order statistics. Sankhya B 30, 59-72.
[13] Bruce, Richard A. (1964). Estimation of the Scale Parameter of the Gamma Distribution by Use of M Order Statistics. Unpublished M. S. thesis, (GRE/Math/64-2), Air Force Institute of Technology. AD 610774.
[14] Cadwell, J. H. (1953). The distribution of quasi-ranges in samples from a normal population. Ann. Math. Statist. 24, 603-613.
[15] Carlton, A. George (1946). Estimating the parameters of a rectangular distribution. Ann. Math. Statist. 17, 355-358.
[16] Carlton, Howard F. (1965). Estimation of the Scale Parameter of the Weibull Probability Density Function by the Use of L-Order Statistics. Unpublished M. S. thesis (GRE/Math/65-2), Air Force Institute of Technology.
[17] Chan, Lai K. (1969). Linear quantile estimates of the location and scale parameters of the logistic distribution. Statist. Hefte 10, 277-282.
[18] Chan, Lai K. (1970). Linear estimation of the location and scale parameters of the Cauchy distribution based on samples quantiles. J. Amer. Statist. Assoc. 65, 851-863.
[19] Chan, Lai K.; Cheng, Smiley W. H. (1971). On the Student's test based on sample percentiles from the normal, logistic, and Cauchy distribution. Technometrics 13, 127-137.
[20] Chan, Lai K.; Kabir, A.B.M. Lutful (1969). Optimum quantiles for the linear estimation of the parameters of the extreme value distribution in complete and censored samples. Naval Res. Logist. Quart. 16, 381-404.
[21] Chernoff, Herman; Gastwirth, Joseph L.;Johns, M. V.,Jr. (1967) Asymptotic distributions of linear combinations of functions of order statistics with applications to estimation. Ann. Math. Statist. 38, 52-72.
[22] Clark, Lester John, Jr. (1964). Estimation of the Scale Parameter of the Weibull Probability Density Function

by Use of One-Order and M-Order Statistics. Unpublished M. S. thesis (GRE/Math/64-3), Air Force Institute of Technology. AD 630809.

[23] Cramer, Harald (1946). Mathematical Methods of Statistics. Princeton University Press, Princeton, N.J.

[24] Dalcher, Andreas (1955). Statistische Schätzungen mit Quantilen. Mitteilungen der Vereinigung Schweizerischer Versicherungsmathematiker. **55**, 475-498.

[25] Dubey, Satya D. (1960). Contributions to Statistical Theory of Life Testing and Reliability. Ph.D. dissertation, Michigan State University. University Microfilms, Ann Arbor, Michigan.

[26] Dubey, Satya D. (1965a). Asymptotic properties of several estimators of Weibull parameters. Technometrics **7**, 423-434.

[27] Dubey, Satya D. (1965b). Asymptotically single observation best estimator of exponential expected life. Sankhya A **27**, 133-142.

[28] Dubey, Satya D. (1967a). Some percentile estimators for Weibull parameters. Technometrics **9**, 119-129.

[29] Dubey, Satya D. (1967b). On some permissible estimators of the location parameter of the Weibull and certain other distribitions. Technometrics **9**, 293-307.

[30] Edgeworth, F. Y. (1886). Observations and statistics: an essay on the theory of errors of observation and the first principles of statistics. Trans. Camb. Philos. Soc. **14** (2), 138-169.

[31] Elwer, Gale E. (1967). Conditional Linear Estimation of the Scale Parameter of the Extreme-Value Distribution by Use of Selected Order Statistics from Doubly Censored Samples. Unpublished M. S. thesis (GRE/Math/67-1), Air Force Institute of Technology.

[32] Epstein, Benjamin (1956). Simple estimators of the parameters of exponential distributions when samples are censored. Ann. Inst. Statist. Math. **8**, 15-26.

[33] Fratzke, William F. (1967). Simultaneous Linear Estimation of the Location and Scale Parameters of the Extreme-Value Distribution Using Selected Order Statistics from Doubly Censored Samples. Unpublished M. S. thesis (GRE/Math/67-2), Air Force Institute of Technology. AD 849869.

[34] Fuller, Weston E. (1914). Flood flows. Trans. Amer. Soc. Civ. Engrs. **77**, 564-617; discussion, 618-694.

[35] Galton, Francis (1875). Statistics by intercomparison, with remarks on the law of frequency of error. Philos.

Mag. (4) **49**, 33-46.
[36] Galton, Francis (1889). Natural Inheritance. Macmillan and Co., London-New York.
[37] Godwin, H. J. (1949). On the estimation of dispersion by linear systematic statistics. Biometrika **36**, 92-100.
[38] Govindarajulu, Zakkula (1966). Best linear estimates under symmetric censoring of the parameters of a double exponential population. J. Amer. Statist. Assoc. **61**, 248-258.
[39] Govindarajulu, Zakkula; Eisenstat, Stan (1965). Best estimates of location and scale parameters of a chi (1 d.f.) distribution, using ordered observations. Rep. Statist. Appl. Res., Un. Japan. Sci. Engrs. **12**, (4), 1-16.
[40] Govindarajulu, Z.; Joshi, M. (1962). Best Linear Unbiased Estimation of Location and Scale Parameters of Weibull Distribution Using Ordered Observations. Report No. 102, Statistical Laboratory, Case Institute of Technology.
[41] Govindarajulu, Zakkula; Joshi, Madhukar (1968). Best linear unbiased estimation of location and scale parameters of Weibull distribution using ordered observations. Rep. Statist. Appl. Res., Un. Japan. Sci. Engrs. **15** (2), 1-14.
[42] Govindarajulu, Z.; Rao, J. N. K. (1968). Asymptotically best estimator of scale based on two symmetric quantiles. Sankhyā A **30**, 335-336.
[43] Gupta, A. K. (1952). Estimation of the mean and standard deviation of a normal population from a censored sample. Biometrika **39**, 260-273.
[44] Gupta, Shanti S. (1960). Order statistics from the Gamma distribution. Technometrics **2**, 243-262; correction, 523-524.
[45] Gupta, S. S.; Qureishi, A. S.; Shah, B. K. (1965). Best Linear Unbiased Estimators of the Parameters of the Logistic Distribution Using Order Statistics. Mimeograph Series No. 52, Department of Statistics, Purdue University.
[46] Gupta, S. S.; Qureishi, A. S.; Shah, B. K. (1967). Best linear estimators of the parameters of the logistic distribution using order statistics. Technometrics **9**, 43-56.
[47] Hager, Harold W.; Bain, Lee J. (1970). Inferential procedures for the generalized Gamma distribution. J. Amer. Statist. Assoc. **65**, 1601-1609.

[48] Harter, H. Leon (1959). The use of sample quasi-ranges in estimating population standard deviation. Ann. Math. Statist. 30, 980-999.
[49] Harter, H. Leon (1961a). Expected values of normal order statistics. Biometrika 48, 151-165; corrigenda, 476.
[50] Harter, H. Leon (1961b). Estimating the parameters of negative exponential populations from one or two order statistics. Ann. Math. Statist. 32, 1078-1090; correction, 35 (1964), 925.
[51] Harter, H. Leon (1964). Expected Values of Exponential, Weibull, and Gamma Order Statistics. ARL 64-31, Aerospace Research Laboratories, Wright-Patterson AFB. AD 436763.
[52] Harter, H. Leon (1966). Asymptotic Variances and Covariances of Maximum-Likelihood Estimators, from Censored Samples, of the Parameters of a Four-Parameter Generalized Gamma Population. ARL 66-0158, Aerospace Research Laboratories, Wright-Patterson AFB. AD 648045.
[53] Harter, H. Leon (1967). Maximum-likelihood estimation of the parameters of a four-parameter generalized Gamma population from complete and censored samples. Technometrics 9, 159-165.
[54] Harter, H. Leon (1970). Order Statistics and their Use in Testing and Estimation, Volume 2: Estimates Based on Order Statistics of Samples from Various Populations. U. S. Government Printing Office, Washington, D. C.
[55] Harter, H. Leon; Moore, Albert H. (1965). Maximum likelihood estimation of the parameters of Gamma and Weibull populations from complete and from censored samples. Technometrics 7, 639-643; errata, 9 (1967), 195.
[56] Harter, H. Leon; Moore, Albert H. (1966a). Iterative maximum-likelihood estimation of the parameters of normal populations from singly and doubly censored samples. Biometrika 53, 205-213; corrections, 56 (1969), 229.
[57] Harter, H. Leon; Moore, Albert H. (1966b). Local-maximum-likelihood estimation of the parameters of three-parameter lognormal populations from complete and censored samples. J. Amer. Statist. Assoc. 61, 842-851; corrigenda, 61 (1966), 1247; 62 (1967), 1519-1520; 63 (1968), 1549.
[58] Harter, H. Leon; Moore, Albert H. (1967a). Asymptotic variances and covariances of maximum-likelihood estimators, from censored samples, of the parameters of Weibull and Gamma populations. Ann. Math. Statist.

38, 557-570.

[59] Harter, H. Leon; Moore, Albert H. (1967b). Maximum-likelihood estimation, from censored samples, of the parameters of a logistic distribution. J. Amer. Stat. Assoc.62, 675-684.

[60] Harter, H. Leon; Moore, Albert H. (1968). Maximum-likelihood estimation, from doubly censored samples, of the parameters of the first asymptotic distribution of extreme values. J. Amer. Statist. Assoc. 63, 889-901.

[61] Hassanein, Khatab M. (1968). Analysis of extreme-value data by sample quantiles for very large samples. J. Amer. Statist. Assoc. 63, 877-888.

[62] Hassanein, Khatab M.(1969a). Estimation of the parameters of the extreme-value distribution by use of two or three order statistics. Biometrika 56, 429-436.

[63] Hassanein, Khatab M. (1969b). Estimation of the parameters of the logistic distribution by sample quantiles. Biometrika 56, 684-687.

[64] Herman, William J. (1968). Least-Squares Conditional Estimation of the Location Parameter of Weibull Populations. Unpublished M. S. thesis (GRE/Math/68-1), Air Force Institute of Technology. AD 684696.

[65] Highfield, Eugene R. (1964). Estimation of the Scale Parameter of the Logistic Distribution by M Order Statistics. Unpublished M. S. thesis (GRE/Math/64-8), Air Force Institute of Technology. AD 610770.

[66] Hill, Bruce M. (1963). The three-parameter lognormal distribution and Bayesian analysis of a point-source epidemic. J. Amer. Statist. Assoc. 58, 72-84.

[67] Hill, Thomas D. (1965). Estimation of the Scale Parameter of the Gamma Distribution by Use of L Order Statistics. Unpublished M.S. thesis (GRE/Math/65-5), Air Force Institute of Technology. AD 630810.

[68] Hoksch, Orville J. (1967). Conditional Estimation of the Standard Deviation of the Normal and Logistic Distributions from Doubly Censored Samples Using Selected Order Statistics. Unpublished M. S. thesis (GRE/Math/67-5), Air Force Institute of Technology.

[69] Huber, Peter J. (1964). Robust estimation of a location parameter. Ann. Math. Statist. 35, 73-101.

[70] Johns, M. V., Jr.; Lieberman, G. J. (1966). An exact asymptotically efficient confidence bound for reliability in the case of the Weibull distribution. Technometrics 8, 135-175.

[71] Jung, Jan (1955). On linear estimates defined by a continuous weight function. Arkiv för Matematik 3(15), 199-209.

[72] Karns, Robert Clay (1963). Scale Parameter Estimation of the Gamma Probability Function Based on One Order Statistic. Unpublished M. S. thesis (GRE/Math/63-8), Air Force Institute of Technology. AD 425223.

[73] Kjelsberg, Marcus O. (1962). Estimation of the Parameters of the Logistic Distribution under Truncation and Censoring. Ph.D. dissertation, University of Minnesota. University Microfilms, Ann Arbor, Michigan.

[74] Kulldorff, Gunnar (1962). On the Asymptotically Optimum Spacings for the Exponential Distribution. Mimeographed report, Univ. of Lund, Sweden.

[75] Kulldorff, Gunnar (1963a). Estimation of one or two parameters of the exponential distribution on the basis of suitably chosen order statistics. Ann. Math. Statist. 34, 1419-1431.

[76] Kulldorff, Gunnar (1963b). Estimation of one or two parameters of the exponential distribution on the basis of suitably chosen order statistics. Part II. Department of Statistics, University of Lund, Sweden.

[77] Kulldorff, Gunnar (1963c). On the optimum spacing of sample quantiles from a normal distribution. I. Skand. Aktuarietidskr. 46, 143-156.

[78] Kulldorff, Gunnar (1964). On the optimum spacing of sample quantiles from a normal distribution. II. Skand. Aktuarietidskr. 47, 71-87.

[79] Le Cam, L. (1970). On the assumptions used to prove asymptotic normality of maximum likelihood estimates. Ann. Math. Statist. 41, 802-828.

[80] Lieblein, Julius (1954). A New Method of Analyzing Extreme-Value Data. NACA Technical Note No. 3053, National Advisory Committee for Aeronautics.

[81] Lieblein, J.; Zelen, M. (1956). Statistical investigation of the fatigue life of deep-groove ball bearings. J. Res. Nat. Bur. Standards 57, 273-316.

[82] Lloyd, E. H. (1952). Least-squares estimation of location and scale parameters using order statistics. Biometrika 39, 88-95.

[83] Mann, Nancy R. (1965). Point and Interval Estimates for Reliability Parameters when Failure Times Have the Two-Parameter Weibull Distribution. Ph.D. dissertation, University of California at Los Angeles. University Microfilms, Ann Arbor, Michigan.

[84] Mann, Nancy R. (1967a). Results on Location and Scale Parameter Estimation with Application to the Extreme-Value Distribution. ARL 67-0023, Aerospace Research Laboratories, Wright-Patterson AFB. AD 653575.
[85] Mann, Nancy R. (1967b). Tables for obtaining the best linear invariant estimates of parameters of the Weibull distribution. Technometrics **9**, 629-645.
[86] Mann, Nancy R. (1968). Point and interval estimation procedures for the two-parameter Weibull and extreme-value distributions. Technometrics **10**, 231-256.
[87] Mann, Nancy R. (1969). Cramér-Rao efficiencies of best linear invariant estimators of parameters of the extreme-value distribution under Type II censoring from above. SIAM J. Appl. Math. **17**, 1150-1162.
[88] Margolin, Barry H.; Mosteller, Frederick (1969). The expected coverage to the left of the i-th order statistic for arbitrary distributions. Ann. Math. Statist. **40**, 644-647.
[89] McCool, John I. (1965). The construction of good linear unbiased estimates from the best linear estimates for a smaller sample size. Technometrics 7, 543-552.
[90] Mosteller, Frederick (1946). On some useful "inefficient" statistics. Ann. Math. Statist. **17**, 377-408.
[91] Musson, Thomas A. (1965). Linear Estimation of the Location and Scale Parameters of the Weibull and Gamma Probability Distributions by the Use of Order Statistics. Unpublished M. S. thesis (GRE/Math/65-9), Air Force Institute of Technology. AD 630811.
[92] Ogawa, Junjiro (1951). Contributions to the theory of systematic statistics, I. Osaka Mathematical Journal **3**, 175-213.
[93] Ogawa, Junjiro (1957). A Further Contribution to the Theory of Systematic Statistics. Mimeograph Series No. 168, Institute of Statistics, University of North Carolina. AD 137834.
[94] Pearson, Karl (1920). On the probable errors of frequency constants. Part III. Biometrika **13**, 113-132.
[95] Plackett, R. L. (1958). Linear estimation from censored data. Ann. Math. Statist. **29**, 131-142.
[96] Polfeldt, Thomas (1969). Asymptotically Best Systematic Statistics in a Non Regular Case. Technical Report No. 62, Department of Statistics, Stanford University.
[97] Prescott, P. (1970). Estimation of the standard deviation of a normal population from doubly censored samples using normal scores. Biometrika **57**, 409-419.

[98] Quayle, Ronald John (1963). Estimation of the Scale Parameter of the Weibull Probability Density Function by Use of One Order Statistic. Unpublished M. S. thesis (GRE/Math/63-10), Air Force Institute of Technology. AD 425228.
[99] Quetelet, A. (1846). Lettres à S.A.R. le Duc Régnant de Saxe-Cobourg et Gotha, sur la Theorie des Probabilités, Appliquée aux Sciences Morales et Politiques. M. Hayez, Bruxelles.
[100] Richards, F.S.G. (1967). On finding local maxima of functions of a real variable. Biometrika 54, 310-311.
[101] Richardson, Everett B. (1966). Simultaneous Linear Estimation of the Location and Scale Parameters of the Extreme-Value and Logistic Distributions by the Use of Selected Order Statistics. Unpublished M. S. thesis (GRE/Math/66-7), Air Force Institute of Technology.
[102] Saleh, A.K. Md. Ehsanes (1966). Estimation of the parameters of the exponential distribution based on optimum order statistics in censored samples. Ann. Math. Statist. 37, 1717-1735.
[103] Salisberry, Rex C. (1966). Simultaneous Linear Estimation of the Location and Scale Parameters of the Gamma and Weibull Distributions by the Use of Selected Order Statistics. Unpublished M. S. thesis (GRE/Math/66-8), Air Force Institute of Technology.
[104] Sarhan, A. E.; Greenberg, B. G. (1956). Estimation of location and scale parameters by order statistics from singly and doubly censored samples. Part I. The normal distribution up to samples of size 10. Ann. Math. Statist. 27, 427-451; correction, 40(1969),325.
[105] Sarhan, A. E.; Greenberg, B. G. (1957). Tables for best linear estimates by order statistics of the parameters of single exponential distributions from singly and doubly censored samples. J. Amer. Statist. Assoc. 52, 58-87.
[106] Sarhan, A. E.; Greenberg, B. G. (1958a). Estimation of location and scale parameters by order statistics from singly and doubly censored samples, Part II. Tables for the normal distribution for samples of size $11 \leq n \leq 15$. Ann. Math. Statist. 29, 79-105.
[107] Sarhan, A. E.; Greenberg, B. G. (1958b). Estimation of Location and Scale Parameters by Order Statistics from Singly and Doubly Censored Samples, Part III. The Normal Distribution for Samples of Size $16 \leq n \leq 20$. Technical Report No. 4, OOR Project 1597, University

of North Carolina. AD 136976.

[108] Sarhan, A. E.; Greenberg, B. G. (1958c). Estimation problems in the exponential distribution using order statistics. Proceedings of the Statistical Techniques in Missile Evaluation Symposium (Blacksburg, Va.), 123-175; discussion, 176-178.

[109] Sarhan, Ahmed E.; Greenberg, Bernard G. (editors) (1962). Contributions to Order Statistics. John Wiley & Sons, New York.

[110] Sarhan, A. E.; Greenberg, B. G.; Ogawa, Junjiro (1963) Simplified estimates for the exponential distribution. Ann. Math. Statist. **34**, 102-116.

[111] Särndal, Carl-Erik (1964). Estimation of the parameters of the Gamma distribution by sample quantiles. Technometrics **6**, 405-414.

[112] Shah, B. K. (1965). On the Bivariate Moments of Order Statistics from a Logistic Distribution and Applications. Mimeograph Series No. 48, Department of Statistics, Purdue University.

[113] Shah, B. K. (1966). On the bivariate moments of order statistics from a logistic distribution. Ann. Math. Statist. **37**, 1002-1010.

[114] Shelnutt, J. William III (1966). Conditional Linear Estimation of the Scale Parameter of the Extreme-Value Distribution by the Use of Selected Order Statistics. Unpublished M. S. thesis (GRE/Math/66-9), Air Force Institute of Technology.

[115] Sheppard, W. F. (1899). On the application of the theory of error to cases of normal distribution and normal correlation. Philos. Trans. Roy. Soc. London A **192**, 101-167.

[116] Siddiqui, M. M. (1963). Optimum estimators of the parameters of negative exponential distributions from one or two order statistics. Ann. Math. Statist. **34**, 117-121.

[117] Simpson, Jackson S. (1967). Simultaneous Linear Estimation of the Mean and Standard Deviation of the Normal and Logistic Distributions by the Use of Selected Order Statistics from Doubly Censored Samples. Unpublished M. S. thesis (GRE/Math/67-10), Air Force Institute of Technology. AD 675544.

[118] Taylor, A. E. (1965). Advanced Calculus (Tenth Printing). Blaisdell Publishing Company, New York.

[119] Teichroew, D. (editor) (1956). Tables of expected values of order statistics and products of order statis-

tics for samples of size twenty and less from the normal distribution. Ann. Math. Statist. 27, 410-426.

[120] Tukey, John W. (1958). A problem of Berkson, and minimum variance orderly estimators. Ann. Math. Statist. 29, 588-592.

[121] Uhlmann, Werner (1963). Ranggrössen als statistische Schätzfunktionen. Z. Angew. Math. Mech. 43, (suppl.), T50-T51.

[122] Weibull, W. (1939). A statistical theory of the strength of materials. Ing. Vetenskaps Akad. Handl., No. 151.

[123] Weiss, Lionel (1963). On the asymptotic distribution of an estimate of a scale parameter. Naval Res. Logist. Quart. 10, 1-9.

[124] White, John S. (1964). Least squares unbiased censored linear estimation for the log Weibull (extreme-value) distribution. J. Indust. Math. Soc. 14, 21-60.

[125] Yamanouchi, Ziro (1949). Estimates of mean and standard deviation of a normal distribution from linear combinations of some chosen order statistics (Japanese. English summary). Bull. Math. Statist. 3(1-2), 52-57.

Supplementary List of Relevant AFIT M. S. Theses
(Furnished by Professor Albert H. Moore)

[S1] Stark, Terry M. (1967). Simultaneous and Conditional Estimation of the Location and Scale Parameters of the Cauchy Distribution by the Use of Selected Order Statistics. (GRE/Math/67-11).

[S2] Bonney, Frederic L. (1968). Nearly Best Unbiased Linear Estimators of the Location and Scale Parameters of the Normal Distribution by Use of Order Statistics. (GRE/Math/68-1).

[S3] Chamberlain, Robert B. (1968). Nearly Best Linear Unbiased Estimators of the Location and Scale Parameters of the Cauchy Distribution by the Use of Censored Order Statistics (GRE/Math/68-2). AD 685112.

[S4] Eickmann, Kenneth E. (1968). Unbiased Nearly Best Linear Estimates of the Scale and Location Parameters of the Extreme Value Distribution by the Use of Order Statistics. (GRE/Math/68-5). AD 693661.

[S5] Morgan, Guy A. (1968). Simultaneous Estimation of the Scale and Location Parameters of the Gamma Probability Distribution by Use of Order Statistics (GRE/Math

/68-8). AD 685114.
[S6] Osborne, Michael C. (1968). Nearly Best Linear Unbiased Estimation of the Location and Scale Parameters of the Type II Extreme Value Distribution by the Use of Order Statistics (GRE/Math/68-10).
[S7] Rodgers, Joe W. (1968). Nearly Best Linear Unbiased Estimation of the Mean and Standard Deviation of the Logistic Distribution (GRE/Math/68-12). AD 685115.
[S8] Stump, Francis B. (1968). Nearly Best Linear Unbiased Estimation of the Location and Scale Parameters of the Weibull Probability Distribution by the Use of Order Statistics (GRE/Math/68-13). AD 685116.
[S9] Davis, Robert I. (1969). Nearly Best Linear Conditional Unbiased Estimation of the Location and Scale Parameters of the Weibull Probability Distribution by the Use of Order Statistics (GRE/Math/69-1).
[S10] Hunter, Stephen A. (1969). Nearly Best Conditional Linear Unbiased Estimation of the Mean and Standard Deviation of the Logistic Distribution (GRE/Math/69-3).
[S11] Jonson, Edward C. (1969). Conditional Nearly Best Linear Estimation of the Location and Scale Parameters of the Cauchy Distribution by the Use of Censored Order Statistics (GRE/Math/69-4).
[S12] Kingcaid, Milton D. (1969). Unbiased Nearly Best Linear Conditional Estimates of the Scale and Location Parameters of the Extreme Value Distributions by Use of Order Statistics (GRE/Math/69-5).
[S13] Socolofsky, James L. (1969). Nearly Best Linear Unbiased Conditional Estimates of the Location and Scale Parameters of the Normal Distribution by Use of Order Statistics (GRE/Math/69-8).
[S14] Criscimagna, Ned H. (1970). Nearly Best Linear Invariant Conditional Estimators of the Location and Scale Parameters of the Normal Distribution by Use of Order Statistics (GRE/Math/70-1).
[S15] Davis, Robert L. (1970). Nearly Best Linear Conditional Unbiased Estimation of the Location and Scale Parameters of the Gamma Probability Distribution by the Use of Order Statistics (GRE/Math/70-2).
[S16] Dunigan, John M. (1970). Nearly Best Linear Invariant Conditional Estimation of the Mean and Standard Deviation of the Logistic Distribution (GRE/Math/70-3).
[S17] Elkins, Robert W. (1970). Nearly Best Linear Invariant Conditional Estimation of the Location and Scale

Parameters of the Weibull Probability Distribution by the Use of Order Statistics (GRE/Math/70-4).

[S18] Gray, Robert M. (1970). Nearly Best Linear Invariant Conditional Estimation of the Scale and Location Parameters of the Gamma Probability Distribution (GRE/Math/70-5).

[S19] Niles, John R. (1970). Best Linear Invariant Estimation of the Location and Scale Parameters of the Weibull Distribution (GRE/Math/70-6).

[S20] Widenhouse, William C. (1970). Conditional Nearly Best Linear Estimation of the Location and Scale Parameters of the First Extreme-Value Distribution (GRE/Math/70-7).

OPTIMAL DESIGNS AND SPLINE REGRESSION

W. J. Studden*

Purdue University

This paper may be divided roughly into two parts. One part consists of an elementary discussion of splines. The second part considers two types of optimal designs: D and $I_\sigma$ optimal designs. Section 1 is of an introductory nature.

The splines are considered in section 2 while sections 3 and 4 are devoted to the D and $I_\sigma$ optimal designs. In sections 5 & 6 we consider some examples and computational procedures for obtaining optimal designs. The paper is mainly expository, however it is restricted to a very small portion of design theory. The applications in section 5 are new.

## §1. Introduction

The design problem under discussion is as follows. Let $f' = (f_1, \ldots, f_m)$ denote an m-vector of continuous functions defined on a compact set X. The points of X are referred to as the possible levels of feasible experiments and the variable $x \in X$ is sometimes called the control variable. For each level $x \in X$ some experiment may be performed whose outcome $Y(x)$ is a random observation with mean value

$$E\,Y(x) = \sum_{i=1}^{m} \theta_i\, f_i(x) \tag{1.1}$$

and variance $\sigma^2$ independent of x. The simplest situation

*This work was supported by the NSF grant GP 20306. Reproduction is permitted in whole or in part for any purposes of the United States Government.

is, say, where X is an interval of the real line, $f_1(x) \equiv 1$, $f_2(x) = x$ and $E\,Y(x) = \theta_1 + \theta_2 x$. The functions $f_1, \ldots, f_m$ are called the regression functions and are known to the experimenter. The regression coefficients or parameters $\theta_1, \ldots, \theta_m$ and $\sigma^2$ are unknown. On the basis of N uncorrelated observations we wish to estimate some function of the parameters $\theta_1, \ldots \theta_m$.

An experimental design specifies a probability measure $\mu$ (usually discrete) on X. The associated experiment involves taking observations at the level x proportional to $\mu$. Thus if $\mu$ assigns mass $p_1, \ldots, p_r$ to $x_1, \ldots, x_r$ and $Np_i = n_i$ are integers the experimenter takes $n_i$ observations at $x_i$. Designs with $Np_i$ not equal to an integer can in practice only be approximated.

If the unknown parameter vector $\theta' = (\theta_1, \ldots, \theta_m)$ is estimated by least squares then the covariance matrix of the estimates $\hat{\theta}$ is given by

$$E(\hat{\theta} - \theta)\ (\hat{\theta} - \theta)' = \frac{\sigma^2}{N} M^{-1}\ (\mu) \tag{1.2}$$

The matrix

$$M(\mu) = \int f(x)\ f'(x)\ d\mu(x)$$

is called the (normalized) information matrix of the design $\mu$. The design problem we consider is to choose $\mu$ so that $M(\mu)$ is large in some sense.

Further discussion of the design problem considered here can be found in Karlin and Studden (1966), Kiefer (1959) and Kiefer and Wolfowitz (1960).

In the following we are interested in two problems 1) maximize the determinant of $M(\mu)$, denoted by $|M(\mu)|$, 2) minimize the quantity $L[M^{-1}(\mu)]$ where L is a non-negative linear functional on the set of positive semidefinite matrices, i.e.

$$L(\alpha D_1 + (1-\alpha)\ D_2) = \alpha L(D_1) + (1-\alpha)\ L(D_2) \tag{1.3}$$

$$0 \leq \alpha \leq 1,\ L(D) \geq 0 \text{ for } D \geq 0.$$

The first of these problems was considered by Kiefer and Wolfowitz (1960); the second by Fedorov (1971).

## §2. Spline Functions

The term spline usually refers to a "piecewise polynomial". Here the space X is an interval [a,b]. The interval [a,b] is divided into k + 1 pieces by k "knots" $\xi_1, \ldots, \xi_k$ where $\xi_0 = a < \xi_1 < \ldots < \xi_k < b = \xi_{k+1}$. A function $s(x)$ on [a,b] is called a spline if $s(x)$ is equal to a polynomial on $(\xi_i, \xi_{i+1})$ (different on each interval) and satisfies certain differentiability conditions at the points $\xi_1, \ldots \xi_k$.

The simplest case stipulates the s(x) is linear on each interval $(\xi_i, \xi_{i+1})$ $i = 0,1, \ldots ,k$ and is continuous at each $\xi_i$, $i = 1, 2, \ldots ,k$. We thus have a polygonal line segment. For the quadratic case we may consider s(x) to be quadratic on each interval with possibly continuity and also differentiability at each $\xi_i$. Discontinuities of the second derivative of g(x) are allowed at each $\xi_i$.

Generally a function which is equal to a polynomial of degree at most n on each interval and has n-1 continuous derivatives at $\xi_i$ can be written in the form

$$\sum_{i=0}^{n} \alpha_i x^i + \sum_{i=1}^{k} \beta_i (x-\xi_i)_+^n \qquad (2.1)$$

where $(x)_+^n = x^n$ for $x \geq 0$ and is zero for $x < 0$. Thus the term $(x-\xi_i)_+^n$ is zero unless $x > \xi_i$. Variable degrees of differentiability can be allowed at $\xi_i$ by using terms $(x-\xi_i)_+^r$ for $r < n$, however we shall not consider those here.

Spline polynomials have received considerable attention from mathematicians, working especially in approximation theory. These functions seem to be extremely suitable for interpolating or approximating data in real world situations since in many cases the underlying functional from s(x) is different on different parts of X. For example if s(x) denotes the distance or path a rocket travels in time

$t = x$, intermittent auxillary rockets may change the form of $s(x)$.

The spline polynomials are the least oscillatory functions for interpolating data. For example, the differentiable function $s(x)$ which interpolates the values $y_i$ at $\xi_i$ (i.e. $s(\xi_i) = y_i$) and minimizes

$$\int (s''(x))^2 dx$$

is a cubic spline of the form (2.1) with knots at $\xi_i$ (and is linear below $\xi_1$ and above $\xi_k$).

In using the functions

$$1\ x_1,\ \ldots\ ,x^n,\ (x-\xi_1)_+^n,\ \ldots\ ,(x-\xi_k)_+^n \tag{2.2}$$

one is sometimes interested in knowing for which set of x values $x_1, x_2, \ldots, x_{n+k+1}$ can one interpolate an arbitrary set of ordinates $y_1, y_2, \ldots, y_{n+k+1}$ using a unique linear combination of the functions (2.2). This is the case if and only if

$$x_i < \xi_i < x_{n+i+1}\quad ,\ i = 1,2,\ \ldots\ ,k \tag{2.3}$$

This result says that we cannot overburden any given interval with two many $x_i$ values. For example, if $n = 1$ and $k = 2$, we use $1,\ x,\ (x-\xi_1)_+,\ (x-\xi_2)_+$ and the inequalities (2.3) say that $x_1, x_2, x_3, x_4$ must satisfy $x_1 < \xi_1 < x_3$ and $x_2 < \xi_2 < x_4$.

For further and more complete discussion of splines we refer the reader to Rice (1968), Rivlin (1969) and Schoenberg (1969).

## §3. Optimal Designs

In this section we discuss two types of problems;

1. Maximize $|M(\mu)|$ with respect to $\mu$.

2. Minimize $L[M^{-1}(\mu)]$ where L is a linear functional

on the set of positive semidefinite matrices such that $L(A) \geq 0$ for $A \geq 0$; i.e. A positive semidefinite.

A solution to problem 1 is called a D-optimal design and a solution to problem 2 an L-optimal design. We are particularly interested in certain special L-optimal designs which we denote by $I_\sigma$-optimal designs. These will be considered near the end of this section. For the D-optimal designs we have the following "Equivalence Theorem" of Kiefer and Wolfowitz (1960) which shows that the D-optimal and minimax designs are equivalent.

Theorem 3.1 The conditions

(i) $\mu^*$ maximizes $|M(\mu)|$

(ii) $\mu^*$ minimizes $\sup_{x \in X} f'(x) M^{-1}(\mu) f(x)$

(iii) $\sup_{x \in X} f'(x) M^{-1}(\mu) f(x) = m$

are equivalent. The set B of all $\mu$ satisfying these conditions is convex and closed and $M(\mu)$ is the same for all $\mu \in B$.

We recall here that $M(\mu)$ is the information matrix and $f'(x) M^{-1}(\mu) f(x)$ is proportional to the variance of the estimate of the regression function at the point x.

Recently V. V. Fedorov has generalized these results to problem 2 and has shown the following.

Theorem 3.2 The conditions

(i) $\mu^*$ minimizes $L[M^{-1}(\mu)]$

(ii) $\mu^*$ minimizes $\sup_{x \in X} L[M^{-1}(\mu) f(x) f'(x) M^{-1}(\mu)]$

(iii) $\sup_{x \in X} L[M^{-1}(\mu) f(x) f'(x) M^{-1}(\mu)] = L[M^{-1}(\mu^*)]$

are equivalent. The set of designs satisfying (i), (ii),

or (iii) is convex.

The second problem is the problem of "quadratic loss". That is, any function L satisfying the required conditions can be put in the form $L(M^{-1}) = \text{tr } M^{-1} C$ where C is positive semidefinite and tr denotes the trace of a matrix. The term quadratic loss is used here since the expected value of $(\hat{\theta} - \theta)' C(\hat{\theta} - \theta)$ is proportional to $\text{tr } M^{-1} C$. The design giving information matrix M is used to obtain the estimates $\hat{\theta}$. Thus, minimizing the expected value of the quadratic form $(\hat{\theta} - \theta)' C(\hat{\theta} - \theta)$ is equivalent to minimizing $L(M^{-1}) = \text{tr } M^{-1} C$.

Both of these theorems may be proven using a variational technique. We consider the design

$$\mu_0 = (1-\alpha)\mu^* + \alpha\mu_x$$

where $\mu_x$ concentrates all of its mass at the single point x. The proof consists of showing that $\mu^*$ minimizes $L(M^{-1}(\mu))$ if and only if the derivative of

$$g(\alpha) = L[M^{-1}((1-\alpha)\mu^* + \alpha\mu_x)]$$

is $\geq 0$ for $\alpha = 0$ for all x. This in turn is equivalent to part (ii) of theorem

We are particularly interested in those L which are invariant under a basis change of the regression functions $f_1, f_2, \ldots, f_m$. Further we would like explicit expressions or characterizations of the optimal designs for the case of spline regression.

One of the invariant functionals L for Theorem 3.2 is the integral of the variance of the response surface estimate. Thus, if $d\sigma$ denotes a measure on X (or possibly on a larger domain) then

$$\int f'(x) M^{-1}(\mu) f(x) d\sigma(x) = \int \text{tr } M^{-1}(\mu) f(x) f'(x) d\sigma(x)$$
$$= \text{tr } M^{-1}(\mu) M(\sigma) = L[M^{-1}(\mu)]$$

For fixed $\sigma$ we wish to minimize this quantity with respect to the design $\mu$.

Note that the expression $\operatorname{tr} M^{-1}(\mu)\, M(\sigma)$ is invariant under basis change of the regression functions. In this case if we know that the minimizing $\mu = \mu^*$ concentrates its mass on m points $x_1^*, x_2^*, \ldots, x_m^*$ (m = the number of regression functions) then we can use as a basis the Lagrange functions $\ell_i(x)$ defined by the conditions $\ell_i(x_j) = \delta_{ij}$, $i,j = 1, \ldots, m$.

Lemma (Fedorov) If for given $\sigma$ the design $\mu^*$ minimizing $L[M^{-1}(\mu)]$ concentrates mass on $x_1^*, \ldots, x_m^*$ then the corresponding weights are proportional to

$$K_{ii}^{\frac{1}{2}} \quad \text{where } K_{ii} = \int \ell_i^2(x)\, d\sigma(x).$$

Proof: This follows by noting that for the Lagrange basis

$$\operatorname{tr} M^{-1}(\mu)\, M(\sigma) \geq \sum_{i\ 1}^{m} \frac{K_{ii}}{p_i}$$

Schwarz's inequality then gives

$$\sum_i \frac{K_{ii}}{p_i} \geq \left( \sum K_{ii}^{\frac{1}{2}} \right)^2$$

with equality only if $p_i = c\, K_{ii}^{\frac{1}{2}}$.

Note that if the $I_\sigma$-optimal design concentrates on m points $x_1^*, \ldots, x_m^*$ then the design problem reduces to minimizing

$$\sum_i \left( \int \ell_i^2(x)\, d\sigma \right)^{\frac{1}{2}}$$

with respect to $x_1^*, x_2^*, \ldots, x_m^*$. For the minimax design the corresponding expression is

$$\max_{x \in X} \sum_i \ell_i^2(x).$$

## §4. Comparison of D and $I_\sigma$ Optimal Designs

In comparing the two types of designs we shall assume that both designs are concentrated on the same set of points $x_1^*, \ldots x_m^*$ and that $X = [-1,1]$. If $\mu$ concentrates mass $p_i$ on $x_i^*$, $i = 1, \ldots ,m$ and $\ell_i(x)$ denotes the Lagrange functions corresponding to $x_1^*, \ldots ,x_m^*$ then

$$f'(x)\, M_f^{-1}(\mu)\, f(x) = \ell'(x)\, M_\ell^{-1}(\mu)\, \ell(x)$$

$$\sum_{i=1}^{m} \frac{\ell_i^2(x)}{p_i}$$

Here we denote the information matrix for the basis $f = (f_1, \ldots ,f_m)$ by $M_f$ and for the Lagrange basis by $M_\ell$. For the Lagrange basis the matrix $M_\ell$ is diagonal with diagonal elements $p_1, \ldots ,p_m$. For the D-optimal design $\mu^*$ on $x_1^*, \ldots ,x_m^*$ the weights are equal and

$$f'(x)\, M_f^{-1}(\mu^*)\, f(x) = m \sum_i \ell_i^2(x).$$

Moreover, this function reaches its maximal value m at the points $x_i^*$.

If the weights $p_i$ at $x_i^*$ are not taken equal then $f'(x)\, M^{-1}(\mu)\, f(x)$ is raised at those $x_i$ for which $p_i < \frac{1}{m}$ and lowered if $p_i > \frac{1}{m}$. For $\sigma$ uniform on $[-1,1]$ the $I_\sigma$-design will produce variances $f'(x)\, M^{-1}\, f(x)$ which are low in the middle and higher at the ends in order to minimize the integral.

For a given $\sigma$ the minimum value of

$$\int f'(x)\, M^{-1}\, f(x)\, d\sigma(x) = \operatorname{tr} M^{-1}(\mu)\, M(\sigma)$$

will be

$$\left(\sum K_{ii}^{\frac{1}{2}}(\sigma)\right)^2 \quad \text{where } K_{ii}(\sigma) = \int \ell_i^2(x)\, d\sigma(x) \quad .$$

The optimal weights are proportional to $K_{ii}^{\frac{1}{2}}$. Using the D-optimal design with equal weights gives a value

$$\mathrm{tr}\, M^{-1}(\mu)\, M(\sigma) = m \sum_i K_{ii}(\sigma) \quad .$$

Thus we should compare

$$\left( \sum K_{ii}^{\frac{1}{2}}(\sigma) \right)^2 \quad \text{with} \quad m \sum_i K_{ii}(\sigma),$$

the former of course being the smaller of the two.

§5. Examples

All of the examples below are concerned with polynomial or polynomial spline regression.

Example 1. Polynomial regression. It has been known for sometime [see Guest (1958)] that if

$$f_i(x) = x^{i-1}, \quad i = 1,2, \ldots ,n+1$$

and $X = [-1,1]$ then the D-optimal design concentrates equal mass on the zeros $x_0* = -1 < x_1* < \ldots < x_{n-1}* < x_n* = 1$ of $(1 - x^2)\, P'_n(x)$ where $P_n$ is the nth Legendre polynomial, orthogonal on $[-1,1]$ to the uniform measure.

Recently, Fedorov has shown that the(I-optimal design for the uniform measure $d\sigma = dx$ is also supported by this same set of points. The weights in this case can be shown to be proportional to $|P_n(x_i*)|^{-1}$. For the case $n = 2$ this gives weight .25, .50, .25 to -1, 0, 1. For $n=3$ the design has weight .154, .346, .346, .154 at the points -1, -.447,+.447, +1. One can choose the measure $\sigma$ with mass outside of X. For example, if $\sigma$ is uniform on $(-a,a)$ and $n = 2$ the resulting design is again on -1, 0, 1 with weights proportional to the square roots of

$$\frac{1}{4}\left(\frac{1}{5} + \frac{1}{3a^2}\right) \qquad \frac{1}{5} - \frac{2}{3a^2} + \frac{1}{a^4} \qquad \frac{1}{4}\left(\frac{1}{5} + \frac{1}{3a^2}\right)$$

Note that the weights are the same for $a = 1$ and $a \to \infty$.

Example 2. Linear-splines. Here are considered the set of regression functions

$$1, x, (x - \xi_1)_+ \ldots (x - \xi_k)_+$$

on the interval $[-1,1]$ where $-1 = \xi_0 < \xi_1 < \ldots < \xi_k < \xi_{k+1} = 1$. This example is fairly easy to work with since we may restrict ourselves to designs concentrating mass only on the $k + 2$ points $\xi_0, \xi_1, \ldots, \xi_k, \xi_{k+1}$. With the aid of the Lemma in §3 it is fairly easy to show that the $I_\sigma$-optimal design for $\sigma$ uniform on $[-1,1]$ has weights proportional to

$$(\xi_1-\xi_0)^{\frac{1}{2}}, (\xi_2-\xi_0)^{\frac{1}{2}}, (\xi_3-\xi_1)^{\frac{1}{2}}, \ldots, (\xi_{k+1}-\xi_{k-1})^{\frac{1}{2}}, (\xi_{k+1}-\xi_k)^{\frac{1}{2}}$$

Referring to the discussion in §4 we note that if we have a pair of intervals close together the regression variance at the middle $\xi_i$ will increase considerably. For the case where the $\xi_i$ are equally spaced the weights are proportional to $1 : \sqrt{2} : \sqrt{2} : \ldots : \sqrt{2} : 1$. For $k = 1$ this gives weights .293, .414, and .293 (approx.) on -1, 0, 1.

Example 3. Quadratic and Cubic Splines. For quadratic splines we consider the regression functions

$$1, x, x^2, (x - \xi_1)_+^2, \ldots, (x - \xi_k)_+^2$$

on the interval $[-1,1]$. It is known (see Studden and Van Arman (1969)) that we can restrict ourselves to designs concentrating mass at $m = k + 3$ points; two of which are the endpoints and one in each of the intervals $(\xi_i, \xi_{i+1})$, $i = 0,1, \ldots, k$. For $k = 1$ the following table indicates the optimal points of allocation for the D-optimal design. The design is on $-1$, $x_1$, $x_2$, 1 and the weights are all equal.

| $\xi$ | 0 | .2 | .4 | .6 | .8 |
|---|---|---|---|---|---|
| $x_1$ | -.390 | -.312 | -.239 | -.166 | -.089 |
| $x_2$ | +.390 | +.476 | .573 | .687 | .825 |

Further calculations show that $x_1$ is always less than zero and approaches zero as $\xi \to 1$.

For k = 1 the I-optimal design for the uniform measure is given below. The design is on points $x_0 = -1$, $x_1, x_2, x_3$ 1 with weights $p_0$, $p_1$, $p_2$, $p_3$.

| $\xi$ | $x_1$ | $x_2$ | $p_0$ | $p_1$ | $p_2$ | $p_3$ |
|---|---|---|---|---|---|---|
| 0 | -.400 | -.400 | .164 | .336 | .336 | .164 |
| .2 | -.325 | .481 | .176 | 356 | 317 | .151 |
| .4 | -.253 | .574 | .187 | .378 | .298 | .137 |
| .6 | -.180 | .684 | .200 | .403 | .280 | .117 |
| .8 | -.099 | .822 | .217 | .435 | .260 | .088 |

Generally speaking the points $x_1$ and $x_2$ are about the same. For $\xi$ less than about .5 they shift slightly toward the endpoints $\pm 1$. The weights $p_i$ however on the interior points $x_1$ and $x_2$ become considerably heavier than on the endpoints.

The results for the quadratic case using $k \geq 1$ are somewhat similar. For example if $\xi_1 = -.3$ and $\xi_2 = .3$ the D-optimal design has equal mass on -1, -.569, 0, 569,1.

For the cubic we take

$$1, x, x^2, x^3, (x - \xi)_+^3 \quad , \quad -1 \leq x \leq 1$$

and consider only the D-optimal design. Our computations again show that the number of points used in the D-optimal design is the same as the number of regresssion functions, in this case m = 5. (This has not been proven and does not follow as in the quadratic case). Here we label the points -1, $x_1$, $x_2$, $x_3$, 1.

| $\xi$ | 0 | .2 | .4 | .6 | .8 |
|---|---|---|---|---|---|
| $x_1$ | -.629 | -.584 | -.547 | -.515 | -.484 |
| $x_2$ | 0 | .104 | .193 | .273 | .352 |
| $x_3$ | .629 | .679 | .733 | .796 | .877 |

We note, as expected, that the middle points $x_2$ is not equal to $\xi$.

## §6. Computational Procedures

Let the design $\mu$ concentrate mass $p_0, p_1, \ldots, p_{r+1}$ at the points $x_0, x_1, \ldots, x_{r+1}$. We shall restrict attention again to polynomial or polynomial spline regression on [-1,1]. In this case we know that the optimal designs have $x_0 = -1$ and $x_{r+1} = +1$. We wish to maximize $|M|$ or minimize $\mathrm{tr}\, M^{-1} C$ with respect to $x_1, x_2, \ldots, x_r$ and $p_0, p_1, \ldots p_{r+1}$ subject to the condition $\Sigma p_i = 1$. If $f'(x) = (f_1(x), \ldots, f_m(x))$ and $g(x)$ denotes the corresponding vector of derivatives one can easily show that for minimizing $\mathrm{tr}\, M^{-1} C$ the $x_i$ and $p_i$ are solutions of the equations

$$\begin{aligned} f'(x_i) M^{-1}(\mu) C M^{-1}(\mu) f(x_i) &= \lambda, \quad i = 0, \ldots, r+1 \\ f'(x_i) M^{-1}(\mu) C M^{-1}(\mu) g(x_i) &= 0, \quad i = 1, \ldots, r \\ \Sigma p_i - 1 &= 0 \end{aligned} \tag{6.1}$$

Here $\lambda$ is an unknown multiplier equal to $\lambda = \mathrm{tr}\, M^{-1}(\mu^*) C$. For maximizing $|M|$ we have the additional information that $f'(x) M^{-1}(\mu^*) f(x) \leq m$ and the corresponding equations are

$$\begin{aligned} f'(x_i) M^{-1}(\mu) f(x_i) &= m, \quad i = 0, 1, \ldots, r+1 \\ f'(x_i) M^{-1}(\mu) g(x_i) &= 0, \quad i = 1, \ldots, r \end{aligned} \tag{6.2}$$

These equations are for the most part impossible to solve by hand, however, standard computer routines seem to give answers fairly quickly at least for r up to four or five.

In obtaining optimal designs the main difficulty is in choosing the points $x_1, \ldots, x_m$. The weights $p_1, \ldots, p_m$ are usually easy to determine. Recently, Federov (1971) (see also Wynn (1970)) has investigated the problem of determining the optimal $\mu^*$ by iterative methods. We shall indicate the procedure for the D-optimal designs. At the kth stage we have a design $\mu_k$. We determine $x_k$ so that

$$f'(x_k) M^{-1}(\mu_k) f(x_k) = \sup_{x \in X} f'(x) M^{-1}(\mu_k) f(x)$$

The new $\mu_{k+1}$ is then

$$\mu_{k+1} = \alpha_k \mu_k + (1 - \alpha_k)\mu_{x_k}$$

where $\mu_{x_k}$ concentrates all of its mass at $x_k$. The scalar $\alpha_k$ is given by

$$\alpha_k = \frac{d_k - m}{[d_k - 1]m} \quad \text{where } d_k = f'(x_k)\, M^{-1}(\mu_k)\, f(x_k).$$

The procedure converges rather slowly at the rate $k^{-1}$. However, in carrying out the procedure one can proceed from

$$M^{-1}(\mu_k) \text{ to } M^{-1}(\mu_{k+1})$$

without recalculating this m x m inverse. Moreover, the problem is reduced essentially to finding

$$\sup_{x \in X} f'(x)\, M^{-1}(\mu_k)\, f(x).$$

References

[1] Fedorov, V.V. (1971), Theory of Optimal Experiments (to appear), Academic Press, New York.

[2] Guest, P. G. (1958), The spacing of observations in polynomial regression, Ann. Math. Statist., Vol. 29, p. 294-299.

[3] Karlin, S. and Studden, W. J. (1966), Optimal experimental designs, Ann. Math. Statist. 37, p. 783-815.

[4] Kiefer, J. (1959), Optimum experimental designs, J. Roy. Statist. Soc. Ser. B, 21, p. 273-319.

[5] Kiefer, J. and Wolfowitz, J. (1960), The equivalence of two extremum problems, Can. J. Math. 12, p. 363-366.

[6] Rice, J. R. (1968), The Approximation of Functions, Vol. 2, Addison-Wesley, Reading, Mass.

[7] Rivlin, T. J. (1969) An Introduction to the Approximation of Functions, Blaisdell Publishing Co., Waltham, Mass.

[8] Schoenberg, I. J. (1969) Approximations with Special

Emphasis on Spline Functions, Academic Press, N.Y.
[9] Studden, W. J. and D. J. Van Arman (1969), Admissible designs for polynomial spline regression, Ann. Math. Statist. 40, 1557-1569.
[10] Wynn, H. P. (1970) The Sequential generation of D-optimal experimental designs, Ann. Math. Statist., Vol. 41, p. 1655-1664.

ISOTONIC APPROXIMATION*

R. E. Barlow and V. Ubhaya

Operations Research Center
University of California, Berkeley

## 1. Introduction

This is a survey of recent work in the field of isotonic regression from a slightly different point of view. H. D. Brunk et al (1955) and C. van Eeden (1956) observed that the solution to certain maximum likelihood estimation problems under an ordering restriction was the same as the solution to the following approximation problem:

$$\text{Minimize } \|g-f\| \text{ , subject to } f \in K \tag{1.1}$$

where the function $g = (g_1, g_2, \ldots, g_k)$ is the given unrestricted maximum likelihood estimate, $K$ is the cone of isotonic or order preserving functions (i.e. $i \leq j$ implies $f_i \leq f_j$) and

$$\|g-f\| = \sqrt{\sum_{i=1}^{k} [g_i - f_i]^2 w_i} \quad . \tag{1.2}$$

The weights $w_1, w_2, \ldots, w_k$ are usually proportional to sample size. The solution, $g^*$, called the <u>isotonic regression</u> can be simply characterized as

*This research was supported in part by the Office of Naval Research under Contract N00014-69-A-0200-1036 and by the National Science Foundation under Grant GP-29123 with the University of California.

$$g_i^* = \underset{t \geq i}{\text{Min}} \quad \underset{s \leq i}{\text{Max}} \quad \frac{\sum_{j=s}^{t} g_j w_j}{\sum_{j=s}^{t} w_j}$$

in the complete (increasing) ordering case. This characterization and other useful properies of g* can be used to obtain asymptotic properties of g* as a function of sample size (see [1]).

Although maximum likelihood was the original motivating principle used, one could start with the approximation problem(for suitable g) and study the properties of estimators obtained in this way. This is, of course, the principle of least squares if we use the norm defined by (1.2). Barlow and van Zwet (1970) exploited this idea to obtain estimators for a nonparametric problem which asymptotically improve on the restricted maximum likelihood estimators.

We consider norms more general than (1.2) and characterize their solution. This was done by van Eeden (1958), Robertson and Waltman (1968) and by Brunk and Johansen (1970) in increasing generality for the $L_p$ $(p > 1)$ norm. The sup norm case was treated by V. Ubhaya (1971) in his Ph. D. thesis. We discuss applications to problems of failure rate estimation. The duality theory is also presented.

## 2. Isotonic Estimation

To motivate the discussion, consider the following problem. Given k normal distributions $N(\mu_i, \sigma_i)$, $(i = 1,2, \ldots ,k)$, let $\bar{x}_i$ be the sample mean from the i-th distribution and $n_i$ the sample size. We suppose $\sigma_i$ $(i = 1,2, \ldots , k)$ are known and wish to maximize the likelihood subject to an order restriction on the means; i.e.

$$\text{Maximize} \quad \prod_{i=1}^{k} \frac{\sqrt{n_i}}{\sqrt{2\pi}\,\sigma_i} \quad \exp\left[-\frac{n_i}{2} \quad (\bar{x}_i - \mu_i)^2/\sigma_i^{\,2}\right]$$

subject to $\mu_i \leq \mu_j$, when $i \leq j$, where " $\leq$ " is a partial ordering on $\Omega = \{1, 2, \ldots, k\}$. The partial ordering is provided by some *a priori* knowledge concerning the means of the distribution. This is equivalent to the problem

$$\text{Minimize} \sum_{i=1}^{k} [\bar{x}_i - \mu_i]^2 w_i \tag{2.1}$$

subject to $\mu_i \leq \mu_j$, when $i \leq j$, where $w_i = n_i/\sigma_i^2 > 0$. To characterize the solution, let $L \subset \Omega$ be a lower set with respect to $\leq$ if $i \in L$ and $j \leq i$ implies $j \in L$. A subset $U$ of $\Omega$ is an upper set if $i \in U$ and $i \leq j$ implies $j \in U$.

The solution to (2.1) can be characterized by

$$\mu_i^* = \underset{\substack{L \\ i \in U \cap L}}{\text{Min}} \; \underset{U}{\text{Max}} \; \sum_{j \in U \cap L} \bar{x}_j w_j \Big/ \sum_{j \in U \cap L} w_j \tag{2.2}$$

In addition to being the maximum likelihood estimator, $\mu^* = (\mu_1^*, \mu_2^*, \ldots, \mu_k^*)$ is consistent and Bayes against the restricted natural conjugate prior [see [1]]. As Brunk [5] has pointed out, $\mu^*$ is the conditional expectation of $\bar{x} = (\bar{x}_1, \bar{x}_2, \ldots, \bar{x}_k)$ with respect to the sigma lattice consisting of the upper sets. Considered as an operator on $L_2$, the solution to (2.1) also has the distance reducing property; i.e. if $\bar{x} = (\bar{x}_1, \bar{x}_2, \ldots, \bar{x}_k)$ and $\bar{y} = (\bar{y}_1, \bar{y}_2, \ldots, \bar{y}_k)$ are given vectors and $\mu^*(\nu^*)$ is the isotonic regression with respect to $\bar{x}$ $(\bar{y})$, then

$$||\mu^* - \nu^*|| \leq ||\bar{x} - \bar{y}|| \tag{2.3}$$

where $||.||$ is the $L_2$ norm with respect to the discrete measure corresponding to the given weights [Brunk(1965)].

Competing estimators to $\mu^*$ can be obtained by considering the $L_p$ $(p \geq 1)$ problem

$$\text{Minimize} \sqrt[p]{\sum_{i=1}^{k} |\bar{x}_i - \mu_i|^p \; w_i} \tag{2.4}$$

subject to $\mu_i \leq \mu_j$, when $i \leq j$. The solution for $p \neq 2$ ($p \geq 1$) does not have an explicit representation. However, an implicit representation similar to (2.2) can be given. Let $M_B$ be the minimizer in a of

$$f_B(a) = \sum_{i \in B} |\bar{x}_i - a|^p w_i .$$

Then

$$\hat{\mu}_i = \underset{\substack{L \\ i \in U \cap L}}{\text{Min}} \; \underset{U}{\text{Max}} \quad M_{U \cap L}$$

solves (2.4). [see van Eeden (1958), p.28] Brunk and Johansen (1970) call this solution a generalized Radon-Nikodym derivative with respect to a signed measure based on the derivative with respect to a of $f_B(a)$. The solution is unique. In contrast with the case $p = 2$, little is known about the properties of this estimator.

Letting $p \to \infty$ we obtain an isotonic problem based on the uniform norm, namely

$$\text{Minimize} \left[ \underset{1 \leq i \leq k}{\text{Max}} |\bar{x}_i - \mu_i| \right] \tag{2.5}$$

subject to $\mu_i \leq \mu_j$, when $i \leq j$. However, the solution is no longer unique.

We consider a slightly more general problem by including weights in the objective function so that our problem becomes

$$\text{Minimize} \left[ \underset{1 \leq i \leq k}{\text{Max}} |\bar{x}_i - \mu_i| \, w_i \right]$$

subject to $\mu_i \leq \mu_j$, when $i \leq j$. V. Ubhaya [11] shows that

$$\underset{\mu \in K}{\text{Min}} \left[ \underset{1 \leq i \leq k}{\text{Max}} |\bar{x}_i - \mu_i| \, w_i \right]$$

$$\underset{i \leq j}{\text{Max}} \frac{w_i \, w_j}{w_i + w_j} [\bar{x}_i - \bar{x}_j] \overset{\text{def}}{=} \theta . \tag{2.6}$$

where K is the cone of isotonic functions. In the complete (increasing) ordering case the solution to (2.5) is any function, $\hat{\mu}$, lying between $\underline{\mu}$ and $\bar{\mu}$ (i.e. $\underline{\mu}_i \quad \hat{\mu}_i \quad \bar{\mu}_i$, i 1,2, ... ,k) where

$$\underline{\mu}_i = \max_{1 \leq s \leq i} [\bar{x}_s - \theta/w_s]$$

and

$$\bar{\mu}_i = \min_{i \leq t \leq k} [\bar{x}_t + \theta/w_t] \quad .$$

Since for purposes of statistical estimation we desire only one solution, the estimator

$$\hat{\mu}_i = \tfrac{1}{2}[\underline{\mu}_i + \bar{\mu}_i] \quad ,$$

i = 1,2, ... ,k, could be used. It has the desirable distance reducing property (2.3) when the weights are equal.

## 3. Duality Theory

Let $(\Omega,\mathfrak{A},\mu)$ be a measure space. Let " $\leq$" be a partial ordering defined on $\Omega$. A real valued function f, defined on $\Omega$ is isotonic if and only if $\omega_1 \leq \omega_2$ implies $f(\omega_1) \leq f(\omega_2)$. Let $(X,\|\cdot\|)$ be a normed linear space of real valued functions defined on $\Omega$ measurable with respect to $\mathfrak{A}$. Let X* be the dual space of continuous linear functionals defined on X. It is easily verified that the set of isotontonic functions, $K \subset X$, is a convex cone. Let $K^* = \{L \mid L \in X^* \text{ and } L(f) \leq 0, \text{ for all } f \in K\}$. Then K* is the dual or polar cone. The following duality theorem is well known [cf. Dieter (1965)].

<u>Theorem.</u> (Duality Theorem). If X is a normed linear space and $K \subset X$ is a convex cone, then given $g \in X$,

$$\inf_{f \in K} \|g-f\| = \max_{L \in K^*} \frac{L(g)}{\|L\|} \tag{3.1}$$

where $K^* \subset X^*$ is the dual cone and

$$\|L\| = \sup_{f \in X} \frac{|L(f)|}{\|f\|} \quad .$$

For the case when K is a subspace of X see Theorem 2

of R. C. Buck (1965).

If the norm is the $L_2$ norm and $X = E^n$, i.e. Euclidean n-space, then (3.1) has an obvious but elegant interpretation. If the primal problem (the left hand side of (3.1)) is to find that point in K closest to g, then the dual problem is to find the hyperplane in K* whose directed distance from g is greatest [cf. Sinden (1963)].

An interesting feature of the solution, g*, of the isotonic regression problem

$$\inf_{f \in K} \|g - f\|$$

when $\|.\|$ is the $L_2$ norm is that it also solves the problem

$$\inf_{g-f \in K^*} \int_\Omega \Theta[f(u)]\, d\mu(u) \tag{3.2}$$

simultaneously for all convex $\Theta$. [cf. Barlow and Brunk (1971), Theorem 3.4 and Veinott [14].] In (3.2), $g - f \in K^*$ means

$$\int_U f\, d\mu \geq \int_U g\, d\mu$$

for all upper sets $U \subset \mathfrak{A}$. In the complete ordering case this corresponds to a stochastic ordering restriction, when f and g are densities.

To interpret the duality theorem for the isotonic regression problem with respect to the uniform norm, let X be the linear space of all bounded real valued functions on [a,b]. Define the norm on X by

$$\|f\| = \sup_{x \in [a,b]} |f(x)|\, w(x)$$

where $w \in X$ and $w > 0$. Generalizing (2.6) we have the duality result

$$\inf_{f \in K} \|g-f\| = \sup_{x \leq y} \frac{w(x)\, w(y)}{w(x) + w(y)} [g(x) - g(y)]. \tag{3.3}$$

[See Ubhaya, (1971).] Let

$$L(g) = g(x) - g(y)$$

for some $x < y$. Clearly, $L(f) \leq 0$, for all $f \in K$ so that $L \in K^*$. Also

$$\| L \| = \frac{w(x) + w(y)}{w(x)w(y)} \quad .$$

The interesting difference between the general duality theorem (3.1) and (3.3) is that we need only take the supremum over a subset of $K^*$.

## 4. Application to Failure Rate Estimation

In many statistical studies involving failure data, biometric mortality data, and actuarial mortality data, the failure rate $r(x) = f(x)/[1 - F(x)]$ for $F(x) < 1$ (corresponding to a lifetime distribution with density f and distribution F, $F(0^-) = 0$ is of prime importance. If F is the life distribution of a person, an airplane engine, a light bulb, a machine tool, etc., then the failure rate function, $r(x)$, describes the way in which the item in question wears out. If the item is aging very rapidly the failure rate function will tend to increase rather steeply. The shape of the failure rate function in the case of equipment often suggests when appropriate maintenance actions should be taken. A problem of considerable interest, therefore, is the estimation of the failure rate function from a sample of n independent identically distributed lifetimes. This was considered in a generalized setting by Barlow and van Zwet (1970). We will follow their approach in describing the sup norm estimator.

Let

$$0 \leq X_{1:n} \leq X_{2:n} \leq \cdots \leq X_{n:n}$$

be an ordered sample from F. Let $F_n$ be the empirical distribution corresponding to our sample. Let $\rho_n$ be the initial or <u>basic</u> <u>estimator</u> for r based on

$$X_{1:n} \leq X_{2:n} \leq \cdots \leq X_{n:n} .$$

For each n, we define a <u>grid</u> on $(-\infty, \infty)$, i.e., a finite or infinite sequence

$$0 = t_{n,0} < t_{n,1} < t_{n,2} < \dots < t_{n,i} < \dots$$

In each *window* $[t_{n,j}, t_{n,j+1})$ we choose a point $x_{n,j}$ such that

$$t_{n,j} < x_{n,j} < t_{n,j+1}$$

and to each point $x_{n,j}$ we assign a nonnegative weight $w(x_{n,j})$. For convenience we abbreviate $x_{n,j}$ to $x_j$. First, we calculate

$$\theta_n = \sup_{x_i \leq x_j} \frac{w(x_i)\, w(x_j)}{w(x_i) + w(x_j)} [\rho_n(x_i) - \rho_n(x_j)]$$

For $t_{n,i} \leq x \leq t_{n,i+1}$ we call

$$\underline{r_n(x)} = \sup_{0 \leq x_s \leq x} \left[ \rho_n(x_s) - \frac{\theta_n}{w(x_s)} \right]$$

and

$$\overline{r_n(x)} = \inf_{x \leq x_t \leq \infty} \left[ \rho_n(x_t) + \frac{\theta_n}{w(x_t)} \right]$$

the lower and upper, respectively, sup norm estimators of $r(x)$. We take as our estimator

$$\hat{r}_n(x) = \tfrac{1}{2}[\underline{r_n(x)} + \overline{r_n(x)}].$$

In practice one often takes the *naive* estimator

$$\rho_n(x) = \frac{F_n(t_{n,i+1}) - F_n(t_{n,i})}{[1 - F_n(x_i)][t_{n,i+1} - t_{n,i}]}$$

for $t_{n,i} \leq x < t_{n,i+1}$ together with the weights

$$w(x_j) = [1 - F_n(x_j)][t_{n,j+1} - t_{n,j}].$$

The properties of $r_n(x)$ relative to the usual estimator

considered in Barlow and van Zwet (1970) are still being investigated.

## References

[1] Barlow, R. E., Bartholomew, D. J., Bremner, J. M., H. D. Brunk, Statistical Inference Under Order Restrictions, J. Wiley and Sons, (to appear).

[2] Barlow, R. E. and H. D. Brunk, "The Isotonic Regression Problem and Its Dual", Journal of the American Statistical Association, (to appear).

[3] Barlow, R. E. and W. R. van Zwet, "Asymptotic Properties of Isotonic Estimators for the Generalized Failure Rate Function," Nonparametric Techniques in Statistical Inference, Cambridge University Press, p. 159-176, 1970.

[4] Brunk, H. D., Ayer, M., Ewing, G. M., W. T. Reid and E. Silverman, "An Empirical Distribution Function for Sampling with Incomplete Information," Ann. Math. Statist., Vol. 26, p. 641-647, 1955.

[5] Brunk, H. D., Conditional Expectation Given a Lattice and Applications, Ann. Math. Statist., Vol. 36, p. 1339-1350, 1965.

[6] Brunk, H. D. and S. Johansen, "A Generalized Radon-Nikodym Derivative and Its Application to Estimation of Ordered Parameters," Pacific J. Math., p. 585-617, 1970.

[7] Buck, R. C., "Applications of Duality in Approximation Theory," Approximation of Functions, Elsevier Pub. Co. New York 1965, Edited by H. L. Garabedian, p. 27-42.

[8] Dieter, U. "Optimierungsaufgaben in Topologischen Vektorraumen I: Dualitatstheorie," Z. Wahrscheinlichkeitstheorie verw. Geb. 5, p. 89-117, 1966.

[9] Robertson, T. P. Waltman, "On estimating Monotone Pameters," Ann. Math. Statist., Vol. 39, p. 1030-1039, 1968.

[10] Sinden, F. W., "Duality in Convex Programming and in Projective Space," J. Soc. Indust. Appl. Math., Vol. 11, p. 535-552, 1963.

[11] Ubhaya, V. "Isotonic and Convex Optimization," Operations Research Center Report ORC 71-13, Univ. of California, Berkeley, 1971.

[12] Van Eeden, C. "Maximum Likelihood Estimation of Ordered Probabilities, " Indag. Math., Volume 18, p. 444-

455, 1956.
[13] Van Eeden, C. "Testing and Estimating Ordered Parameters of Probability distributions," (Thesis at Univ. of Amsterdam), 1958.
[14] Veinott, A. F., Jr., Least d-Majorized Network Flows with Inventory and Statistical Applications," Management Science, Vol. 17, p. 547-567, May 1971.

ASYMPTOTICALLY EFFICIENT ESTIMATION OF NONPARAMETRIC REGRESSION COEFFICIENTS

L. Weiss and J. Wolfowitz

Cornell University and University of Illinois

Abstract

For each n, $X_1(n), \ldots, X_n(n)$ are independent random variables, the density function for $X_i(n)$ being $f(x - \theta_1 - \theta_2 t_i(n))$. $\theta_1$, $\theta_2$ are unknown parameters to be estimated. $t_i(n)$ are known, and $-\infty < A \leq t_i(n) \leq B < \infty$ for all i and n.

$$\lim_{n \to \infty} \frac{1}{n} \sum_i t_i(n) = T,$$

say, and

$$\lim_{n \to \infty} \frac{1}{n} \sum_i (t_i(n) - T)^2 = S,$$

say, where $S > 0$. $f(y)$ is symmetric about zero,

$$\int_{-\infty}^{\infty} (f'(y)/f(y))^2 f(y)\, dy = J,$$

say, where $0 < J < \infty$, and $f(y)$ satisfies some straightforward regularity conditions. If $f(y)$ is known, and $Q_1(n)$, $Q_2(n)$ denote the maximum probability estimators of $\theta_1$, $\theta_2$ respectively, so that $Q_1(n)$, $Q_2(n)$ are asymptotically efficient, then the asymptotic joint distribution of

$$n^{\frac{1}{2}}(Q_1(n) - \theta_1) \quad , \quad n^{\frac{1}{2}}(Q_2(n) - \theta_2)$$

is bivariate normal, with zero means and covariance matrix

$$\begin{pmatrix} 1/J + T^2/JS & -T/JS \\ -T/JS & 1/JS \end{pmatrix}.$$

Now suppose $f(y)$ is not known, but satisfies the conditions above. Let $F(y)$ denote

$$\int_{-\infty}^{y} f(t)\, dt.$$

For a given $p$ in the open interval $(0,\frac{1}{2})$, let $H(p)$ denote

$$\frac{2}{p} f^2(F^{-1}(p)) + \int_{F^{-1}(p)}^{F^{-1}(1-p)} (f'(y)/f(y))^2 f(y)\, dy.$$

Estimators $\bar{Q}_1(n)$, $\bar{Q}_2(n)$ of $\theta_1$, $\theta_2$ respectively, are constructed, such that the joint distribution of

$$n^{\frac{1}{2}}(\bar{Q}_1(n) - \theta_1), \qquad n^{\frac{1}{2}}(\bar{Q}_2(n) - \theta_2)$$

is bivariate normal, with zero means and covariance matrix

$$\begin{pmatrix} 1/H(p) + T^2/SH(p) & -T/SH(p) \\ -T/SH(p) & 1/SH(p) \end{pmatrix}.$$

The construction uses methods described in "Asymptotically Efficient Non-Parametric Estimators of Location and Scale Parameters," by L. Weiss and J. Wolfowitz, Z. Wahr. verw. Geb., 1970.

# COMPARISONS OF ORDER STATISTICS AND OF SPACINGS FROM HETEROGENEOUS DISTRIBUTIONS[1]

Gordon Pledger and Frank Proschan

The Florida State University

## Abstract

Given two sets of independent, possibly unlike, components, conditions involving majorization are given which insure that any k-out-of-n system constructed of components in the first set will have reliability at least as great as that of a corresponding system constructed of components in the second set. Since the ordered failure times of the components represent order statistics from heterogeneous distributions, we obtain stochastic comparisons between the order statistics from one set of underlying distributions $\{F_1, \ldots, F_n\}$ and those from another set $\{F_1^*, \ldots, F_n^*\}$ under both parametric and nonparametric assumptions. As a sample result, if one vector of component hazards $(-\log[1 - F_1(t)], \ldots, -\log[1 - F_n(t)])$ majorizes a second such vector $(-\log[1 - F_1^*(t)], \ldots, -\log[1 - F_n^*(t)])$ for each $t \geq 0$, then for $k = 1, \ldots, n$, the k-th order statistic from the set $\{F_1, \ldots, F_n\}$ is stochastically larger than the k-th order statistic from the set $\{F_1^*, \ldots, F_n^*\}$. Results of this type can be used to find bounds for the reliability of a k-out-of-n system of unlike components in terms of a k-out-of-n system of like components.

Under more restrictive hypotheses, among them the as-

---

[1]Research sponsored by the Air Force Office of Scientific Research, AFSC, USAF, under Grant No. AFOSR-71-2058 and the National Science Foundation under Grant GU 2612. The United States Government is authorized to reproduce and distribute reprints for Governmental purposes notwithstanding any copyright notation hereon.

sumption that $F_1^* = \cdots = F_n^*$, both stochastic and expected value comparisons for the spacings between order statistics are obtained. It is also shown that in the case of proportional concave hazards, the successive normalized spacings from one set of heterogeneous distributions are stochastically increasing.

## 1. Introduction

A great body of theory and methods exists for order statistics and their spacings from a single underlying distribution. See for example David (1970), Sarhan and Greenberg (1962), and Pyke (1965,1970), and the references contained therein.

A far smaller set of results is available for the case of order statistics from underlying heterogeneous distributions. See for example Sen (1970). One motivation for considering underlying heterogeneous distributions arises in reliability theory, where so-called "k-out-of-n" systems are studied.

A system of n components is called a k-out-of-n system if it functions if and only if at least k components function. See Barlow and Proschan (1965), Chapter 7, and the references contained therein. Note that the time of failure of a k-out-of-n system of independent components with respective life distributions $F_1, \ldots, F_n$ corresponds to the $(n - k + 1)$th order statistic from the set of underlying heterogeneous distributions $\{F_1, \ldots, F_n\}$. With this fact in mind, we may wish to approximate the life distribution of a k-out-of-n system of stochastically unlike components by the life distribution of a k-out-of-n system consisting of stochastically like components. Thus we would be comparing order statistics from underlying heterogeneous populations with corresponding order statistics from some appropriate underlying homogeneous population.

Sen (1970) derives such comparisons, taking as the homogeneous population the equally weighted mixture of the original $F_1, \ldots, F_n$. In the present paper, we present additional comparisons involving not only order statistics but also spacings between order statistics. In some of our comparisons, the underlying heterogeneous distributions are compared against a single underlying homogeneous distribution, while in others, they are compared against another set of distributions, less heterogeneous in the sense of majorization.

## 2. Comparisons for k-out-of-n Systems and Corresponding Order Statistic Implications

The simplest comparisons of k-out-of-n systems may be made by taking fixed component reliabilities, $p_1, \ldots, p_n$, rather than time-dependent component reliabilities $\bar{F}_1(t), \ldots, \bar{F}_n(t)$. For component reliability $p_i$ we define the corresponding component hazard $R_i$ by:

$$R_i = -\log p_i \quad . \tag{2.1}$$

Many of our comparisons involve the notion of majorization. Vector $\underline{a}$ majorizes vector $\underline{b}$ (written $\underline{a} \overset{m}{>} \underline{b}$) if, possibly after reordering components, $a_1 \geq \cdots \geq a_n$, $b_1 \geq \cdots \geq b_n$,

$$\sum_{i=1}^{k} a_i \geq \sum_{i=1}^{k} b_i \quad \text{for } k = 1, \ldots, n-1,$$

and

$$\sum_{i=1}^{n} a_i = \sum_{i=1}^{n} b_i \; .$$

Throughout this paper we shall assume, without loss of generality, that each vector $\underline{a}$ is already ordered in accordance with this definition (so that $a_1 \geq \cdots \geq a_n$).

To obtain our comparisons, we also need the notion of a Schur function. A function $f(\underline{a})$ with continuous partial derivatives of first order is a Schur function if for all $i \neq j$,

$$(a_i - a_j)\left(\frac{\partial f}{\partial a_i} - \frac{\partial f}{\partial a_j}\right) \geq 0.$$

The notions of majorization and Schur function are linked by the following result of Ostrowski (1952). This result, as well as others relating to majorization, can be found in Mitrinović (1970), pp. 162-170.

Theorem 2.1. (Ostrowski) Let f be a symmetric function with continuous partial derivatives of first order. Then a necessary and sufficient condition for $f(\underline{a}) \geq f(\underline{b})$ whenever $\underline{a} \overset{m}{>} \underline{b}$ is that f be a Schur function.

Using these notions, we may state a comparison for a k-out-of-n system; we denote its system reliability $h_k(p_1, \ldots, p_n)$, as a function of component reliabilities $p_1, \ldots, p_n$.

Theorem 2.2. Let $\underline{R} = (R_1, \ldots, R_n)$ be a vector of component hazards which majorizes $\underline{R}^* = (R_1^*, \ldots, R_n^*)$, a second vector of component hazards. Then the corresponding reliabilities for a k-out-of-n system satisfy

$$h_k(\underline{p}) \geq h_k(\underline{p}^*) \quad \text{for } k = 1, \ldots, n-1, \tag{2.2a}$$

and

$$h_n(\underline{p}) = h_n(\underline{p}^*) \tag{2.2b}$$

Proof. (2.2a) It will suffice to prove for k = 1, ... n - 1, that

$$\log h_k(e^{-R_1}, \ldots, e^{-R_n})$$

is a Schur function of $(R_1, \ldots, R_n)$, since by Ostrowski's Theorem, this would imply

$$\log h_k(e^{-R_1}, \ldots, e^{-R_n}) \geq \log h_k(e^{-R_1^*}, \ldots, e^{-R_n^*}),$$

which implies (2.2a). (The function $\eta_k(\underline{R}) = -\log h_k(\underline{p})$ is called the hazard transform of the k-out-of-n system; it represents the system hazard as a function of component hazards $R_1, \ldots, R_n$. See Esary, Marshall, and Proschan (1970) for a systematic treatment of hazard transforms and their applications in reliability theory.)

Write for k = 1, ... , n,

$$\eta_k(\underline{R}) = -\log h_k(\underline{p}) = -\log \left\{ e^{-\sum_1^n R_i} \left[1 + \sum_1^n (e^{R_i} - 1) + \cdots \right.\right.$$

$$\left.\left. + \sum_{i_1 < \cdots < i_{n-k}} (e^{R_{i_1}} - 1) \cdots (e^{R_{i_{n-k}}} - 1)\right]\right\} =$$

$$\sum_1^n R_i - \log C_k, \text{ where } C_k = 1 + \sum_1^n (e^{R_i} - 1) + \cdots$$

$$+ \sum_{i_1 < \cdots < i_{n-k}} (e^{R_{i_1}} - 1) \cdots (e^{R_{i_{n-k}}} - 1) > 0$$

for $k = 1, \ldots, n-1$, and $C_k = 1$ for $k = n$.

For $k = 1, \ldots, n-1$, to show that $-\eta_k(\underline{R})$ is a Schur function, we determine

$$\frac{\partial \eta_k}{\partial R_j} = 1 - \frac{1}{C_k} \left[e^{R_j} + e^{R_j} \sum_{i \neq j} (e^{R_i} - 1) + \cdots \right.$$

$$\left. + e^{R_j} \sum_{\substack{i_1 < \cdots < i_{n-k-1} \\ i_\alpha \neq j}} (e^{R_{i_1}} - 1) \cdots (e^{R_{i_{n-k-1}}} - 1)\right].$$

Suppose $j > m$, so that $R_j \leq R_m$. Define

$$\Delta_k = \frac{\partial \eta_k}{\partial R_j} - \frac{\partial \eta_k}{\partial R_m} .$$

Then

$$\Delta_{n-1} \overset{\text{sgn}}{=} e^{R_m} - e^{R_j} \geq 0,$$

where $x \overset{\text{sgn}}{=} y$ means x and y have the same sign. A straightforward induction shows that

$$\Delta_k \overset{\text{sgn}}{=} (e^{R_m} - e^{R_j}) \sum_{\substack{i_1 < \cdots < i_{n-k-1} \\ i_\alpha \neq j,m}} (e^{R_{i_1}} - 1) \cdots (e^{R_{i_{n-k-1}}} - 1) \geq 0,$$

for k = 1, ... , n - 2. Hence $-\eta_k(\underline{R})$ is a Schur function for k = 1, ... ,n - 1, so that (2.2a) holds.

$$\text{(b)}\quad h_n(\underline{p}) = e^{-\sum_1^n R_i} = e^{-\sum_1^n R_i^*} = h_n(\underline{p}^*).$$

(We note incidentally that $-\eta_n(\underline{R})$ is a Schur function.) ||

A case of special interest occurs when

$$R_1^* = \cdots = R_n^* = \frac{1}{n} \sum_1^n R_i ,$$

or equivalently, when

$$p_1^* = \cdots = p_n^* = \left( \prod_1^n p_i \right)^{\frac{1}{n}},$$

the geometric mean of the $p_1, \ldots ,p_n$. Since we are assuming $R_1 \geq \cdots \geq R_n$, it follows that $\underline{R} \overset{m}{>} \underline{R}^*$. Thus we may apply Theorem 2.2 to obtain

<u>Corollary 2.3.</u> Let $\underline{p}$ be a vector of component reliabilities for a k-out-of-n system, and let

$$p_G = \left(\prod_1^n p_i\right)^{\frac{1}{n}} .$$

Then

$$h_k(p_1, \dots, p_n) \geq h_k(p_G, \dots, p_G) \quad \text{for } k = 1, \dots, n. \tag{2.3}$$

This bound is sharp.

Conclusion (2.3) gives a lower bound for the reliability of k-out-of-n systems of unlike components in terms of the corresponding reliability in the case of like components. The bound is sharp in the following sense. If any value

$$p > p_G = \left(\prod_1^n p_i\right)^{\frac{1}{n}}$$

is used for common component reliability, then (2.3) cannot hold for $k = 1, \dots, n$. This is a consequence of the fact that equality holds in (2.3) for $k = n$.

The following lemma provides a means of deriving new Schur functions.

Lemma 2.4. Let $f(\underline{y})$ be a Schur function, with $\frac{\partial f}{\partial y_i} \leq (\geq)\ 0$ for $i = 1, \dots, n$. Let $y_i = g(x_i)$, where g is increasing, differentiable, and concave (convex). Then $h(\underline{x}) = f(g(x_1), \dots, g(x_n))$ is a Schur function.

Proof. The conclusion follows from inspection of

$$\frac{\partial h}{\partial x_j} - \frac{\partial h}{\partial x_i} = \frac{\partial f}{\partial y_j}\frac{dy_j}{dx_j} - \frac{\partial f}{\partial y_i}\frac{dy_i}{dx_i} . \;\|$$

Using Lemma 2.4 we may make comparisons of k-out-of-n systems when components have comparable odds ratios. We define the odds ratio $r_i$ in favor of failure of component i by

$$r_i = \frac{1 - p_i}{p_i} . \tag{2.4}$$

It follows from (2.4) that

$$r_i = e^{R_i} - 1, \text{ or } R_i(r_i) = \log(1 + r_i),$$

a differentiable, increasing, concave function of $r_i$. This last identity, in conjunction with Lemma 2.4 and Theorem 2.2, yields

Theorem 2.5. If $\underline{r} \overset{m}{>} \underline{r}^*$, then

$$h_k(\underline{p}) \geq h_k(\underline{p}^*) \quad \text{for } k = 1, \dots, n.$$

Note that the conclusions of Theorem 2.2 and 2.5 are the same for $k = 1, \dots, n-1$, while the hypotheses differ. A careful examination shows that neither of the hypotheses $\underline{R} \overset{m}{>} \underline{R}^*$ and $\underline{r} \overset{m}{>} \underline{r}^*$ implies the other. Thus there may arise cases where Theorem 2.2 applies while Theorem 2.5 does not, and vice versa.

However in one important case, Theorem 2.2 yields a stronger result than does Theorem 2.5. If in Theorem 2.5 we take

$$r_1^* = \cdots = r_n^* = \frac{1}{n}\sum_1^n r_i,$$

then the corresponding common component reliability is

$$p_H = [\frac{1}{n}\sum_1^n \frac{1}{p_i}]^{-1},$$

the harmonic mean of the $p_i$. Recall that the harmonic mean $p_H \leq p_G$, the geometric mean. Thus a k-out-of-n system of components each with reliability $p_H$ is less reliable than a k-out-of-n system of components each with reliability $p_G$.

Our results thus far have involved comparisons for a fixed k-out-of-n system. Next we present a set of inequalities among system reliabilities $h_k(\underline{p})$, $k = 1, \dots, n$, for a fixed vector $\underline{p}$ of component reliabilities.

Theorem 2.6. Let $h_k(\underline{p})$ be the reliability of a k-out-of-n system of independent components with respective reliabilities $p_1, \ldots, p_n$. Then

$$h_k^2(\underline{p}) \geq h_{k-1}(\underline{p})h_{k+1}(\underline{p}) \quad \text{for } k = 2, \ldots, n-1, \qquad (2.5)'$$

i.e., for fixed $\underline{p}$, $h_k(\underline{p})$ is a Pólya frequency sequence of order 2 ($PF_2$) in the index k.

Proof. Write

$$h_k(\underline{p}) = \sum_{j=-\infty}^{\infty} e_{n-k-j}H_j,$$

where

$$e_j = \begin{cases} 0 & \text{for } j \leq -1 \\ 1 & \text{for } j = 0 \\ \displaystyle\sum_{i_1 < \cdots < i_j} \frac{(1-p_{i_1})\cdots(1-p_{i_j})}{p_{i_1}\cdots p_{i_j}} & \text{for } j \geq 1, \end{cases}$$

and

$$H_j = \begin{cases} 0 & \text{for } j < 0 \\ \displaystyle\prod_1^n p_i & \text{for } j \geq 0. \end{cases}$$

$\{H_j\}$ is clearly $PF_2$ and $\{e_j\}$ is $PF_2$ by Theorem 2, p. 95, of Mitronović (1970). Therefore, their convolution $h_k$ is $PF_2$ (Karlin,1968. p. 394). ||

Up to now we have been studying the reliability of k-out-of-n systems of components having fixed reliabilities (i.e., non time-dependent). Next we consider time-dependent models and present stochastic comparisons for order statistics. We say X is stochastically greater than Y ($X \overset{st}{>} Y$ or $Y \overset{st}{<} X$) if $P[X > x] \geq P[Y > x]$ for each real x. We say X is stocastically equal to Y ($X \overset{st}{=} Y$) if $P[X > x] = P[Y > x]$ for each real x.

In some cases the results follow immediately by setting $p_i = \bar{F}_i(t)$ (and $p_i^* = \bar{F}_i^*(t)$), where $\bar{F}_i(t) = 1 - F_i(t)$ represents the survival probability corresponding to life distribution $F_i(t)$, $i = 1, \dots, n$. Such results are numbered in a way which indicates their connection with earlier theorems.

Our model is as follows. We assume independent observations, one observation from distribution $F_i(F_i^*)$, $i = 1, \dots, n$. The ordered observations are denoted by $Y_1 \le \cdots \le Y_n$ $(Y_1^* \le \cdots \le Y_n^*)$. From Theorem 2.2 we immediately obtain:

Theorem 2.2'. (a) Let $(-\log \bar{F}_1(t), \dots, -\log \bar{F}_n(t)) \overset{m}{>} (-\log \bar{F}_1^*(t), \dots, -\log \bar{F}_n^*(t))$ for each $t \ge 0$. Then

$$Y_1 \overset{st}{=} Y_1^* \text{ and } Y_k \overset{st}{\ge} Y_k^* \quad \text{for } k = 2, \dots, n.$$

(b) Let $(-\log F_1(t), \dots, -\log F_n(t)) \overset{m}{>} (-\log F_1^*(t), \dots, -\log F_n^*(t))$ for each $t \ge 0$. Then

$$Y_k \overset{st}{\le} Y_k^* \text{ for } k = 1, \dots, n-1, \text{ and } Y_n \overset{st}{=} Y_n^*.$$

Note that (b) is the dual of (a), obtained by interchanging $F_i$ and $\bar{F}_i$, and $P[Y_k > t]$ and $P[Y_{n-k+1} \le t]$.

Just as Corollary 2.3 represents a case of special interest obtained from Theorem 2.2, so Corollary 2.3' below represents a case of special interest obtained from Theorem 2.2'.

Corollary 2.3'. (a) Let

$$\bar{F}_1^*(t) = \cdots = \bar{F}_n^*(t) = \left[\prod_1^n \bar{F}_i(t)\right]^{\frac{1}{n}}$$

for each $t \ge 0$. Then

$$Y_k \overset{st}{\ge} Y_k^* \text{ for } k \quad 1, \dots, n; \text{ in particular, } Y_1 \overset{st}{=} Y_1^*. \quad (2.6a)$$

(b) Let

$$F_1^*(t) = \cdots = F_n^*(t) = \left[\prod_1^n F_i(t)\right]^{\frac{1}{n}} \text{ for } t \geq 0.$$

Then

$$Y_k \overset{st}{\leq} Y_k^* \text{ for } k = 1, \ldots, n; \text{ in particular, } Y_n \overset{st}{=} Y_n^*. \tag{2.6b}$$

These bounds are sharp.

As above, the sharpness of the bounds is a consequence of the fact that if in (a) we replace the common value

$$\left[\prod_1^n \bar{F}_i(t)\right]^{\frac{1}{n}}$$

by a larger value for some t, then the conclusion $Y_1 \overset{st}{\geq} Y_1^*$ would no longer hold; similarly for (b).

In a similar fashion we may obtain the analogue of Theorem 2.5 above, with stochastic comparisons of order statistics, rather than reliability comparisons for k-out-of-n systems.

<u>Theorem 2.5'.</u> (a) Let

$$\left(\frac{F_1(t)}{\bar{F}_1(t)}, \ldots, \frac{F_n(t)}{\bar{F}_n(t)}\right) \overset{m}{>} \left(\frac{F_1^*(t)}{\bar{F}_1^*(t)}, \ldots, \frac{F_n^*(t)}{\bar{F}_n^*(t)}\right)$$

for each $t \geq 0$. Then $Y_k \overset{st}{\geq} Y_k^*$ for $k = 1, \ldots, n$.
(b) Let

$$\left(\frac{\bar{F}_1(t)}{F_1(t)}, \ldots, \frac{\bar{F}_n(t)}{F_n(t)}\right) \overset{m}{>} \left(\frac{\bar{F}_1^*(t)}{F_1^*(t)}, \ldots, \frac{\bar{F}_n^*(t)}{F_n^*(t)}\right)$$

for each $t \geq 0$. Then $Y_k \overset{st}{\leq} Y_k^*$ for $k = 1, \ldots, n$.

Next we apply these theorems to several broad parametric families of distributions. We consider the case of proportional hazards, discussed in Esary, Marshall, and Proschan (1970). In keeping with the definition (2.1) of hazard in the non time-dependent case, we may define the <u>hazard function</u> (or more simply, <u>hazard</u>) R(t) corresponding to survival probability $\bar{F}(t)$ in the time-dependent case by

$$R(t) = -\log \bar{F}(t) \quad \text{for } t \geq 0. \tag{2.7}$$

Thus survival probability may be expressed in terms of the hazard function by

$$\bar{F}(t) = e^{-R(t)} \quad \text{for } t \geq 0. \tag{2.8}$$

Hazards are proportional if hazard $R_i(t)$ may be expressed as

$$R_i(t) = \lambda_i R(t) \quad \text{for } t \geq 0,\ \lambda_i > 0,\ i = 1,\ldots,n, \tag{2.9}$$

where $R(t)$ is a hazard function.

Corollary 2.7. Assume proportional hazards $\lambda_i R(t)$ (alternately $\lambda_i^* R(t)$), $i = 1, \ldots, n$, where $R(t)$ is a hazard function. Let $\underline{\lambda} \overset{m}{>} \underline{\lambda}^*$. Then $Y_1 \overset{st}{=} Y_1^*$ and $Y_k \overset{st}{\geq} Y_k$ for $k = 2,\ldots,n$.

Proof. Observe that $\underline{\lambda} \overset{m}{>} \underline{\lambda}^*$ implies that

$$(-\log \bar{F}_1(t), \ldots, -\log \bar{F}_n(t)) \overset{m}{>}$$

$$(-\log \bar{F}_1^*(t), \ldots, -\log \bar{F}_n^*(t))$$

for each $t \geq 0$. The result follows by Theorem 2.2'. ||

Important special cases of proportional hazards are:
(i) Weibull survival probability

$$\bar{F}_i(t) = e^{-\lambda_i t^{\alpha}}, \quad \alpha > 0,\ i = 1, \ldots, n.$$

(ii) Specializing further, with $\alpha = 1$, exponential survival probability

$$\bar{F}_i(t) = e^{-\lambda_i t}, \quad i = 1, \ldots, n.$$

Another interesting comparison can be obtained when survival probability is logarithmically convex in the parameter $\lambda$.

Theorem 2.8. For $t \geq 0$ let $\bar{F}(t,\lambda)$ $(F(t,\lambda))$ be differentiable, monotone, and log convex in $\lambda \geq 0$.

If $\underline{\lambda} \overset{m}{>} \underline{\lambda}^*$, then $Y_k \overset{st}{\geq} (\overset{st}{\leq})\ Y_k^*$ for $k = 1, \ldots, n$.

Proof. Assume $\bar{F}(t,\lambda)$ is differentiable, monotone, and log convex in $\lambda$. For $t \geq 0$, define $\phi_k(\underline{\lambda}) = P[Y_k > t]$. Then

$$\phi_1 = \prod_1^n \bar{F}(t,\lambda_i)$$

and

$$\frac{\partial \phi_1}{\partial \lambda_j} = \left( \prod_{i=1}^n \bar{F}(t,\lambda_i) \right) \frac{\dfrac{\partial \bar{F}(t,\lambda_j)}{\partial \lambda}}{\bar{F}(t,\lambda_j)} .$$

Since $\bar{F}(t,\lambda)$ is log convex in $\lambda$, it follows that $\phi_1(\underline{\lambda})$ is a Schur function. Also

$$\phi_2 = \prod_1^n \bar{F}(t,\lambda_i) \left[ 1 + \sum_1^n \frac{F(t,\lambda_i)}{\bar{F}(t,\lambda_i)} \right] ,$$

and

$$\frac{\partial \phi_2}{\partial \lambda_j} = \prod_{i=1}^n \bar{F}(t,\lambda_i) \cdot \frac{\dfrac{\partial \bar{F}(t,\lambda_j)}{\partial \lambda}}{\bar{F}(t,\lambda_j)} \cdot$$

$$\left[ 1 + \sum_{i=1}^n \frac{F(t,\lambda_i)}{\bar{F}(t,\lambda_i)} - \frac{1}{\bar{F}(t,\lambda_j)} \right]$$

$$= \prod_{i=1}^n \bar{F}(t,\lambda_i) \frac{\dfrac{\partial \bar{F}(t,\lambda_j)}{\partial \lambda}}{\bar{F}(t,\lambda_j)} \sum_{i \neq j} \frac{F(t,\lambda_i)}{\bar{F}(t,\lambda_i)} .$$

Since $\bar{F}(t,\lambda)$ is log convex and $\bar{F}(t,\lambda)$ is monotone in $\lambda$, it follows that $\phi_2(\underline{\lambda})$ is a Schur function. Finally for $k = 2, \ldots, n-1$,

$$\frac{\partial\phi_{k+1}}{\partial\lambda_j} = \prod_{i=1}^{n} \bar{F}(t,\lambda_i) \frac{\frac{\partial \bar{F}(t,\lambda_j)}{\partial\lambda}}{\bar{F}(t,\lambda_j)} \cdot$$

$$\left[ \sum_{i_1 < \cdots < i_k} \frac{F(t,\lambda_{i_1})\cdots F(t,\lambda_{i_k})}{\bar{F}(t,\lambda_{i_1})\cdots \bar{F}(t,\lambda_{i_k})} - \frac{F(t,\lambda_j)}{\bar{F}(t,\lambda_j)} \sum_{\substack{i_1 < \cdots < i_{k-1} \\ i_\alpha \neq j}} \frac{F(t,\lambda_{i_1})\cdots F(t,\lambda_{i_{k-1}})}{\bar{F}(t,\lambda_{i_1})\cdots \bar{F}(t,\lambda_{i_{k-1}})} \right] \cdot$$

Suppose $\bar{F}(t,\lambda)$ is decreasing in $\lambda$. Since $\log \bar{F}(t,\lambda)$ is convex in $\lambda$, then for $\lambda_j \geq \lambda_m$,

$$\frac{\frac{\partial \bar{F}(t,\lambda_j)}{\partial\lambda}}{\bar{F}(t,\lambda_j)} \geq \frac{\frac{\bar{F}(t,\lambda_m)}{\partial\lambda}}{\bar{F}(t,\lambda_m)} ,$$

with both ratios negative. Also

$$0 \leq \sum_{i_1 < \cdots < i_k} \frac{F(t,\lambda_{i_1})\cdots F(t,\lambda_{i_k})}{\bar{F}(t,\lambda_{i_1})\cdots \bar{F}(t,\lambda_{i_k})} - \frac{F(t,\lambda_j)}{\bar{F}(t,\lambda_j)} \sum_{\substack{i_1 < \cdots < i_{k-1} \\ i_\alpha \neq j}} \frac{F(t,\lambda_{i_1})\cdots F(t,\lambda_{i_{k-1}})}{\bar{F}(t,\lambda_{i_1})\cdots \bar{F}(t,\lambda_{i_{k-1}})}$$

$$\leq \sum_{i_1 < \cdots < i_k} \frac{F(t,\lambda_{i_1}) \cdots F(t,\lambda_{i_k})}{\bar{F}(t,\lambda_{i_1}) \cdots \bar{F}(t,\lambda_{i_k})}$$

$$- \frac{F(t,\lambda_m)}{\bar{F}(t,\lambda_m)} \sum_{\substack{i_1 < \cdots < i_{k-1} \\ i_\alpha \neq m}} \frac{F(t,\lambda_{i_1}) \cdots F(t,\lambda_{i_{k-1}})}{\bar{F}(t,\lambda_{i_1}) \cdots \bar{F}(t,\lambda_{i_{k-1}})} .$$

It follows that $\phi_{k+1}$ is a Schur function for $k = 2, \ldots, n-1$. A similar argument applies if $\bar{F}(t,\lambda)$ is increasing in $\lambda$. It follows by Theorem 2.1 that for $k = 1, \ldots, n$, $\phi_k(\underline{\lambda}) \geq \phi_k(\underline{\lambda}^*)$, and so, $Y_k \overset{st}{\geq} Y_k^*$.

A similar proof holds for the case $F(t,\lambda)$ differentiable, monotone, and log convex in $\lambda$. For this case, we show that $P[Y_{n-k} < t]$ is a Schur function of $\underline{\lambda}$, since it has the same form as $\phi_{k+1}$ with each $\bar{F}_i$ replaced by $F_i$.||

A familiar example of a class of distributions satisfying the hypothesis of Theorem 2.8 is obtained by considering survival probability $\bar{F}(t,\lambda) = \bar{G}(\lambda t)$, where distribution G has <u>decreasing failure rate</u> (DFR) and $\lambda$ occurs as a scale factor. The hypothesis is satisfied since a DFR survival probability is log convex (Barlow and Proschan, 1965, Chapter 2.)

We now give some examples of survival probabilities $\bar{F}(t,\lambda)$ which are log convex in $\lambda$, so that Theorem 2.6 holds.

(i) $\bar{F}(t,\lambda) = e^{-(\lambda t)^\alpha}$ for $0 < \alpha \leq 1$. This parameterization of the Weibull distribution is different from that following Corollary 2.7. In the latter case, $\log \bar{F}$ is linear in $\lambda$, and so the restriction to $\alpha \leq 1$ is not necessary.

(ii) $F(t,\lambda) = \int_0^t \frac{\lambda^\alpha x^{\alpha-1}}{\Gamma(\alpha)} e^{-\lambda x} dx$ for $0 < \alpha \leq 1$, a gamma distribution with shape parameter $\alpha \leq 1$.

$$\text{(iii)} \quad F(t,\lambda) = \sum_1^n p_i(1 - e^{-a_i\lambda t}), \; a_i \geq 0, \; p_i \geq 0,$$

$$\sum_1^n p_i = 1.$$

A mixture of exponential distributions has decreasing failure rate; see Barlow and Proschan (1965), Chapter 2.

(iv) More generally, let $\bar{F}_i(t,\lambda)$ be log convex in $\lambda$, $p_i \geq 0$,

and $\quad \sum p_i = 1.$ Then $\bar{F}(t,\lambda) = \sum_1^n p_i F_i(t,\lambda)$

is log convex in $\lambda$. See Barlow and Proschan (1965), Chapter 2.

The last result of this section holds for the smallest and largest order statistics, or equivalently, for series and parallel systems.

<u>Theorem 2.9.</u> Let $F(t,\lambda)$ $(\bar{F}(t,\lambda))$ be log concave and differentiable in $\lambda \geq 0$ for each fixed $t \geq 0$. If $\underline{\lambda} \overset{m}{>} \underline{\lambda}^*$, then

$$Y_n \overset{st}{\geq} Y_n^* \; (Y_1 \overset{st}{\leq} Y_1^*).$$

<u>Proof.</u> Assume $F(t,\lambda)$ is log concave and differentiable in $\lambda \geq 0$. Then

$$\frac{\partial}{\partial\lambda_j} P[Y_n > t] \quad \frac{\partial}{\partial\lambda_j} \left[ 1 - \prod_1^n F(t,\lambda_i) \right]$$

$$= - \prod_1^n F(t,\lambda_i) \frac{\frac{\partial F(t,\lambda_j)}{\partial\lambda}}{F(t,\lambda_j)} \; .$$

If $\lambda_j \geq \lambda_m$, then

$$\frac{\partial}{\partial\lambda_j} P[Y_n > t] - \frac{\partial}{\partial\lambda_m} P[Y_n > t] \overset{\text{sgn}}{=}$$

$$\frac{\dfrac{\partial F(t,\lambda_m)}{\partial\lambda}}{F(t,\lambda_m)} - \frac{\dfrac{\partial F(t,\lambda_j)}{\partial\lambda}}{F(t,\lambda_j)} \geq 0 \quad ,$$

since $\log F(t,\lambda)$ is concave in $\lambda \geq 0$.

The dual part of the theorem is proved in a similar fashion. ||

Note that if the density is log concave, or equivalently, $PF_2$, in t, then both F and $\bar{F}$ are log concave in t. If, in addition, F is a function of $\lambda t$, then $F(\lambda t)$ and $\bar{F}(\lambda t)$ are log concave in $\lambda$ and thus both conclusions of Theorem 2.9 hold. Examples of such $PF_2$ densities are:

(i) Weibull. $f(t,\lambda) = \alpha\lambda^{\alpha}t^{\alpha-1}\, e^{-(\lambda t)^{\alpha}}$ for $\alpha > 1$.

(ii) Gamma. $f(t,\lambda) = \dfrac{\lambda^{\alpha}t^{\alpha-1}}{\Gamma(\alpha)}\, e^{-\lambda t}$ for $\alpha \geq 1$.

(iii) Truncated Normal. $f(t,\lambda) = a\lambda e^{-\frac{1}{2}\lambda^2(t-\mu)^2}$, $t \geq 0$, $\lambda > 0$, where $a > 0$ is such that

$$\int_0^{\infty} f(t,\lambda)dt = 1.$$

(Here the density is log concave in $t - \mu$.)

## 3. Spacings

In this section we present comparisons under majorization for spacings between order statistics. In addition, we present stochastic comparisons among the spacings arising from a single set of order statistics. In each case we assume the underlying heterogeneous distributions have proportional hazards $\lambda_1 R(t), \ldots ,\lambda_n R(t)$.

Let $D_{1,n} = Y_1$, $D_{2,n} = Y_2 - Y_1$, ... , $D_{n,n} = Y_n - Y_{n-1}$ denote the spacings between order statistics $Y_1 \leq \cdots \leq Y_n$ computed from independent observations, one from each of $n$ heterogeneous distributions. Recall that when the observations come from a single underlying distribution F, an exponential, then the normalized spacings $nD_{1,n}$, $(n-1)D_{2,n}$, ... ,$D_{n,n}$ are independently distributed according to the same exponential distribution F. More generally, we can show that when the underlying distributions have concave proportional hazards, the normalized spacings increase stochastically.

<u>Theorem 3.1.</u> Let $\bar{F}_i(t) = e^{-\lambda_i R(t)}$, $\lambda_i > 0$, for i = 1, ... ,n, where R(t) is a concave, differentiable hazard function. Then

$$nD_{1,n} \overset{st}{\leq} (n-1)D_{2,n} \overset{st}{\leq} \cdots \overset{st}{\leq} D_{n.n} .$$

<u>Proof.</u> First we show $nD_{1,n} \overset{st}{\leq} (n-1)D_{2,n}$ for each n; then we use induction on n. Let

$$\lambda = \sum_1^n \lambda_i \,, \; \bar{\lambda} = \lambda/n, \; r = \frac{dR}{dt} . \quad \text{For } x > 0,$$

$$P[(n-1)D_{2,n} > x]$$

$$= \sum_{i=1}^{n} \int_0^\infty \lambda_i r(t) e^{-\lambda_i R(t) - (\lambda - \lambda_i) R(t + \frac{x}{n-1})} dt$$

$$= \int_0^\infty \lambda \sum_{i=1}^{n} \frac{\lambda_i}{\lambda} r(t) e^{\lambda_i [R(t + \frac{x}{n-1}) - R(t)]} e^{-\lambda R(t + \frac{x}{n-1})} dt.$$

Since $e^{ay}$ is a convex function, by Jensen's inequality we have

$$P[(n-1)D_{2,n} > x]$$

$$\geq \int_0^\infty \lambda r(t) e^{\bar{\lambda}[R(t + \frac{x}{n-1}) - R(t)] - \lambda R(t + \frac{x}{n-1})} dt$$

$$= \int_0^\infty \lambda r(t) e^{-(n-1)\bar{\lambda} R(t + \frac{x}{n-1}) - \bar{\lambda} R(t)} dt$$

$$\geq \int_0^\infty \lambda r(t) e^{-(n-1)\bar{\lambda} R(t) - (n-1)\bar{\lambda} R(\frac{x}{n-1}) - \bar{\lambda} R(t)} dt,$$

since R is subadditive. Thus

$$P[(n-1)D_{2,n} > x] \geq e^{-(n-1)\bar{\lambda}R(\frac{x}{n-1})} \int_0^\infty \lambda r(t) e^{-\lambda R(t)} dt$$

$$= e^{-(n-1)\bar{\lambda}R(\frac{x}{n-1})} \geq e^{-\lambda R(\frac{x}{n})}, \text{ since R is concave.}$$

But the last expression gives $P[nD_{1,n} > x]$. Thus we have proved

$$nD_{1,n} \overset{st}{\leq} (n-1)\, D_{2,n}$$

for each n.

Next suppose that for fixed i,

$$(n - i + 1)D_{i,n} \overset{st}{\geq} (n - i + 2)D_{i-1,n} \text{ for all } n > i.$$

Write

$$P[(n-i)\, D_{i+1,n} > x] = \int_0^\infty \sum_{j=1}^{n} \lambda_j\, r(t) e^{-\lambda R(t)}$$

$$P[(n-i)D_{i,n-1} > x | \lambda_j \text{ deleted}, R_t],$$

where

$$P[(n-i)D_{i,n-1} > x|\lambda_j \text{ deleted}, R_t]$$

indicates that $F_j$ is no longer present and that hazard $R(x)$ is to be replaced by $R_t(x) = R(t + x) - R(t)$. Using the inductive hypothesis, we have

$$P[(n-i)D_{i+1,n} > x] \geq \int_0^\infty \sum_{j=1}^n \lambda_j r(t) e^{-\lambda R(t)} \cdot$$

$$P[(n-i+1)D_{i-1,n-1} > x|\lambda_j \text{ deleted}, R_t] \, dt$$

$$= \quad P[(n - i + 1)D_{i,n} > x].$$

Thus

$$(n-i)D_{i+1,n} \overset{st}{\geq} (n-i+1)D_{i,n} \quad ,$$

and the proof is complete by induction. ||

Next we compare spacings from two sets of underlying exponential distributions. We drop the second subscript in the spacings $D_{i,n}$, and define $D_i^*$, $i = 1, \ldots, n$, to be the spacings arising from $\{F_1^*, \ldots, F_n^*\}$, the second set of exponential distributions.

<u>Theorem 3.2.</u> Let $\bar{F}_i(t) = e^{-\lambda_i t}$ and $\bar{F}_i^*(t) = e^{-\bar{\lambda} t}$ for $i = 1, \ldots, n$, where

$$\bar{\lambda} = \frac{1}{n} \sum_1^n \lambda_i \, .$$

Then $D_1 \overset{st}{=} D_1^*$ and $D_i \overset{st}{\geq} D_i^*$ for $i = 2, \ldots, n$.

<u>Proof.</u> By Theorem 3.1, we know that

$$nD_1 \overset{st}{\leq} (n-1)D_2 \overset{st}{\leq} \cdots \overset{st}{\leq} D_n \, . \tag{3.1}$$

It is also well known that the normalized spacings from a single exponential distribution are stochastically alike, i.e.,

$$nD_1^* \overset{st}{=} (n-1)D_2^* \overset{st}{=} \cdots \overset{st}{=} D_n^*. \tag{3.2}$$

See Epstein and Sobel (1953 ,1954) or Sukhatme (1936,1937). Finally, we may verify directly that for $x > 0$,

$$P[nD_1 > x] = \prod_1^n e^{-\lambda_i \frac{x}{n}} = e^{-\bar{\lambda}x} = P[nD_1^* > x],$$

so that

$$nD_1 \overset{st}{=} nD_1^* \quad . \tag{3.3}$$

From (3.1), (3.2), and (3.3), we obtain the conclusion of the theorem. ||

Note that the mean $\bar{\lambda}$ of the $\lambda_1, \ldots, \lambda_n$ is used for the comparison. Is the same conclusion possible using a vector $\underline{\lambda}^*$, where $\underline{\lambda} \overset{m}{>} \underline{\lambda}^*$ ? That the conclusion need not hold when $\underline{\lambda} \overset{m}{>} \underline{\lambda}^*$ may be shown by the counterexample: Let $n = 3$, with $\lambda_1 = 10$, $\lambda_2 = 4$, $\lambda_3 = 1$, and $\lambda_1^* = 7$, $\lambda_2^* = 7$, $\lambda_3^* = 1$. Then, letting $\lambda = \lambda_1 + \lambda_2 + \lambda_3$,

$$P(D_3 > 1) = \left(\frac{\lambda_1}{\lambda}\frac{\lambda_2}{\lambda-\lambda_1} + \frac{\lambda_2}{\lambda}\frac{\lambda_1}{\lambda-\lambda_2}\right)e^{-\lambda_3} +$$

$$\left(\frac{\lambda_1}{\lambda}\frac{\lambda_3}{\lambda-\lambda_1} + \frac{\lambda_3}{\lambda}\frac{\lambda_1}{\lambda-\lambda_3}\right)e^{-\lambda_2} + \left(\frac{\lambda_2}{\lambda}\frac{\lambda_3}{\lambda-\lambda_2} + \frac{\lambda_3}{\lambda}\frac{\lambda_2}{\lambda-\lambda_3}\right)e^{-\lambda_1}$$

$$< P(D_3^* > 1) \, .$$

In Theorem 3.2 we averaged the exponential failure rates to obtain a stochastic comparison. Next we average the exponential means to obtain a comparison; the comparison is not stochastic, but rather is for expectations. We will first need to show

<u>Lemma 3.3.</u> Let $\underline{x} \overset{m}{>} \underline{x}^*$ and $\underline{y} \overset{m}{>} \underline{y}^*$. Then

$$\sum_1^k x_i y_i \geq \sum_1^k x_i^* y_i^* \text{ for } k = 1, \ldots, n.$$

Proof. Define

$$\phi(\underline{x}) = \sum_{i=1}^k x_i y_i$$

for fixed $k$, $1 \leq k \leq n$, and fixed $y_1 \geq \cdots \geq y_k$. Then

$$\frac{\partial \phi}{\partial x_i} = \begin{cases} y_i & \text{for } i \leq k \\ 0 & \text{for } i > k. \end{cases}$$

If $x_i \geq x_j$, then $y_i \geq y_j$, and thus $\phi$ is a Schur function. By Theorem 2.1,

$$\sum_1^k x_i y_i \geq \sum_1^k x_i^* y_i \quad .$$

Using the same argument with fixed $\underline{x}^*$ and $\underline{y} \overset{m}{>} y^*$, we obtain the desired conclusion. ||

Theorem 3.4. Let $0 \leq a_1 \leq \cdots \leq a_n$, $\bar{F}_i(t) = e^{-t/\mu_i}$,

$$\bar{F}_i^*(t) = e^{-t/\mu_i^*}, \quad \text{where } \mu_i^* = \frac{1}{n} \sum_1^n \mu_i ,$$

$i = 1, \ldots, n$. Then

$$E \sum_{i=1}^n a_i(n - i + 1) D_i \geq E \sum_{i=1}^n a_i(n - i + 1) D_i^* .$$

Proof. $D_1^*, \ldots, D_n^*$ are spacings from a common exponential distribution, and thus by (3.2),

$$EnD_1^* = E(n-1)D_2^* = \cdots = ED_n^* \quad .$$

By Theorem 3.1, $ED_n \geq E2D_{n-1} \geq \cdots \geq EnD_1$. Since

$$E(nD_1 + \cdots + D_n) = E\sum_1^n Y_i = \sum_1^n \mu_i$$

$$= E\sum_1^n Y_i^* = E(nD_1^* + \cdots + D_n^*),$$

it follows that

$$(ED_n, \ldots, EnD_1) \overset{m}{>} (ED_n^*, \ldots, EnD_1^*).$$

By Lemma 3.3 the desired conclusion follows. ||

<u>Remark.</u> Let $\tau(Y_i) = nD_1 + (n-1)D_2 + \cdots + (n-i+1)D_i$, the "total time on test" accumulated by the time of the i-th failure; $\tau$ is used commonly in life testing under the exponential assumption. Note that

$$E\tau(Y_i) \leq E\tau(Y_i^*) \quad \text{for } i = 1, \ldots, n,$$

since

$$(ED_n, E2D_{n-1}, \ldots, EnD_1) \overset{m}{>} (ED_n^*, E2D_{n-1}^*, \ldots, EnD_1^*),$$

as shown in the proof of Theorem 3.4.

<u>Corollary 3.5.</u> Let $\bar{F}_i(t) = e^{-t/\mu_i}$, $\bar{F}_i^* = e^{-t/\mu_i^*}$, where

$$\mu_i^* = \frac{1}{n}\sum_1^n \mu_i, \quad i = 1, \ldots, n.$$

Then $E(Y_n - Y_i) \geq E(Y_n^* - Y_i^*)$ for $i = 0, \ldots, n-1$, where $Y_0 \equiv 0 \equiv Y_0^*$.

<u>Proof.</u> In Theorem 3.4, choose $a_1 = \cdots = a_{i-1} = 0$,

$$a_i = \frac{1}{n-i+1}, \; a_{i+1} = \frac{1}{n-i}, \; \dots, a_n = 1.$$

Then

$$\sum_{j=1}^{n} a_j (n-j+1) D_j = Y_n - Y_i \, . \quad \|$$

## 4. Applications

In this section we describe some applications of the results of Sections 2 and 3 to reliability problems.

Typically, a system (say a relay network) is to be designed using like units. However, due to random fluctuations in the units, the individual unit reliabilities vary. From a knowledge of the average (not necessary arithmetic mean) reliability of components, we wish to predict conservatively the reliability of k-out-of-n systems of the unlike components. The various theorems in Section 2 describe conditions under which we may obtain such conservative predictions.

A statistical application in the same spirit may be made as follows. Suppose for a sample of m systems, the first r successive failure times are observed for the components. Suppose further that the underlying component distributions are exponential with <u>differing</u> failure rates. From this data, we may obtain a conservative estimate of the life distribution of a k-out-of-n system, where k may be greater than r. Theorems and applications illustrating how conservative confidence bounds may be derived are presented in Barlow, Madansky, Proschan, and Scheuer (1968).

Statistical applications of this type as well as other statistical applications of the results of Sections 2 and 3 are being developed; a forthcoming report will give details.

## References

Barlow, R. E., Madansky, A., Proschan, F., and Scheuer, E. M. (1968). Statistical estimation procedures for the "burn-in" process, <u>Technometrics</u> <u>10</u>, 51-62.

Barlow, R. E. and Proschan, F. (1965). Mathematical Theory of Reliability, John Wiley & Sons, Inc.

David, H. A. (1970). Order Statistics, John Wiley & Sons, Inc.

Epstein, B. and Sobel, M. (1953). Life Testing, J. Amer. Statist. Assoc. 48 486-502.

Epstein, B. and Sobel, M. (1954). Some theorems relevant to life testing from an exponential distribution, Ann. Math. Statist. 25, 373-381.

Esary, J. D., Marshall, A. W., and Proschan, F. (1970). Some reliability applications of the hazard transform. SIAM J. Appl. Math. 18, 849-860.

Karlin, S. (1968). Total Positivity, Vol. I, Stanford University Press.

Mitrinović, D. S. (1970). Analytic Inequalities. Springer- Verlag.

Pyke, R. (1965). Spacings, J. R. Statist. Soc. B 27, 395-436. Discussion: 437-449.

Pyke, R. (1970). Spacings revisited, Technical Report No. 25, Department of Mathematics, University of Washington.

Sarhan, A. E. and Greenberg, B. G. (Eds.) (1962). Contributions to Order Statistics, John Wiley & Sons, Inc.

Sen, P. K. (1970). A note on order statistics for heterogeneous distributions, Ann. Math. Statist. 41, 2137-2139.

Sukhatme, P. V. (1936). On the analysis of k samples from exponential populations with especial reference to the problem of random intervals, Statist. Res. Mem. 1, 94-112.

Sukhatme, P. V. (1937). Tests of significance for samples of the $\chi^2$- population with two degrees of freedom, Ann. Eugen., Lond. 8, 52-56.

# MOMENT PROBLEMS WITH CONVEXITY CONDITIONS I[1]

by

J. H. B. Kemperman

University of Rochester

## 1. Introduction

Many moment problems arising in practice involve the condition that the measure $\mu$ on hand satisfies certain convexity conditions. Often such convexity conditions can be expressed as the requirement that the measure $\mu$ admits a representation of the form

$$\mu(A) = \int P(y, A)\, \nu(dy) \; . \tag{1.1}$$

Here, $P(y, A)$ denotes a given kernel while $\nu$ is a rather unrestricted measure. Frequently, a representation (1.1) arises from an application of Choquet's theorem and then the measure $\mu_y(A) = P(y, A)$ is usually extremal in some sense.

Section 2 describes the type of moment problem we want to consider. In the Sections 3 and 4 we describe in detail in what manner a representation (1.1) enables us to transfer the original moment problem concerning $\mu$ to an often more convenient moment problem concerning $\nu$ .

The basic idea of transferring a moment problem from one measure $\mu$ to a related measure $\nu$ has been used in many situations, see Johnson and Rogers [9] p. 435; Royden [22] p. 371; Mallows [15] p. 225; [16] p. 163; [17] p. 410; Mulholland and Rogers [20] p. 215; Karlin and Studden [10] pp. 483-504 and others.

---

1) This research was supported by the National Science Foundation.

The main purpose of the present paper is to make a more systematic study of explicit integral representations relating certain classes of probability measures to simpler classes and their use in transferring moment problems, first on the general level and afterwards, in more detail, for each specific application.

In Section 5, as a first but very central illustration, we make a detailed study of the class $\mathcal{M}_s[x_o]$ consisting of all probability measures $\mu$ on $R$ such that the corresponding cumulative distribution function $F$ has in the interval $(x_o, +\infty)$ an $(s-1)$-th derivative $F^{(s-1)}(x)$ which in $(x_o, +\infty)$ is convex or concave depending on whether $s$ is even or odd. One has $\mathcal{M}_s[x_o] \subset \mathcal{M}_{s-1}[x_o]$ .

It follows from Theorem 5.28 that each measure $\mu$ of class $\mathcal{M}_s[x_o]$ admits a representation (1.1) with $\nu$ as an arbitrary probability measure on $R$ , while the kernel $P(y, A)$ is given by (5.23) and (5.27) . The resulting correspondence $\mu = T\nu$ between $\mu \in \mathcal{M}_s[x_o]$ and $\nu$ is one to one. In order that $\mu$ be supported by a closed interval containing $x_o$ it is necessary and sufficient that $\nu$ be supported by that interval. It would not be very hard to establish analogous integral representations for the case that the condition $\mu \in \mathcal{M}_s[x_o]$ is replaced by certain more general convexity conditions, for instance, those of the form $L(\mu) \geq 0$ with $L$ as an appropriate linear differential operator.

In Section 6 these ideas are applied to a number of moment problems involving the condition $\mu \in \mathcal{M}_s[x_o]$ , in particular, an explicit method for obtaining the best bounds on a moment $\int h(x)\, \mu(dx)$ or the measure $\mu(I)$ of a given interval $I$ , when it is known that $\mu$ belongs to $\mathcal{M}_s[x_o]$ and further that

$$\int x^{r_j}\, \mu(dx) = \alpha_{r_j} , \qquad (j = 1, \ldots, n) ,$$

with the $r_j$ and $\alpha_{r_j}$ as given real numbers, $r_j > 0$ .

It is also assumed that the measure $\mu$ is supported by a prescribed interval $[0, b]$ containing $x_o$ where we allow that $b = +\infty$ .

Instead of working out some special problems in full detail, I have tried to show that many such problems can be handled very well by more or less classical techniques, namely, after a transformation (1.1) from $\mu$ to $\nu$. Other applications of this idea may be found in the papers referred to above. It should be noted that many of the so-called classical moment problems not involving a convexity condition have not been solved yet either.

In Section 7 we give the full details for the special case where $\mu \in \mathcal{M}_s[x_o]$ is supported by a given interval $[0, b]$ and only one power moment $\alpha_r$ is prescribed; $r$ need not be an integer. We are interested in the probability content $\mu(I)$ of a given interval $I = [z, b]$. In the case $x_0 = 0$, the precise upper bound on $\mu(I)$ is given by (7.30), (7.31) and (7.32), while the precise lower bound is given by (7.41) and (7.43). In the case $x_o > 0$, the best upper bound on $\mu(I)$ is given by (7.46) and (7.47); for $s = 1$ these are due to von Mises [19].

In Section 7 there is also some discussion of another method due to van Dantzig [5] which could have been used equally well in deriving the latter optimal bounds, see Theorem 7.64.

Also note that there would be no difficulty in replacing the prescribed power moment $\alpha_r$ by a prescribed moment of the form $\int \phi(x)\, \mu(dx) = c$, provided the given function $\phi$ is non-decreasing and $\phi(0) = 0$.

In a subsequent paper, the methods of Sections 3 and 4 will be applied to the class $\mathcal{M} = \mathcal{M}_s[x_1, \ldots, x_s]$ of so-called bell-shaped distributions $\mu$ on $R$. Our results generalize previous work of Johnson and Rogers [9], Mallows [15], [16], [17], Mulholland and Rogers [20], Karlin and Studden [10] and others.

Here, a measure $\mu \in \mathcal{M}_s[x_1, \ldots, x_s]$ is a probability measure on $R$ characterized by the property that its distribution function $F$ has in each interval $(x_{r-1}, x_r)$ an $(s - 1)$-th derivative $F^{(s-1)}$, which is convex or concave depending on whether $r$ is odd or even, $(r = 1, \ldots, s + 1;\ x_0 = -\infty,\ x_{s+1} = +\infty)$. In the special case $s = 1$ one would speak of a distribution which is unimodal about

$x_1$ . It is allowed that $F^{(s-1)}$ has a jump at $x_r$ but this jump should then be upward or downward, depending on whether r is odd or even, respectively.

It will be shown that also this class admits a representation of the type (1.1) with $\nu$ as an unrestricted probability measure on R . The resulting correspondence between $\mu$ and $\nu$ is again one to one. The kernel P(A, y) can be explicitly given and has many beautiful properties, such as total positivity.

It is also possible to replace the ordinary convexity of $\pm F^{(s-1)}$ by a generalized type of convexity of F relative to a fixed Tchebycheff system. The corresponding representations (1.1) allow us to handle a large number of moment problems involving the side condition that the measure $\mu$ be bell-shaped in some specified sense.

## 2. Statement of the moment problem

In the present Section we describe the general type of moment problem we shall be interested in. In the sequel, $X = (X, \mathcal{A}_X)$ will denote a given measurable space while $\mathcal{M}$ is a given collection of probability measures on X . It will be convenient to introduce a random variable Z taking values in X and having a distribution

$$\mu(A) = \Pr(Z \in A), \qquad (A \in \mathcal{A}_X) ,$$

which belongs to the given class $\mathcal{M}$ .

Let further $g_j : X \to R$ $(j = 0, 1, \ldots, n)$ be given real-valued measurable functions on X . Here, $g_o$ will always denote the special function with $g_o(x) = 1$ for all $x \in X$ .

The moment space M , corresponding to $\mathcal{M}$ and the functions $g_j$, is defined as the set of all points $c = (c_1, \ldots, c_n)$ in $R^n$ such that there exists at least one $\mu \in \mathcal{M}$ with

$$(2.1) \qquad Eg_j(Z) = \int g_j(x)\mu(dx) = c_j, \quad \text{for all } j = 1,\ldots,n;$$

(also for $j = 0$ provided $c_o = 1$) . Here, we naturally include the tacit assumption that $g_j$ is integrable relative to $\mu$ , that is, $\int |g_j| d\mu < \infty$, $(j = 1, \ldots, n)$.

It will be convenient to introduce the function $g: X \to R^n$ defined by

(2.2) $$g(x) = (g_1(x), g_2(x), \ldots, g_n(x)) .$$

Thus, (2.1) can also be written as $\int g \, d\mu = c$. Here, $\mu$ is a probability measure, therefore,

(2.3) $$M = \{c = \int g \, d\mu : \mu \in \mathcal{M}\} \subset \text{conv } g(X) ,$$

(namely, by a result of Richter [23] p. 151 and Rogosinsky [24] p. 4, see also [13] p. 95). Here, $g(X) = \{g(x) : x \in X\}$ denotes the range of $g: X \to R^n$, while conv(S) with $S \subset R^n$ denotes the convex hull of S.

One would even have $M = \text{conv } g(X)$ in the special case that $\mathcal{M}$ contains each probability measure $\mu$ on X with a finite support consisting of at most $n + 1$ points. In that case, it is often easy to describe the moment space M or to minimize $c_n$, given that $\mu \in \mathcal{M}$ and given the moments $c_1, \ldots, c_{n-1}$.

(2.4) Let $h: X \to R$ denote a given measurable function. Often one is interested in determining the best lower bound

(2.5) $$L(c) = \inf\{\int h \, d\mu : \mu \in \mathcal{M}(c)\} ,$$

where

(2.6) $$\mathcal{M}(c) = \{\mu : \mu \in \mathcal{M}, \int g \, d\mu = c\} .$$

Thus $Eh(Z) \geq L(c)$ whenever $Z \in X$ has a distribution $\mu \in \mathcal{M}$ and satisfies (2.1); moreover, this lower bound cannot be improved.

In (2.5) and (2.6), we regard $c \in R^n$ as fixed. We put $L(c) = +\infty$ if $\mathcal{M}(c)$ is empty, that is, if $c \notin M$. As is natural, we agree to take in (2.5) the integral $\int h \, d\mu = -\infty$ as soon as $\int h^- \, d\mu = +\infty$, whether or not $\int h^+ \, d\mu < +\infty$; (here, $h^+ = \max(0, h)$ and $h^- = \max(0, -h)$). In view of this convention, we may (and will) as well restrict ourselves to those $c \in R^n$ which satisfy

(2.7) $$\int h^- \, d\mu < +\infty \quad \text{for all} \quad \mu \in \mathcal{M}(c) .$$

For, if (2.7) were not satisfied we would have $L(c) = -\infty$. A sufficient condition in order that (2.7) holds for all c would be that

(2.8) $\qquad h(x) + \sum_{j=0}^{n} \gamma_j |g_j(x)| \geq 0$ , for all $x \in X$ ,

and for some choice of the positive constants $\gamma_0, \gamma_1, \ldots, \gamma_n$ .

## 3. Transforming a moment problem

We now consider another measurable space $Y = (Y, \mathcal{A}_Y)$ and further a given function $P(y, A)$ on $Y \times \mathcal{A}_X$ which is a Markov kernel. That is, for each fixed $y \in Y$ it is a probability measure on $X$, while for each fixed $A \in \mathcal{A}_X$ it is a measurable function on $Y$. For each probability measure $\nu$ on $Y$, let $\mu = T\nu$ denote the probability measure on $X$ given by

(3.1) $\qquad (T\nu)(A) = \int P(y, A)\, \nu(dy)$ .

Let us now introduce the assumption that the given collection $\mathcal{M}$ of probability measures on $X$ (as in Section 2) is of the form

(3.2) $\qquad \mathcal{M} = \{\mu = T\nu : \nu \in \mathcal{M}^*\}$ ,

where $\mathcal{M}^*$ denotes a given collection of probability measures on $Y$. Observe that $\mathcal{M}$ is convex as soon as $\mathcal{M}^*$ is convex. In the applications we have in mind, the structure of $\mathcal{M}^*$ is much simpler than the structure of $\mathcal{M}$ and it becomes advantageous to transform the moment problem posed in Section 2 from the space $X$ to the space $Y$.

REMARK. It was shown by Mulholland and Rogers [20] (see also [10] p. 103) that many classes $\mathcal{M}$ of probability measures on $X = R$ admit a representation as in (3.1) and (3.2), but they usually do not specify the kernel $P(y, A)$ on hand.

More precisely, they show that for many such classes $\mathcal{M}$ there exists a distinguished and much smaller subset $\mathcal{M}_1$ of $\mathcal{M}$ (usually consisting of extreme elements in some sense) such that each $\mu \in \mathcal{M}$ can be represented as an integral

$$\mu(A) = \int_0^1 \eta_t(A)\, dt , \qquad \text{for all } A \in \mathcal{A}_X ,$$

and where $\eta_t \in \mathcal{M}_1$ for all $0 \leq t \leq 1$ . Note that the mapping $t \to \eta_t$ depends on $\mu$ , but in an unspecified way.

Suppose we enumerate $\mathcal{M}_1$ as $\mathcal{M}_1 = \{P(y, .), y \in Y]$ . Let further the above measure $\eta_t$ in $\mathcal{M}_1$ depending on $\mu$ be denoted by $P(\phi_\mu(t), .)$. Finally, letting

$$\nu(B) = \lambda(\{t: \phi_\mu(t) \in B\}) ,$$

(with $B \subset Y$ and $\lambda$ as Lebesgue measure), it follows that $\mu = T\nu$ with $T\nu$ as in (3.1). In such a situation, the class $\mathcal{M}^*$ may often be taken as the class of all probability measures on $Y$ .

(3.3) In transferring the moment problem to $Y$ , we will need the following (adjoint) transformation from functions to functions. Namely, for each measurable function $\phi: X \to R$, let us define

$$\text{(3.4)} \qquad \phi^*(y) = (T\phi)(y) = \int \phi(x)\, P(y, dx), \qquad (y \in Y) ,$$

allowing the values $\phi^*(y) = \pm \infty$ . Note that $\phi^*(y)$ is equal to a weighted average of the values $\phi(x)$ with $x \in X$ .

The <u>domain</u> of $\phi^*$ is defined as the set of all $y \in Y$ for which the integral (3.4) is well defined, in the sense that either

$$\int \phi^-(x)\, P(y, dx) < \infty , \quad \text{or} \quad \int \phi^+(x)\, P(y, dx) < \infty .$$

Consider any measure $\mu \in \mathcal{M}$ such that $\mu(\phi) = \int \phi(x)\mu(dx)$ is well-defined. Let further $\nu \in \mathcal{M}^*$ be such that $T\nu = \mu$ . It follows from Fubini's theorem that $\phi^*(y)$ is well-defined for almost $[\nu]$ all values $y \in Y$; in other words, the domain of $\phi^*$ has full $\nu$-measure. Moreover, the integral $\nu(\phi^*) = \int \phi^*(y)\, \nu(dy)$ is well-defined and, finally,

$$\text{(3.5)} \qquad \int \phi(x)\, \mu(dx) = \int \phi^*(y)\, \nu(dy).$$

In other words, $\int \phi\, d(T\nu) = \int (T\phi) d\nu$ .

Observe that for a <u>nonnegative</u> measurable function $\phi$ the relation (3.5) always hold while $\phi^*(y)$ is well-defined for each $y \in Y$ . Consequently, if $\phi: X \to R$ is an arbitrary measurable function then

(3.6) $\int \phi^+ d\mu < \infty$ if and only if $\int T\phi^+ d\nu < \infty$ ,

and

(3.7) $\int \phi^- d\mu < \infty$ if and only if $\int T\phi^- d\nu < \infty$ ;

here, $\mu = T\nu$ with $\mu \in \mathcal{M}$ and $\nu \in \mathcal{M}^*$ . Jointly, the two properties (3.6) and (3.7) are equivalent to the $\mu$-integrability of $\phi$ , (that is, $\mu(|\phi|) < \infty$) . They in turn imply that $\nu(\phi^*) = \nu(T\phi)$ is well-defined and finite, equivalently, that $\nu(|T\phi|) < \infty$ , equivalently, that $\nu((T\phi)^-) < \infty$ and $\nu((T\phi)^+) < \infty$ .

In other words, if $\phi$ is integrable relative to $\mu = T\nu$ then $\phi^* = T\phi$ is integrable relative to $\nu$ . However, <u>the converse is false</u>, see (5.46) for a counterexample. This is related to the fact that the obvious inequalities

$$(T\phi)^- \leq T\phi^- \ ; \ (T\phi)^+ \leq T\phi^+ ; |T\phi| \leq T|\phi|$$

need not be inequalities.

(3.8) In view of (3.5), the transferred moment problem will involve the functions $g_j^* = Tg_j$ $(j = 0, 1, \ldots, n)$ , $g_0^*(y) = 1$ for all $y \in Y$ . An important role will be played by the intersection

$$Y_o = \mathcal{D}_1 \cap \ldots \cap \mathcal{D}_n ,$$

where

$$\mathcal{D}_j = \{y \in Y: \int |g_j(x)| P(y, dx) < \infty\}$$

denotes the set of values $y$ for which $g_j^*(y)$ is well-defined and finite. We shall also need the set

$$\mathcal{D}(h^*) = \{y \in Y: \int h^-(x) \ P(y, dx) < \infty\}$$

of values $y$ for which $h^*(y) = (Th)(y)$ is well-defined with a value $h^*(y) > -\infty$ . Note that these domains $\mathcal{D}_j$ , $Y_o$ and $\mathcal{D}(h^*)$ depend only on the kernel $P$ and the functions $g_j$ and $h$ (not on any measure $\mu$ or $\nu$) .

(3.9) DEFINITION. Consider any point $c = (c_1, \ldots, c_n) \in R^n$. In the transferred moment problem, the role of $\mathcal{M}(c)$ defined by (2.6) will be taken over by the collection $\mathcal{M}^*(c)$

of all measures $\nu \in \mathfrak{M}^*$ which satisfy

(3.10) $$\int (T|g_j|)(y)\ \nu(dy) < \infty\ , \quad \text{for all} \quad j = 1, \ldots, n,$$

and

(3.11) $$\int g_j^*(y)\ \nu(dy) = c_j\ , \quad \text{for all} \quad j = 1, \ldots, n\ .$$

Observe that $\nu(\mathfrak{D}_j) = 1$ for all $j$, thus, $\nu(Y_o) = 1$. Condition (3.10) is sufficient but not necessary for the absolute convergence of the integrals in (3.11). In many applications, the absolute convergence of the integrals in (3.11) (which normally would be a tacit assumption in (3.11)) happens to be equivalent to (3.10) because of a relation of the type $T|g_j| \leq a|Tg_j| = a|g_j^*|$ (for all $y \in Y$ ; $j = 1, \ldots, n$) where $a$ denotes a positive constant.

Also note that $\mathfrak{M}^*(c)$ contains all measures $\nu \in M^*$ which satisfy (3.11) and have a finite support contined in $Y_o$. For, then (3.10) is trivially satisfied, compare the definition of $Y_o$.

It follows from the definitions of $\mathfrak{M}(c)$ and $\mathfrak{M}^*(c)$ and the comments in (3.3) that

(3.12) $$\mathfrak{M}(c) = \{\mu\colon\ \mu = T\nu \quad \text{with} \quad \nu \in \mathfrak{M}^*(c)\}\ .$$

Thus, for $c \in R^n$ we have $c \in M$ if and only if $\mathfrak{M}(c)$ is non-empty if and only if $\mathfrak{M}^*(c)$ is non-empty.

If desired, one may as well assume that the $n + 1$ functions $g_o^*(y)$ , $g_1^*(y), \ldots, g_n^*(y)$ on $Y_o$ are linearly independent (on $Y_o$) , since, otherwise, some of the moment conditions (3.11) would be redundant. In order that the $g_j^*$ ($j = 0, 1, \ldots, n$) are linearly independent on $Y_o$ it is necessary but not sufficient that the functions $g_j$($j = 0, 1, \ldots, n$) on $X$ are linearly independent.

(3.13) Let us finally transfer the best lower bound $L(c)$ defined by (2.5) to the space $Y$ . Thus consider a point $c \in R^n$ such that (2.7) holds; (if not then we would have $L(c) = -\infty$) . Recall that $0 \leq (h^*)^- = (Th)^- \leq Th^-$ . By (2.7), we have for each $\mu \in \mathfrak{M}(c)$ that $\int h^-\, d\mu < \infty$ . Hence, by (3.7) and (3.12), we have $\int Th^- d\nu < \infty$ , for all $\nu \in \mathfrak{M}^*(c)$ , in particular, $\nu(\mathfrak{D}(h^*)) = 1$ for all

$\nu \in \mathfrak{M}^*(c)$ .

Finally, we have for each $\nu \in \mathfrak{M}^*(c)$ that

$$\int (h^*)^- \, d\nu < \infty \quad \text{and} \quad \int h^* \, d\nu = \int h \, d\mu > -\infty ,$$

where $\mu = T\nu$ and each integral is well-defined. We conclude that the definition (2.5) of $L(c)$ may be rewritten as

$$L(c) = \inf \left\{ \int h^* \, d\nu : \nu \in \mathfrak{M}^*(c) \right\} . \tag{3.14}$$

Naturally, $L(c) = \infty$ if $\mathfrak{M}^*(c)$ happens to be empty.

The transferred moment problem now consists of the relations $\nu \in \mathfrak{M}^*$; (3.10), (3.11) (defining $\mathfrak{M}^*(c)$); and (3.14). One can even transfer the condition (2.7) on $c$ , by writing it as

$$\int Th^- \, d\nu < \infty \quad \text{for all} \quad \nu \in \mathfrak{M}^*(c) . \tag{3.15}$$

If the latter condition is not satisfied then $L(c) = -\infty$ .

## 4. Auxiliary results

In the present Section, we record a few more or less known results which will often be helpful in solving the transferred moment problem.

In the sequel, $g^* : Y_o \to R^n$ will denote the function defined by

$$g^*(y) = (g_1^*(y), \ldots, g_n^*(y)) . \tag{4.1}$$

We know, see (3.9), that for each $c \in R^n$ and each $\nu \in \mathfrak{M}^*(c)$ the measure $\nu$ is supported by $Y_o$ while $\int g^*(y) \, \nu(dy) = c$ . We conclude that $c \in \text{conv}[g^*(Y_o)]$ as soon as $\mathfrak{M}^*(c)$ is non-empty; the latter is equivalent to $\mathfrak{M}(c)$ being non-empty and also to $c \in M$ . Therefore, (2.3) can be improved to

$$M \subset \text{conv } g^*(Y_o) \subset \text{conv } g(X) . \tag{4.2}$$

The latter inclusion can be seen as follows. Consider a

point $c \in \text{conv}[g^*(Y_o)]$ . Equivalently, $c = \int g^* \, d\nu$ with $\nu$ as a probability measure on $Y_o$ of finite support $\{y_1, \ldots, y_k\} \subset Y_o$; (it is not necessary that $\nu$ belongs to $\mathfrak{M}^*$ ; one can always attain that $k \leq n + 1$ ; if the set $g^*(Y_o)$ in $R^n$ happens to be connected one can even attain that $k \leq n$) . Next, let $\mu = T\nu$ denote the corresponding probability measure on $X$ . Using the definition of $Y_o$ , see (3.8), it follows that $\int |g| \, d\mu < \infty$ and we conclude that

$$c = \int g^* \, d\nu = \int Tg \, d\nu = \int g \, d\mu \in \text{conv } g(X) .$$

This establishes (4.2).

Observe that the convex set $\text{conv}[g^*(Y_o)]$ in $R^n$ has a non-empty interior if and only if the $n + 1$ functions $g_o^* \equiv 1, g_1^*, \ldots, g_n^*$ on $Y_o$ are linearly independent.

(4.3) In the remaining part of the present Section we will restrict ourselves to the following special case. For a more general treatment, compare [10], [12], [14] and [25].

ASSUMPTION: the class $\mathfrak{M}^*$ contains all probability measures $\nu$ on $Y$ which have a finite support $\{y_1, \ldots, y_k\}$ which is contained in $Y_o$ .

It follows from (3.2) that the original class $\mathfrak{M}$ contains all probability measures $\mu$ on $X$ of the special form

$$\mu(A) = p_1 \, \mu_{y_1}(A) + \ldots + p_k \, \mu_{y_k}(A) . \tag{4.4}$$

Here, $\mu_y(A) = (T\epsilon_y)(A) = P(y, A)$ and $y_i \in Y_o$, $p_i \geq 0$, $\Sigma \, p_i = 1$ .

In the present case, the moment space $M$ is given by

$$M = \text{conv } g^*(Y_o) , \tag{4.5}$$

compare the remarks following (4.2). This implies, for instance, that $M$ is contained in a half space $a_o + \sum_{j=1}^{n} a_j c_j \geq 0$ if and only if $g^*(Y_o)$ is contained in that half space, equivalently,

$$\sum_{j=o}^{n} a_j g_j^*(y) = \int (\sum_{j=o}^{n} a_j g_j(x))\, P(y, dx) \geq 0 \quad \text{for all} \quad y \in Y_o \,.$$

DEFINITION. For each $c \in R^n$, let $\mathcal{M}_f^*(c)$ denote the collection of all probability measures $\nu$ on $Y$ having a finite support which is contained in $Y_o$ and such that $\int g^* \, d\nu = c$, (equivalently, (3.11) holds).

We have from (4.5) that $\mathcal{M}_f^*$ is non-empty if and only if $c \in M$. By the above assumption and a remark preceding (3.12), we further have

(4.6) $\mathcal{M}_f^*(c) \subset \mathcal{M}^*(c)$, for each $c \in R^n$.

(4.7) LEMMA. <u>Under the above assumptions, we have</u>

(4.8) $L(c) = \inf \{ \int h^* \, d\nu : \nu \in \mathcal{M}_f^*(c) \}$,

<u>for each</u> $c \in R^n$ <u>which satisfies</u> (2.7) <u>which is equivalent to</u> (3.15). <u>If this is not satisfied then</u> $L(c) = -\infty$.

(4.9) <u>Proof</u>. Let the right hand side of (4.8) be denoted by $L_f(c)$. It follows from (4.6) that $L(c) \leq L_f(c)$. We may as well assume that $L(c) < \infty$.

Let $\gamma$ denote any finite number with $L(c) < \gamma$. Choose $\nu \in \mathcal{M}^*(c)$ such that $\int h^* d\nu < \gamma$. By the definition (3.9) of $\mathcal{M}^*(c)$, we have $\int |g^*| \, d\nu < \infty$ and $\int g^* \, d\nu = c$. Moreover, using the assumption (3.15), $\int h^*)^- \, d\nu \leq \int (Th^-)\, d\nu < \infty$, hence, the integral $\int h^* \, d\nu < \gamma$ is well-defined and in fact absolutely convergent.

But then there always exists a probability measure $\tilde{\nu}$ on $Y_o$ of finite support and having exactly the same moments as $\nu$ relative to the functions $g_1^*, \ldots, g_n^*$ and $h^*$. It follows that

$$\tilde{\nu} \in \mathcal{M}_f^*(c) \quad \text{and} \quad \int h^* \, d\tilde{\nu} = \int h^* \, d\nu < \gamma,$$

consequently, $L_f(c) < \gamma$. Since $\gamma > L(c)$ is arbitrary, we have $L_f(c) \leq L(c)$ , thus, $L_f(c) = L(c)$ .

(4.10) Let us review some of the known techniques for computing a quantity $L(c)$ of the type (4.8). For convenience, assume that $h: X \to R$ satisfies a condition of the type (2.8) in which case each $c \in R^n$ satisfies (2.7) and hence (4.8). It also follows from (2.8) that $L(c) > -\infty$ for all $c \in R^n$ . Further, $L(c) < \infty$ if and only if $c \in M = \text{conv } g^*(Y_o)$ . In view of the remarks preceding (3.13) and (4.3), we may as well assume that $M$ has a non-empty interior.

A convenient way for determining $L(c)$ when $c \in \text{int}(M)$ is often as follows, (see [13] for further details). Select constants $d_o, d_1, \ldots, d_n$ such that

$$(4.11) \qquad d_o + \sum_{j=1}^{n} d_j g_j^*(y) \leq h^*(y) \quad \text{for all} \quad y \in Y_o .$$

Such constants can always be found. Clearly, (4.11) implies that $L(c) \geq \sum_{j=o}^{n} d_j c_j$ for all $c$ ; here, $c_o = 1$ .

(4.12) LEMMA. <u>Under the above assumptions we have for each</u> $c \in \text{int}(M)$ <u>that</u>

$$(4.13) \qquad L(c) = \max \sum_{j=o}^{n} d_j c_j ,$$

<u>where</u> $d = (d_o, \ldots, d_n)$ <u>ranges through the</u> $(n+1)$<u>-tuples of real numbers satisfying</u> (4.11). <u>Moreover, for almost all</u> $c \in \text{int}(M)$ <u>the above maximum is assumed by a unique</u> $(n+1)$<u>-tuple</u> $d$ .

(4.14) LEMMA. <u>Let</u> $c \in \text{int}(M)$ <u>be fixed such that the infimum</u> $L(c)$ <u>in</u> (4.8) <u>is assumed by at least one</u> $\nu \in \mathcal{M}_f^*(c)$ . <u>Then one can find an</u> $(n+1)$<u>-tuple</u> $d$ <u>satisfying</u> (4.11) <u>such that the infimum</u> (4.8) <u>is even assumed by a measure</u> $\nu \in \mathcal{M}_f^*(c)$ <u>carried by the "contact set"</u> $B(d)$ <u>of all points</u> $y \in Y_o$ <u>for which</u> (4.11) <u>holds with the equality sign</u>.

<u>For almost all</u> $c \in \text{int}(M)$ <u>for which</u> (4.8) <u>is assumed the latter</u> $(n+1)$<u>-tuple</u> $d$ <u>is unique. One can even attain that the support</u> $\{y_1, \ldots, y_k\} \subset B(d)$ <u>of</u> $\nu$ <u>(which achieves</u> $L(c)$<u>)</u> <u>contains</u> $k \leq n+1$ <u>points; (even</u> $k \leq n$ <u>if</u> $g^*(B(d))$ <u>is connected</u>).

(4.15) In many applications the set $Y_o$ happens to be a compact topological space (made into a measurable space by the $\sigma$-field of all Borel subsets), while the functions $g_j^*$ are continuous on $Y_o$ and $h^*$ is at least lower semi-continuous on $Y_o$. In that case we have for each $c \in M$ that the infimum in (4.8) is assumed (by some $\nu \in \mathcal{M}_f^*(c)$). It follows from the above Lemmas that then one needs to consider only admissible measures $\nu$ in the computation of $L(c)$ for $c \in \text{int}(M)$. Here, a measure $\nu$ is said to be admissible if it is a probability measure on $Y_o$ of finite support which is carried by the contact set $B(d)$ corresponding to some $(n+1)$-tuple $d$ satisfying (4.11). Observe that for <u>each</u> admissible measure $\nu$ we have

$$(4.16) \qquad L(c) = \int h^* \, d\nu \, , \quad \text{where} \quad c = \int g^* \, d\nu \, .$$

Letting $\nu$ run through all admissible measures, (4.13) yields a parametric representation of the function $L(c)$, at least as long as $c \in \text{int}(M)$.

## 5. A class of generalized convex distributions

Let $Z$ be a real random variable and consider the problem of minimizing $Eh(Z)$ subject to the moment conditions $Eg_j(Z) = c_j$ $(j = 1, \ldots, n)$ and further certain convexity conditions imposed on the distribution function $F$ of $Z$. In applying the general principles outlined in the previous sections, our first effort would be to remove these convexity conditions by a Markov transformation $\mu = T\nu$ as in (3.1) which relates the distribution $\mu$ of $Z$ to a rather unrestricted measure $\nu$. In fact, the present section is largely devoted to a detailed study of such a transformation relative to the following class $\mathcal{M}_s[x_o]$ of measures.

(5.1) DEFINITION. Let $s \geq 1$ be a fixed integer and let $x_o$ be a fixed real number. Then $\mathcal{M}_s[x_o]$ will denote the class of probability measures $\mu$ on $R$ such that the corresponding cumulative distribution function $F$ possesses an $(s-1)$-th derivative $F^{(s-1)}(x)$ throughout the interval $(x_o, \infty)$ and that further $(-1)^s F^{(s-1)}(x)$ is convex in $(x_o, \infty)$.

Sometimes (rarely and only when stated) we shall allow

that $s = 0$ in which case $\mathcal{M}_0[x_0]$ will stand for the class of all probability measures on $R$.

The above class $\mathcal{M}_s[x_0]$ was first used in this connection by van Dantzig [5] p. 38 in his generalization of the inequality of Camp and Meidell, see Section 7.

(5.2) Consider a measure $\mu \in \mathcal{M}_s[x_0]$ with c.d.f. $F$. Since $(-1)^s F^{(s-1)}$ is convex in $(x_0, \infty)$ it must be continuous there. Moreover, the right hand derivative $F^{(s)}(x+0)$ will exist and be finite for all $x > x_0$ in such a way that $(-1)^s F^{(s)}(x+0)$ is continuous to the right and non-decreasing. Further, $F$ is bounded, hence, $F^{(s)}(+\infty) = 0$, thus,

$$(-1)^s F^{(s)}(x) \leq 0 \quad \text{for all} \quad x > x_0;$$

(the s-th derivative of $F$ will always be taken as the right hand derivative). This in turn implies that $(-1)^{s-1} F^{(s-1)}(x)$ is non-decreasing so that $(-1)^{s-1} F^{(s-2)}(x)$ is convex (if $s \geq 2$). In this way, we have proved that

$$(5.3) \qquad \mathcal{M}_s[x_0] \subset \mathcal{M}_{s-1}[x_0], \qquad (s \geq 2).$$

Therefore,

$$(5.4) \qquad (-1)^{j-1} F^{(j)}(x) \geq 0 \quad \text{for all} \quad x > x_0;\ j = 1, \ldots, s,$$

as soon as $F$ corresponds to $\mu \in \mathcal{M}_s[x_0]$.

(5.5) The class $\mathcal{M}_s[x_0]$ is closed under convergence in distribution (and thus compact in the weak* topology). For, (compare [10] p. 376), it is not difficult to see that $\mu \in \mathcal{M}_s[x_0]$ if and only

$$(5.6) \qquad (-1)^s \Delta_h^{s+1} F(x) = (-1)^s \sum_{j=0}^{s+1} (-1)^{s+1-j} \binom{s+1}{j} F(x+jh) \geq 0$$

holds for each choice of $x_0 > 0$ and $h > 0$.

A measure $\mu \in \mathcal{M}_s[x_0]$ is completely arbitrary on the interval $(-\infty, x_0]$ and may, for instance, have a positive mass at $x_0$; but on the open interval $(x_0, \infty)$ it is always absolutely continuous with density $f(x) = F'(x+0)$.

Since $\mu$ also belongs to $\mathcal{M}_1[x_o]$ this density is even non-increasing.

It is useful to consider the class $\mathcal{F}_s[x_o]$ of all non-negative scalar multiples $\lambda f$ $(\lambda \geq 0)$ of such density functions $f$ corresponding to measures $\mu \in \mathcal{M}_s[x_o]$.

In view of (5.3) and (5.6), it is easily seen that a function $f$ on $(x_o, \infty)$ belongs to $\mathcal{F}_s[x_o]$ if and only if $(-1)^j \Delta_h^j f(x) \geq 0$ $(j=0,1,\ldots,s)$, for all $x > x_o$, all $h > 0$, and further $f$ is integrable on $(x_o, \infty)$. Thus $\mathcal{F}_s[x_o]$ is a convex cone in an obvious sense. Moreover, if $f_1$ and $f_2$ are in $\mathcal{F}_s[x_o]$ then so is their product $f_1 f_2$.

Next, by (5.3), we have that $\mathcal{F}_s[x_o] \subset \mathcal{F}_{s-1}[x_o]$. Let $\mathcal{F}_\infty[x_o]$ denote their intersection. Thus $f \in \mathcal{F}_\infty[x_o]$ if and only if it is integrable and completely monotone on $(x_o, \infty)$, (that is, $f(-x)$ has all its differences non-negative). It follows from a well-known theorem of Bernstein (compare [7] p. 416) that $f \in \mathcal{F}_\infty[x_o]$ if and only if on $(x_o, \infty)$ there is a representation of the form

$$f(x) = \int_0^\infty e^{-xy}(ye^{x_o y})\rho(dy), \qquad (x > x_o), \tag{5.7}$$

with $\rho$ as a finite nonnegative measure on $(0, \infty)$; (the finiteness of $\rho$ guarantees that $f$ will be integrable on $(x_o, \infty)$). For example, if $x_o = 0$ and $\epsilon > 0$ then

$$f(x) = \int_1^\infty e^{-xy}\, y^{-\epsilon}\, dy = x^{-1+\epsilon}\int_x^\infty e^{-z}\, x^{-\epsilon}\, dz$$

corresponds to a function in $\mathcal{F}_\infty[x_o]$; $f(x) \sim \Gamma(1-\epsilon)x^{-1+\epsilon}$ if $x$ is small and $0 < \epsilon < 1$.

Observe that (5.7) sets up a Markov transformation as in (3.2) between the class $\mathcal{M}^*$ of all probability measures $\rho$ on R and the class $\mathcal{M}_\infty[x_o]$; if $x_o = 0$ one takes $\mu(A) = \rho(A)$ when $A \subset (-\infty, x_o]$. This in turn can be used to remove a condition of the form $\mu \in \mathcal{M}_\infty[x_o]$, namely, simply by transferring the moment problem on hand to $\mathcal{M}^*$ as indicated in Section 3.

(5.8) Often one would be interested in measures $\mu \in \mathcal{M}_s[x_o]$ which are supported by an interval $(-\infty, x_1]$

where $x_1 < \infty$. In that case we have $F(x) = 1$ for all $x \geq x_1$ thus $F^{(j)}(x + 0) = 0$ for all $x \geq x_1$ and $j = 1, \ldots, s$. In particular, since $F^{(j)}(x)$ is continuous throughout $(x_0, \infty)$ for all $j = 1, \ldots, s - 1$, we have in this case that

$$(5.9) \qquad F^{(j)}(x_1) = 0 \quad \text{for} \quad j = 1, \ldots, s - 1 .$$

For $j = s$ we have that $(-1)F^{(s)}(x + 0)$ is non-decreasing, thus

$$(5.10) \qquad (-1)^s F^{(s)}(x_1 - 0) \leq 0 = (-1)^s F^{(s)}(x_1 + 0) .$$

(5.11) Let us now start with the promised Markov transformation of our class $\mathcal{M}_s[x_0]$.

To each $\mu \in \mathcal{M}_s[x_0]$ (with corresponding c.d.f. $F$ and density $f$ on $x > x_0$), we associate the following <u>nonnegative</u> measures

$$(5.12) \qquad \eta(A) = (-1)^s \int_A dF^{(s)}(u)$$

and

$$(5.13) \qquad \nu(A) = \int_A (u - x_0)^s / s! \; \eta(du) .$$

Here, $A$ runs through the Borel subsets of $(x_0, \infty)$. We will extend $\nu$ to a measure on all of $R$ by defining

$$(5.14) \qquad \nu(A) = \mu(A) \qquad \text{for each} \qquad A \subset (-\infty, x_0] .$$

Since $F^{(s)}(\infty) = 0$, the above measure $\eta$ is actually finite on each interval $[x_0 + \varepsilon, +\infty)$ with $\varepsilon > 0$. The measure $\nu$ on $R$ will turn out to be a probability measure.

If $\mu$ happens to be supported by an interval $(-\infty, x_1]$ then $F^{(s)}(x) = 0$ for all $x > \max(x_0, x_1)$. Consequently, $\mu$ is supported by $(-\infty, x_1]$ if and only if $\nu$ is supported by $(-\infty, x_1]$.

(5.15) LEMMA. <u>Let</u> $\mu \in \mathcal{M}_s[x_0]$ <u>and let</u> $\eta$ <u>and</u> $\nu$ <u>be as above</u>. <u>Then</u>

$$(5.16) \qquad (-1)^{s-1-j} F^{(s-j)}(x) = \int_x^\infty (u - x)^j / j! \; \eta(du) ,$$

*for all* $x > x_o$ *and all* $j = 0, 1, \ldots, s-1$; *(here, we interpret* $F^{(s)}(x)$ *as a right hand derivative). In particular,*

(5.17) $$f(x) = \int_x^{\infty} (u-x)^{s-1}/(s-1)!\ \eta(du) .$$

*Hence,*

(5.18) $$\bar{F}(x) = 1 - F(x) = \int_x^{\infty} (u-x)^s/s!\ \eta(du) \quad \textit{if} \quad x \geq x_o .$$

*In particular,*

(5.19) $$\nu((x_o, +\infty)) = \int_{x_o}^{\infty} (u-x_o)^s/s!\ \eta(du) = \bar{F}(x_o) ,$$

*showing that* $\nu$ *is in fact a probability measure.*

(5.20) *Proof.* From the definition (5.12) of $\eta$ and using that $F^{(s)}(+\infty) = 0$, we see that

$$(-1)^{s-1} F^{(s)}(x+0) = \int_{x+0}^{\infty} \eta(du) , \quad \text{if} \quad x > x_o .$$

Using (5.4) and Fubini, it follows by induction that (5.16) holds for $j = 0, 1, \ldots, s-1$ and all $x > x_o$. The special case $j = s-1$ of (5.16) easily implies the remaining assertions.

(5.21) In view of the definition (5.13) of $\nu$, (5.17) may be rewritten as

(5.22) $$f(x) = \int K_s(u, x)\ \nu(du), \qquad \text{if} \quad x > x_o ,$$

where the kernel $K_s(u, x)$ is defined by

(5.23) $$K_s(u, x) = \frac{s(u-x)^{s-1}}{(u-x_o)^s} \qquad \text{if} \quad x_o < x < u ,$$

$$= 0 , \qquad \text{otherwise.}$$

Note that $K_s(u, x) \geq 0$ and further

(5.24) $$\int K_s(u, x)\, dx = 1 \qquad \text{for all} \quad u > x_o .$$

(5.25) In view of (5.14) and (5.22), it is natural that to

each probability measure $\nu$ on R we associate another measure $\mu = T\nu$ defined as follows:

(i) if $A \subset (-\infty, x_o]$ then $\mu(A) = \nu(A)$ ;

(ii) if $A \subset (x_o, +\infty)$ then $\mu(A) = \int_A f(x)\,dx$;

here, f is defined by (5.22).

Using (5.24), we see that the resulting measure $\mu$ is also a probability measure. In fact, the present transformation T is precisely a Markov transformation

$$(5.26) \qquad (T\nu)(A) = \int P_s(u, A)\,\nu(du)$$

as discussed in Section 3, provided we take

$$(5.27) \qquad \begin{aligned} P_s(u, A) &= \epsilon_u(A) && \text{if } u \le x_o; \\ &= \int_A K_s(u, x)\,dx && \text{if } u > x_o . \end{aligned}$$

Here, and in the sequel, $\epsilon_u$ will denote the probability measure which is entirely concentrated at the point u . Let further $\mu_u = T\epsilon_u$ thus $\mu_u(A) = P_s(u, A)$ . If $u \le x_o$ then $\mu_u = \epsilon_u$ . If $u > x_o$ then $\mu_u$ is absolutely continuous (relative to Lebesgue measure) having a support $[x_o, u]$ and there a density $a(u - x)^{s-1}$ , $(x_o \le x \le u$; a is a constant depending on u).

If $s = 0$ were allowed we would take $P_o(u,A) = \epsilon_u(A)$ for all u and T would be the identity transformation, both for measures and for functions.

(5.28) THEOREM. Let the real number $x_o$ and the integer $s \ge 1$ be fixed. Define the transformation $\mu = T\nu$ as in (5.26), (5.27) with $K_s(u, x)$ as in (5.23).

We assert that the correspondence $\nu \to \mu = T\nu$ is a 1:1 correspondence between the class $\mathcal{M}^*$ of all probability measures on R on the one hand and the class $\mathcal{M}_s[x_o]$ on the other hand.

Thus, a probability measure $\mu$ on R belongs to $\mathcal{M}_s[x_o]$ if and only if its restriction to the interval

$(x_o, \infty)$ is absolutely continuous with a density $f$ admitting a representation as in (5.22) with $\nu$ as a nonnegative measure.

(5.29) Proof. It follows from Lemma 5.15 that each $\mu \in \mathcal{M}_s[x_o]$ has a representation as $\mu = T\nu$ with $\nu \in \mathcal{M}^*$.

Conversely, let $\nu$ be any probability measure on $R$ and put $\mu = T\nu$. We must show that $\mu \in \mathcal{M}_s[x_o]$ and further that $\nu$ is uniquely determined by $\mu$.

Since $T$ is a Markov transformation, the measure $\mu$ is clearly a probability measure. From the definition of $\mu$, we have $\mu(A) = \nu(A)$ for $A \subset (-\infty, x_o]$, while the restriction of $\mu$ to $(x_o, \infty)$ is absolutely continuous with a density defined by (5.22). The latter is equivalent to

$$(5.30) \qquad F'(x+0) = \frac{1}{(s-1)!}\int_x^{\infty} (u-x)^{s-1}\eta(du), \quad (x > x_o).$$

Here, $\eta$ is the nonnegative measure on $(x_o, \infty)$ defined by

$$\eta(du) = s!(u - x_o)^{-s}\,\nu(du).$$

Also observe that

$$(5.31) \qquad \int (u - x_o)_+^s\,\eta(du) < \infty.$$

It remains to show that the derivative $F^{(s-1)}(x)$ exists for $x > x_o$; that $(-1)^s F^{(s-1)}(x)$ is convex for $x > x_o$; and finally that $\eta$ can be retrieved from $F$, (so that $\nu$ is uniquely determined by $\mu$).

It suffices to prove, for $j = 1, \ldots, s$, that (*) the derivative $F^{(j)}(x)$ exists for all $x > x_o$ in such a way that

$$(5.32) \qquad (-1)^{j-1}F^{(j)}(x) = \frac{1}{(s-j)!}\int_{x+0}^{\infty} (u-x)^{s-j}\,\eta(du),$$

for all $x > x_o$; (for $j = s$ we mean the right hand derivative). For, clearly this implies the stated assertions, in particular, $\eta$ can be retrieved from $F$ as in (5.12).

For $j = 1$ the assertion (*) holds by (5.30). Let $1 \leq j \leq s - 1$ be fixed such that (*) holds. Let $x > x_o$, and consider the difference quotient

$$\frac{1}{h}[F^{(j)}(x + h) - F^{(j)}(x)]$$

$$= \frac{(-1)^j}{(s - j - 1)!} \int_{x+o}^{\infty} (u - x - \theta h)^{s-j-1} \eta(du)$$

$$+ \frac{(-1)^j}{(s - j)!} \int_{x+o}^{x+h+o} \frac{1}{h}(u - x - h)^{s-j} \eta(du)$$

(if $h > 0$ ; a similar formula holds for $h < 0$) . Here, we used the mean value theorem so that $\theta$ describes a quantity satisfying $0 < \theta < 1$ . If $j \leq s - 2$ then the latter integral tends to 0 as $h \to 0$ and (5.32) with $j$ replaced by $j + 1$ follows from the dominated convergence theorem, compare (5.31).

If $j = s - 1$ then the latter integral tends to 0 as $h$ tends to 0 , provided $x$ is a continuity point of $\eta$ , $(\eta(\{x\}) = 0)$ . Moreover, it tends to 0 as $h \downarrow 0$ for each $x > x_o$ since $\eta((x, x + h]) \downarrow 0$ as $h \downarrow 0$ ; (for an analogous proof, see [11] p. 331). This completes the proof of the Theorem.

(5.33) COROLLARY. If $\mu \in \mathcal{M}_s[x_o]$ and $F$ is the corresponding distribution function then

(5.34) $$\lim_{x \to \infty} x^j F^{(j)}(x) = 0 , \qquad (j = 1, \ldots, s) ,$$

and also

(5.35) $$\lim_{x \downarrow x_o} (x - x_o)^j F^{(j)}(x) = 0 , \ (j = 1, \ldots, s) .$$

An analogous result was obtained by Mulholland and Rogers [20] p. 209.

(5.36) Proof. In view of (5.3), it suffices to establish (5.34) and (5.35) for the case $j = s$ . It follows from

(5.12) and (5.19) that

$$\int_{x_o+0}^{\infty} (u - x_o)^s/s! \; d[(-1)^s F^{(s)}(u + 0)] < \infty ,$$

where $(-1)^s F^{(s)}(u)$ is non-decreasing toward 0 as $u \to \infty$.

Let $\epsilon > 0$ be given, and choose $a > x_o$ such that $\int_a^{\infty} < \epsilon$ ; (here, we refer to a truncation of the latter integral). Then we have for all $x > a$ that

$$\epsilon > \int_x^{\infty} \geq \frac{(x - x_o)^s}{s!} |F^{(s)}(x + 0)| ,$$

which implies (5.34). Next, let $a > x_o$ be such that $\int_{x_o}^{a} < \epsilon$ . Then we have for all $x_o < x < a$ that

$$\epsilon > \int_x^{a} \geq \frac{(x - x_o)^s}{s!} |F^{(s)}(x) - F^{(s)}(a)| ,$$

which implies (5.35); (it is possible for $(-1)^s F^{(s)}(x)$ to tend to $-\infty$ as $x \downarrow x_o$) .

(5.37) <u>Further comments</u>. Theorem 5.28 asserts that each $\mu \in \mathcal{M}_s[x_o]$ can be represented in a unique way as a (convex) mixture

$$\mu = \int \mu_u \, \nu(du)$$

of the special measures $\mu_u = T\epsilon_u$ , where $\nu$ is a unique probability measure. It follows that $\mu_u (u \in R)$ are precisely the extreme points of the convex set of probability measures $\mathcal{M}_s[x_o]$ .

Naturally, this relates to the fact that the mapping T from $\mathcal{M}^*$ onto $\mathcal{M}_s[x_o]$ is 1:1 and affine (linear). It may be of interest to note that this mapping is even a homeomorphism (provided we supply $\mathcal{M}^*$ and $\mathcal{M}_s[x_o]$ with the weak* topology).

Since T is 1:1 and $\mathcal{M}^*$ is compact, it suffices to verify that T is continuous. Thus, let $\nu_j \in \mathcal{M}^*$ converge to $\nu \in \mathcal{M}^*$ (in the weak* topology). We must prove that $\mu_j = T\nu_j$ converges to $\mu = T\nu$ .

Let $\phi$ be any bounded continuous function on R . We must prove that $\int \phi \, d\mu_j \to \int \phi \, d\mu$ . Introducing

(5.38) $$\phi^*(u) = (T\phi)(u) = \int \phi(x) \, P_s(u, dx) ,$$

(with $P_s$ as in (5.27)), we have

$$\int \phi \, d\mu_j = \int \phi \, d(T\nu_j) = \int (T\phi) \, d\nu_j = \int \phi^* \, d\nu_j .$$

Similarly, $\int \phi \, d\mu = \int \phi^* \, d\nu$ , hence, it suffices to verify that $\phi^*$ is continuous; (it is obviously bounded). But this follows immediately from the following more explicit formula for $T\phi = \phi^*$ , obtained from (5.23) and (5.27).

(5.39)
$$\begin{aligned} \phi^*(u) &= \phi(u) && \text{if } u < x_o; \\ &= \int_{x_o}^{u} K_s(u,x)\phi(x) \, dx \\ &= \frac{s}{(u - x_o)^s} \int_{x_o}^{u} (u - x)^{s-1} \phi(x) \, dx \\ &= \int_0^1 \phi((1 - t)u + tx_o) \, st^{s-1} \, dt && \text{if } u > x_o. \end{aligned}$$

(5.40) Let us now make a closer study of the present transformation $\phi^* = T\phi$ as given in (5.39). From now on, let $\phi$ denote a measurable function on $[x_o, \infty)$ which is locally integrable (relative to Lebesgue measure), so that $\phi^*(u)$ is well-defined and finite for all $u \in R$ . If $u > x_o$ we have from (5.39) that

(5.41) $$\frac{1}{s!}(x - x_o)^s \phi^*(u) = \frac{1}{(s - 1)!} \int_{x_o}^{u} (u - x)^{s-1} \phi(x) \, dx .$$

In particular, if $r > -1$ then

(5.42) $$\phi(x) = (u - x_o)^r \text{ implies } \phi^*(u) = \binom{r+s}{s}^{-1} (u-x_o)^r ,$$

for all $u > x_o$ .

It further follows from (5.41) that the image function $\phi^* = T\phi$ is not entirely arbitrary, in the sense that it always has certain special properties. Namely, the right hand side of (5.41) is precisely equal to the s-fold integral $I^s\phi$, where

$$(I\psi)(u) = \int_{x_o}^{u} \psi(x)\,dx ;$$

(this yields another way for deriving (5.42)). It follows that the right hand side of (5.41) admits in $(x_o, \infty)$ an (s - 1)-th derivative which itself is absolutely continuous having almost everywhere a derivative equal to the locally integrable function $\phi(u)$. Similarly, $\phi^*(u)$ itself has in $(x_o, \infty)$ an absolutely continuous (s - 1)-th derivative in such a way that

$$(5.43) \qquad \phi(x) = \frac{1}{s!}\left(\frac{d}{dx}\right)^s [(x-x_o)^s \phi^*(x)] = \sum_{j=0}^{s} \binom{s}{j} \frac{(x-x_o)^j}{j!} \left(\frac{d}{dx}\right)^j \phi^*(x) ,$$

for almost all $x > x_o$.

Observe that knowing $\phi$ the differential equation $(5.43)_s$ already determines $\phi^*$ up to a function of the form $\sum_{j=1}^{s} a_j (x - x_o)^{-j}$, where the $a_j$ are constants. Hence, (5.43) admits at most one solution $\phi^*$ which is locally integrable at $x_o$.

A further conclusion one may draw from (5.41) is that at each fixed value $x_1 > x_o$ the function $\phi^*$ tends to be more regular than $\phi$ by having there s - 1 more derivatives; and even s more derivatives if $\phi$ is continuous at $x_1$. At the point $x_o$ itself $\phi^*$ is at least regular as $\phi$.

(5.44) In transferring a moment problem (see Section 3), a condition $\int |\phi|\, d\mu < \infty$ imposed on $\mu \in \mathcal{M}_s[x_o]$ becomes the condition $\int (T|\phi|)\, d\nu < \infty$ for the corresponding probability measure $\nu$ on R, (compare (3.6)). Since the restrictions of $\mu$ and $\nu$ to $(-\infty, x_o]$ coincide, we only need to study the condition

$$\int_{x_o}^{\infty} |\phi(x)|\mu(dx) = \int_{x_o}^{\infty} |\phi(x)| f(x)\, dx < \infty$$

as compared to the condition $\int_{x_o}^{\infty} |\phi^*(u)|\ \nu\ (du) < \infty$ .

To judge the magnitude of the function $\phi^*(u)$ on $(x_o, \infty)$, observe that in (5.41) we have $|u - x| \leq |u - x_o|$ . In this way we arrive at the useful inequality

(5.45) $$|\phi^*(u)| \leq (T|\phi|)(u) \leq s(u - x_o)^{-1} \int_{x_o}^{u} |\phi(x)|\ dx ,$$

holding for all $u > x_o$ . Using (5.45) and Fubini, it is easily seen that

$$\int_{x_o}^{a} |\phi^*(u)| du < \infty \text{ as soon as } \int_{x_o}^{a} |\phi(x) \log(x - x_o)| dx < \infty .$$

If $\phi$ satisfies this condition (which is only slightly stronger than integrability) then the differential equation (5.43) has a unique solution $\phi^*$ which is integrable near $x_o$ .

In any case, we have from (5.45) that $(u - x_o)\phi^*(u) \to 0$ as $u \downarrow x_o$ . Moreover, $T|\phi|$ is bounded in any neighborhood of $x_o$ where $\phi$ is bounded.

(5.46) Counterexample. In connection with the comments in Section 3 following (3.7), let us now present an example of a locally integrable function $\phi$ on $(x_o, \infty)$, (taking $\phi(x) = 0$ for $x \leq x_o$), such that $\phi^* = T\phi$ is integrable relative to all $\nu \in \mathcal{M}^*$ , (in fact, the function $\phi^*$ is bounded), while nevertheless the original function $\phi$ is not integrable relative to all $\mu \in \mathcal{M}_s[x_o]$ . Hence, for such a function $\phi$ the condition $\int |\phi|\ d\mu < \infty$ imposed on $\mu \in \mathcal{M}_s[x_o]$ is not equivalent to the condition $\int |\phi^*|\ d\nu < \infty$ imposed on the corresponding measure $\nu \in \mathcal{M}^*$ , $(\mu = T\nu)$ .

For convenience, choose $x_o = 0$ . Let $\beta$ be any constant with $0 < \beta < \frac{1}{s}$ and take

(5.47) $$\phi(x) = \frac{1}{s!} \left(\frac{d}{dx}\right)^s [x^s \sin x^{-\beta}] .$$

Thus,

$$(5.48) \qquad \phi(x) \sim x^{-s\beta}[c_1 \begin{smallmatrix}\sin\\ \cos\end{smallmatrix}(x^{-\beta}) + 0(x^{\beta})] \qquad \text{as} \quad x \downarrow 0 ,$$

where $c_1$ is a non-zero constant. Since $s\beta < 1$ we have that $\phi$ is locally integrable. From (5.43) and (5.47) (and the remark following (5.45)), we have that

$$\phi^*(u) = \sin u^{-\beta} ,$$

which is a bounded function. On the other hand, we have that $\int_{x_0}^{\infty} |\phi(x)| \mu(dx) = \infty$ for many $\mu \in \mathcal{M}_s[x_o]$. For instance, choose $\mu \in \mathcal{M}_s[x_o] = \mathcal{M}_s[0]$ such that its density function $f$ satisfies $f(x) \sim c_2 x^{-1+\epsilon}$ as $x \downarrow 0$, where $c_2 > 0$ and $0 < \epsilon \leq s\beta$. This is always possible, compare the remark following (5.7).

## 6. Some concrete moment problems with convexity conditions

Let $s \geq 1$ be a fixed integer and let $x_o \geq 0$ be a given real number. In the present Section, we will consider a nonnegative random variable $Z$ having a distribution $\mu(A) = \Pr(Z \in A)$ of class $\mathcal{M}_s[x_o]$ as defined in (5.1). Thus the cumulative distribution function $F$ of $Z$ ($F(x) = \Pr(Z \leq x)$ ; $F(0-) = 0$) has in the interval $(x_o, \infty)$ a derivative $F^{(s-1)}$ such that $(-1)^s F^{(s-1)}$ is convex. It follows (see (5.3)) that for all $0 \leq j \leq s - 1$ the function $(-1)^{j+1} F^{(j)}(x)$ exists and is convex throughout $(x_o, \infty)$. Moreover, for all $0 \leq j \leq s$, the derivative $(-1)^j F^{(j)}(x + 0)$ exists and is non-decreasing in $(x_o, \infty)$.

In particular, $F$ is concave for $x > x_o$ with a non-increasing density $f(x) = F'(x + 0)$. If $s \geq 2$ then $f$ is even convex on $(x_o, \infty)$ and so on.

In many applications, it is more or less necessary to take $x_o > 0$. For, the density $f$ of $Z$ even if it exists everywhere will usually only decrease from a certain point $x_o$ on, compare the remarks in [3] p. 431. In order that $f$ be convex in $(x_o, \infty)$ one may have to choose a still larger value $x_o$.

It may be of interest to note that one can also handle

the moment problem where it is given that $F^{(s-1)}$ exists and is convex in an interval $(-\infty, x_1)$. For, this happens if and only if $\tilde{F}(x) = 1 - F(-x)$ corresponds to a measure in $\mathcal{M}_s[-x_1]$.

In the present Section, starting with (6.13), we assume actually that $x_o = 0$. The case $x_o > 0$ will be taken up again in Section 7.

(6.1) We shall now consider the problem of determining the best possible upperbound of the form

(6.2) $$\Pr(Z \geq z) \leq U = U(c) \text{ ,}$$

(with $z > 0$ as a given number), or the best possible lower bound of the form

(6.3) $$Eh(Z) \geq L(c) \text{ ,}$$

given that

(6.4) $$Z \geq 0 \text{ ; } \mu \in \mathcal{M}_s[x_o]$$

and that further

(6.5) $$Eg_j(Z) = c_j \quad \text{for } j = 1, \ldots, n \text{ .}$$

Here, the $g_j$ and $h$ are given Borel measurable functions on $[0, \infty)$ such that

(6.6) $$h(x) + \sum_{j=o}^{n} \gamma_j |g_j(x)| \geq 0 \text{ ,} \qquad \text{for all } x \geq 0 \text{ ,}$$

and at least one choice of the constants $\gamma_j$. Here, as usual, $g_o(x) = 1$ for all $x$. We will also assume that the functions $g_j$ are locally integrable on $[x_o, \infty)$, (relative to Lebesgue measure).

The set of all possible points $c = (c_1, \ldots, c_n)$ in $R^n$ (subject to (6.4) and (6.5)) will be denoted by $M$. Most of the time it will be allowed to take $s = 0$, in which case the condition $\mu \in \mathcal{M}_s[x_o]$ is void.

(6.7) In applying the results of the Sections 3 and 4, we take there $X = [0, \infty)$, $Y = [0, \infty)$, $\mathcal{M} = \mathcal{M}_s[x_o]$ while $\mathcal{M}^*$

is the collection of all probability measures on $[0, \infty)$. The Markov kernel $P(u, A)$ is presently defined by (5.23) and (5.27). Since the $g_j$ were assumed to be locally integrable on $[x_o, \infty)$, we have that the domain $\mathcal{D}_j$ of each transform $g_j^* = Tg_j$ and also their intersection $Y_o$ is all of $Y = [0, \infty)$, (compare (3.8)). Here, $\phi^* = T\phi$ is given by (5.39). In view of (4.5), since $\mathfrak{M}^*$ contains all measures of finite support on $X = [0, \infty)$, we have that

$$(6.8) \qquad M = \text{conv}\, g^*(X) .$$

Here, $g^*(u) = (g_1^*(u), \ldots, g_n^*(u))$. Finally, we conclude from Lemma 4.7 that for each $c \in M$ the best lower bound $L(c)$ in (6.3) coincides with the best lower bound in

$$(6.9) \qquad Eh^*(W) \geq L(c) .$$

Here, $W$ will stand for a nonnegative random variable taking only finitely many different values which satisfies the moment conditions

$$(6.10) \qquad Eg_j^*(W) = \int g_j^*(u)\nu(du) = c_j \qquad \text{for } j = 1, \ldots, n ,$$

where $\nu$ stands for the distribution of $W$. Provided $c \in \text{int}(M)$, the Lemmas 4.12 and 4.14 will often be very useful in determining the best lower bound $L(c)$ in (6.9). In particular, often one only needs to consider distributions $\nu$ of $Z$ of a finite support which is contained in a contact set $B(d)$. Here, $d$ stands for any $(n + 1)$-tuple of real numbers satisfying (4.11) (with $Y_o = [0, \infty)$) and $B(d)$ is precisely the set of points where (4.11) holds with the equality sign.

By the way, if necessary one could easily add conditions of the type $f(\xi) = q$ ($\xi > x_o$ and $q > 0$ given) to the original moment problem; for, by (5.22) this would only amount to an additional condition $EK_s(\xi, W) = q$ on $W$ which could easily be added to (6.10).

(6.11) The fact that in (6.9) and (6.10) one only needs to consider measures $\nu$ of finite support means that in the original moment problem it would be no loss of generality (as to the computation of $L(c)$) if one would assume that the distribution $\mu \in \mathcal{M}_s[x_0]$ of $Z$ has the special form

$$(6.12) \qquad \mu(A) = \sum_{i=1}^{k} p_i \, \mu_{y_i}(A) \, ,$$

where $p_i \geq 0$, $\Sigma p_i = 1$. Further, $\mu_y = T\epsilon_y$ equals $\epsilon_y$ when $y \leq x_0$. Otherwise (when $s \geq 1$), $\mu_y$ is an absolutely continuous measures with density $a(y - x)^{s-1}$ on $[x_0, y]$, (rectangular if $s = 1$; triangular if $s = 2$, etc.). Restricting the original moment problem to measures of the form (6.12) is more or less the same as solving the transformed moment (6.9), (6.10) with $W$ of finite support.

(6.13) The remaining part of the present Section will be concerned with the easier case $x_0 = 0$; (moment problems with $x_0 > 0$ may be found in Section 7). We now have that the distribution $\mu$ of $Z \geq 0$ belongs to the class $\mathcal{M}_s[0]$. Presently, $(-1)^s F^{(s-1)}(x)$ is convex for all $x > 0$ (if $s \geq 1$) but $\mu(\{0\}) = \nu(\{0\})$ can still be positive. The case $s = 1$ often arises in applications as $Z = |S - m|$, where $S$ is a random variable which is unimodal about $m$; similarly, for $s \geq 2$.

By (5.39), the transform $\phi^* = T\phi$ of $\phi$ is given by

$$(6.14) \qquad \phi^*(u) = su^{-s} \int_0^u (u - x)^{s-1} \phi(x)\, dx \, ,$$

(if $s \geq 1$, $u > 0$). Further, $\phi^*(0) = \phi(0)$. In particular,

$$\text{if} \quad \phi(x) = x^r \quad \text{then} \quad \phi^*(u) = \binom{r+s}{s}^{-1} u^r, \quad (r \geq 0) \, .$$

Therefore, the moment

$$(6.15) \qquad \alpha_r = EZ^r = \int_0^\infty x^r \, \mu(dx)$$

can also be expressed as

$$(6.16) \qquad \alpha_r = \binom{r+s}{s}^{-1} \beta_r \, ,$$

where

$$(6.17) \qquad \beta_r = \int_0^\infty u^r \, \nu(du) \, , \qquad (T\nu = \mu) \, .$$

Here, $\nu$ can be any probability measure on $[0, \infty)$. Hence, given the numbers $0 < r_1 < r_2 < \ldots < r_n$, the set of all possible n-tuples $(\beta_{r_1}, \beta_{r_2}, \ldots, \beta_{r_n})$ is exactly the convex hull of the curve $\{(u^{r_1}, \ldots, u^{r_n}): 0 \leq u < \infty\}$ in $R^n$. Using (6.16), we then also know the collection of possible n-tuples $(\alpha_{r_1}, \ldots, \alpha_{r_n})$. If the measure $\mu$ is assumed to have its support in $[0, b]$ then so has $\nu$ and one must restrict $u$ to $0 \leq u \leq b$.

In the special case $r_j = j$ $(j = 1, \ldots, n)$ it is known that the convex hull of the curve $\{(u, u^2, \ldots, u^n): 0 \leq u < \infty\}$ can also be described as the set of points $(\beta_1, \ldots, \beta_n)$ for which a number of simple determinants in terms of the $\beta_j$ are nonnegative, see [10] p. 173 (and **[10] p. 107 for** the case of a finite support $[0, b]$). Using (6.16), these become simple determinants in terms of the $\alpha_r$, $(r = 1, \ldots, n)$.

REMARK. The above results can be applied to the expected values $E(Z_j)$ of the order statistics $Z_1 \leq Z_2 \leq \ldots \leq Z_n$ based on a sample of $n$ independent observations on a random variable $0 \leq Z \leq 1$ having a <u>convex</u> distribution function $F$; we allow $F$ to have a discontinuity $F(1-) < F(1) = 1$ but, otherwise, $F$ must be continuous. We have

$$E(Z_j) = \int_0^1 \phi_j(F(x))\, dx,$$

where $\phi_j$ denotes the polynomial

$$\phi_j(w) = \sum_{i=o}^{j-1} \binom{n}{i} w^i (1-w)^{n-i} = \sum_{k=i}^{n} a_{jk} w^k, \qquad \text{(say)}.$$

The inverse function $x = G(y)$ of $y = F(x)$ is a completely arbitrary <u>concave</u> distribution function on $[0, 1]$, while

$$E(Z_j) = \int_0^1 \phi_j(y)\, dG(y) = \sum_{k=i}^{n} a_{jk}\, \alpha_k,$$

where

$$\alpha_k = \int_0^1 y^k\, dG(y).$$

Therefore, the set $D$ of all possible n-tuples

$(E(Z_1), \ldots, E(Z_n))$ is a simple linear transformation of the known convex set of all possible n-tuples $(\alpha_1, \ldots, \alpha_n)$. Namely, the latter is precisely the convex hull of the curve $\{(\frac{1}{2}u, \frac{1}{3}u^2, \ldots, \frac{1}{n+1}u^n): 0 \le u \le 1\}$. The extreme points of D correspond to the distribution functions F which are linear on [0, 1) but may have a jump at 1. The boundary points of D correspond to piecewise linear F with about n/2 discontinuities for $f = F'$; again there may be a jump at 1.

(6.18) As an illustration of the general problem (6.3), (6.4), (6.5), let us now study the best lower bound L(c) for Eh(Z) given that $Z \ge 0$ has its distribution $\mu$ in $\mathcal{M}_s[0]$ and given the power moments

$$(6.19) \qquad EZ^{r_j} = \alpha_{r_j}, \quad \text{for } j = 1, \ldots, n.$$

Here, $0 < r_1 < \ldots < r_n$ are fixed. Defining $\beta_r$ as in (6.16), this becomes the best lower bound (6.9), given that $W \ge 0$ takes only finitely many values and satisfies the moment conditions

$$(6.20) \qquad EW^{r_j} = \beta_{r_j}, \quad \text{for } j = 1, \ldots, n;$$

(by the way, much of what follows can be extended to Tchebycheff systems other than $g_j(x) = x^{r_j}$, $j = 1, \ldots, n$).

We shall initially assume that Z is also subject to the condition $0 \le Z \le b$ with b finite; the case $b = \infty$ may afterwards be derived as a monotone limit. Assuming that h is integrable on [0, b] and continuous at 0, we see from (6.14) that h* is continuous on [0, b] in which case the infimum (6.9) (with the extra condition $0 \le W \le b$) is even assumed. It follows from Lemma 4.14, and the comments in (4.15), that in computing the best lower bound L(c) in (6.9) one needs to consider only measures $\nu(A) = \Pr(W \in A)$ of finite support which are carried by a contact set B(d), (provided at least the moment point c is in the interior of the moment space M of all possible moment points; this moment space depends on b).

In the present case, $d = (d_0, d_1, \ldots, d_n)$ ranges through the (n + 1)-tuples such that

$$(6.21) \qquad d_o + \sum_{j=1}^{n} d_j u^{r_j} \leq h^*(u) \quad \text{for all} \quad 0 \leq u \leq b \ .$$

The corresponding set $B(d)$ is defined as the set of points $0 \leq u \leq b$ such that (6.21) holds with the equality sign. Probability measures carried by such contact sets always achieve the infimum in (6.9). Consequently, the study of the function $L(c)$ on $\text{int}(M)$ can more or less be *reduced to a study of possible contact sets*.

(6.22) The *size* of a contact set $B(d)$ may be defined either as the number $[B(d)]$ of points in $B(d)$ or as the sum $|B(d)|$ of the multiplicities of the roots $0 \leq u \leq b$ of the equation $\psi^*(u) = 0$. Here, $\psi^*$ will be defined by

$$(6.23) \qquad \psi^*(u) = h^*(u) - d_o - \sum_{j=1}^{n} d_j u^{r_j} \ .$$

In the case on hand, $\psi^*$ depends on $d$ and satisfies $\psi^*(u) \geq 0$ for all $0 \leq u \leq b$. The multiplicity of $\psi^*$ at a root $0 \leq u_o \leq b$ of the equation $\psi^*(u) = 0$ is meant as the multiplicity of $u_o$ as a zero of $\psi^*$ but the latter notion needs to be defined. For concreteness, we shall adopt the following definition (though others would be possible).

Consider a real function $\theta$ on the finite interval $[0, b]$, and let $0 \leq u_o \leq b$. The multiplicity of $u_o$ as a zero of $\theta$ will take to be zero unless the function $\theta$ is continuous in a neighborhood of $u_o$ and, moreover, $\theta(u_o) = 0$ in which case the multiplicity of $\theta$ at $u_o$ is $\geq 1$; (for $u_o = 0$ or $u_o = b$ one takes a one-sided neighborhood). The multiplicity of $\theta$ at $u_o$ is $\geq k$ (with $k$ as a positive integer) if in some neighborhood of $u_o$ the function $\theta$ admits $k - 1$ continuous derivatives and, moreover, $\theta^{(j)}(u_o) = 0$ for $j = 0,1,\ldots,k-1$.

In most applications there would be a low upperbound $N$ (such as $N = 3$) on the number of points $u_o$ at which $\psi^* \geq 0$ can have a zero of multiplicity $1$. Then one has the following relation between $[B(d)]$ and $|B(d)|$:

$$(6.24) \qquad [B(d)] \leq \frac{1}{2}(N + |B(d)|) \ .$$

Next, observe that we have $\psi^* = T\psi$, where

(6.25) $$\psi(x) = h(x) - d_o - \sum_{j=1}^{n} d_j \binom{r_j+s}{s} x^{r_j}, \quad (0 \le x \le b) .$$

One may ask in how far the multiplicity sum $|B(d)|$ for $\psi^*$ is related to the behavior of the function $\psi$. In this direction, we have the following result which turns out to be quite useful in many moment problems.

(6.26) LEMMA. Let $\phi$ be a piecewise continuous function on $[0, b]$ which is continuous at $0$. Suppose that $\phi^* = T\phi$ has $k$ zeros at least in $[0, b]$ counted according to their multiplicity. Then $\phi$ itself must have at least $k$ zeros counted according to their multiplicity provided a change of sign of $\phi$ at a discontinuity $z$ ($\phi(z - 0) \le 0 \le \phi(z+0)$ or vice versa) is counted as a single zero.

(6.27) Proof. Recall (compare the comment preceding (5.44)) that at each point $u_o > 0$ the function $\phi^*$ has $s$ more derivatives than $\phi$. As to $u_o = 0$, if $\phi(x) = a + 0(x^{\gamma})$ for $x$ small then $\phi^*(u) = a + 0(u^{\gamma})$ for $u$ small (where $a$ and $\gamma$ denote constants).

Since $\phi^*$ has $k$ zeros in $[0, b]$ counting multiplicities, the function $x^s\phi^*(x)$ has $k + s$ zeros in $[0, b]$ counting multiplicities. It follows by induction (using Rolle's theorem) that the function $(\frac{d}{dx})^j(x^s\phi^*(x))$ has at least $k + s - j$ zeros in $[0, b]$ counting multiplicities $(j = 0, 1, \ldots s)$; note that for $j = s - 1$ this function is continuous and piecewise of class $C^1$ so that for $j = s$ a jump across zero must be counted also.

Taking $j = s$, one obtains the stated result. After all, we have from (5.43) (with $x_o = 0$ this time) that

$$s!\,\phi(x) = (\frac{d}{dx})^s (x^s\phi^*(x)) .$$

Actually, a careful analysis will even show that $\phi$ has even $k$ distinct zeros provided $\phi^*$ has a zero of multiplicity $\le 1$ at $x_o = 0$ and a zero of multiplicity $\le s + 1$ at any other point.

(6.28) As a more concrete application of the above ideas,

let us consider the problem of computing the best lower bound $L(c)$ on $Eh(Z)$ in the situation (6.19) for the case where the given function $h$ is piecewise constant. In other words, let

$$(6.29) \qquad h(x) = \sigma_j \quad \text{when} \quad z_j \le x < z_j, \qquad (j = 0, 1, \ldots, q) ,$$

where $0 = z_0 < z_1 < \ldots < z_q < z_{q+1} = b$ $(q \geq 1)$ and the $\sigma_j$ denote constants with $\sigma_{j-1} \neq \sigma_j$ ; (by the way, this moment problem has not even been solved for the special case $s = 0$, $q = 2$, $n \geq 3$ to which we will return in a separate paper).

Let $c \in \text{int}(M)$ be fixed and let $B(d)$ be a contact set carrying the random variable $W$ for which the infimum $Eh^*(W) \geq L(c)$ is assumed subject to the moment conditions (6.20). Here,

$$c_j = \alpha_{r_j} = \binom{r_j + s}{s}^{-1} \beta_{r_j} .$$

An upperbound on the number of points $[B(d)]$ in $B(d)$ is given by (6.24). If $s \geq 2$ then the function $\psi^*$ is quite regular, (namely, it has a continuous derivative and is piecewise analytic) so that $\psi^* \geq 0$ can have multiplicity 1 only at the endpoints 0 and $b$ . Thus (6.24) holds with $N = 2$ , yielding that

$$(6.30) \qquad [B(d)] \leq \frac{1}{2} |B(d)| + 1 \qquad \text{if} \quad s \geq 2 .$$

Similarly,

$$(6.31) \qquad [B(d)] \leq \frac{1}{2} |B(d)| + 1 + \frac{1}{2} q', \qquad \text{of} \quad s = 1 .$$

Here, $q'$ denotes the number of indices $j = 1, \ldots, q$ with $\sigma_{j-1} < \sigma_j$ $(0 \leq q' \leq q)$ . Namely, if $s = 1$ then $\psi^*$ is continuous and piecewise analytic. Besides 0 and $b$ the only candidates for a zero of multiplicity 1 of the function $\psi^* \geq 0$ are the points $z_1, \ldots, z_q$ . However, if $\sigma_{j-1} > \sigma_j$ then the slope of $\psi^*$ decreases jumpwise at $z_j$ and $z_j$ cannot be a zero of $\psi^*$ at all.

(6.32) LEMMA. <u>Let</u> $B(d)$ <u>be a contact set relative to the function</u> (6.29). <u>Then we have</u> $|B(d)| \leq n + 2q$ , <u>unless</u>

$d_1 = d_2 = \dots = d_n = 0$ .

(6.33) Proof. Recall that $k = |B(d)|$ denotes the number of zeros of $\psi^*$ in $[0, b]$ counting multiplicities. Here, $\psi^* = T\psi$ with $\psi$ as in (6.25). By Lemma 6.26, also $\psi$ itself has a least $k$ zeros counting multiplicities and also counting any change of sign of $\psi$ at a jump point $z_j$, $(j = 1, \dots, q)$ . Disregarding the latter, there are at least $k - q$ zeros.

Let $k_j$ denote the number of zeros of $\psi$ in the interval $z_{j-1} \leq x \leq z_j$ , counting multiplicities. From what we have just proved, $\sum_{j=0}^{q} k_j \geq k - q$ . But then the derivative $\psi'$ of $\psi$ will have at least $\Sigma(k_j - 1) \geq k - 2q - 1$ zeros counting multiplicities. But from (6.25), since $h$ is piecewise constant

$$\psi'(x) = \sum_{j=1}^{n} d_j \binom{r_j + s}{s} r_j x^{r_j - 1} .$$

Provided not all $d_1, \dots, d_n$ are equal to zero, $\psi'$ can have at most $n - 1$ zeros counting multiplicities. Hence, $k - 2q - 1 \leq n - 1$ thus $k \leq n + 2q$ which was to be proved.

(6.34) REMARKS. If $s = 0$ we can do better. Namely, here the above proof shows that $|B(d)| \leq n + q$ . Moreover, we have (6.24) with $N = q + 2$ , hence, $[B(d)] \leq \frac{1}{2}n + q + 1$ , a bound which is often sharp.

Observe that in the above proof we never used property (6.21). Hence, we really proved that for $h$ as in (6.29) and $d_1, \dots, d_n$ not all zero, the function $\psi^*$ defined by (6.23) can have at most $n + 2q$ zeros in the interval $[0, b]$ counting multiplicities.

Let $s \geq 0$ . It will be convenient to call a contact set each set $B$ which is contained in some contact set of the form $B(d)$ . We already observed (compare (6.21)) that the present moment problem can be reduced to a study of contact sets.

Observe that from (6.14), since $h$ is of the form (6.29), we have

(6.35) $$h^*(u) = \sigma_o + \sum_{j=1}^{q} (\sigma_j - \sigma_{j-1})(1 - z_j/u)_+^s ,$$

where $v_+ = v^+ = \max(0, v)$.

Let us briefly consider the contact set $B(0)$ corresponding to the choice $d_1 = \ldots = d_n = 0$ (and which was excluded in Lemma 6.32). In order that it be non-empty we must (compare (6.21)) take $d_o$ equal to the minimal value $\inf h^* = \min\{h^*(u) : 0 \leq u \leq b\}$. In other words, a set $B$ is a contact set contained in $B(0)$ if and only if we have $h^*(u) = \inf h^*$ for all $u \in B$. As is easily seen from (6.35), $B(0)$ is a (non-empty) union of single point sets $\{z_j\}$ and intervals $[z_{j-1}, z_j]$. Depending on the $z_j$ and $\sigma_j$, it is quite possible for $B(0)$ to contain infinitely many points.

On the other hand, a contact set $B$ not contained in $B(0)$ must always be finite and, in fact, the number $[B]$ of points in such a contact set cannot exceed a certain bound $K$ furnished by (6.30), (6.31) and (6.32); a related result with $q = 1$ for bell-shaped distributions was obtained by Mallows [17] p. 410.

Not each set $B$ with $[B] \leq K$ will be a contact set, however. One must also verify the condition that, for some $(n + 1)$-tuple $d = (d_o, d_1, \ldots, d_n) \neq (d_o, 0, \ldots, 0)$ satisfying (6.21), the equality sign in (6.21) holds for all $u \in B$. This verification may be involved, but is more or less unavoidable. In any case, we now have a definite program for handling the moment problem on hand.

(6.36) REMARK. Consider for the moment the limiting case $b = +\infty$. Here, it becomes interesting to see what happens if $s$ is large. Since $\beta_r = \alpha_r \binom{r+s}{r} \sim \alpha_r s^r/r!$, it may be more revealing to transform $u$ to the variable $y = s/u$. A typical term in (6.35) becomes

$$(1 - z_j/u)^s = (1 - z_j y/s)^s \to e^{-z_j y} = \int_{z_j}^{\infty} e^{-xy}\, y\, dx .$$

Thus, in the limit our transformed moment problem becomes the transformed moment problem relative to the class $\mathcal{M}_\infty[0]$, which is governed by the transformation (5.7), (with $x_o = 0$) .

(6.37) In order to have some more concrete answers, let us first consider the relatively easy case where $n = 1$ . Here, it is given that

$$\mu \in \mathcal{M}_s[0] \ ; \quad 0 \leq Z \leq b \ ; \qquad E(Z^r) = \alpha_r \ ,$$

where $s \geq 1$ , $b > 0$ and $r > 0$ are fixed. Further, $\mu$ denotes the distribution of $Z$ . As a standing notation, $\beta_r = \binom{r+s}{r} \alpha_r$ .

Consider a given integrable function $h$ on $[0, b]$ which is continuous at $0$ . Let $h^* = Th$ and $\mu = T\nu$ and note that

$$Eh(Z) = \int h \, d\mu = \mu(h) = \nu(h^*) = \int h^* \, d\nu \ .$$

Letting $\alpha_r$ be free, we have that the set of all possible pairs $(\alpha_r = \mu(x^r), \mu(h))$ coincides with the set of all possible pairs $(\binom{r+s}{s}^{-1} \nu(u^r), \nu(h^*))$, where $\nu$ runs through all probability measures on $[0, b]$ . In other words, the set of all possible pairs $(\beta_r , \mu(h))$ is __precisely the convex hull__ of the curve

$$\Gamma = \{(u^r , h^*(u)) : \ 0 \leq u \leq b\} \ .$$

In order to determine the smallest possible value $L(\alpha_r)$ of $\mu(h)$ , one must determine the lowest point in this convex hull which is on the vertical through $(\beta_r , 0)$ . This in turn leads in a natural way to a consideration of the tangent line $\xi_2 = d_o + d_1 \xi_1$ through $(\beta_r, L(\alpha_r))$ and supporting the curve $\Gamma$ from below. The set of values $u$ corresponding to the points where this tangent line touches the curve $\Gamma$ would be precisely the contact set $B(d)$ needed for the computation of $L(\alpha_r)$ . If $d_1 = 0$ this contact set may be infinite. If $d_1 \neq 0$ and $h$ is as in (6.29) (and in many other applications) the contact set would be finite.

(6.38) As a concrete illustration, suppose that

$$h(x) = 0 \quad \text{if} \quad 0 \leq x < z_1 \quad \text{or} \quad z_2 < x \leq b ;$$
$$= 1 \quad \text{if} \quad z_1 \leq x \leq z_2 ,$$

where $0 < z_1 < z_2 \leq b$. Observe that in this case $L(\alpha_r)$ would be equal to the best lower bound on $Pr(Z \notin [z_1, z_2])$. Thus,

$$(6.39) \qquad Pr(z_1 \leq Z \leq z_2) \leq 1 - L(\alpha_r)$$

would be a sharp upperbound (in terms of $s$, $b$, $z_1$, $z_2$ and $\alpha_r$). The function $h^*(u)$ is presently given by (6.35) with $q = 2$ ; $\sigma_o = 1$ ; $\sigma_1 - \sigma_o = -1$ ; $\sigma_2 - \sigma_1 = +1$. For example, in the special case $z_2 = b$ we have $h^*(u) = 1 - (1 - z_1/u)_+^s$ for $0 \leq u \leq b$.

The curve $\Gamma$ begins with a horizontal line segment from the point $(0, 1)$ (corresponding to $u = 0$) to the point $(z_1^r, 1)$. This is followed by a downward bulge from $(z_1^r, 1)$ to $(b^r, h^*(b))$; there is a discontinuity in the s-th derivative at the point $(z_2^r, h^*(z_2))$.

As is easily seen, the bottom part of conv($\Gamma$) has no horizontal line segments so that the contact set $B(0)$ can be neglected. Further, each contact point consists either of a single point $u$; or of a pair $\{0, u\}$; or of a pair $\{u, b\}$.

In the limiting case $b = \infty$, the contact set consists either of the unique point $u_1$ where $h^*$ is minimal, that is, with a horizontal supporting line, (namely, when $\beta_r \geq u_1^r$) ; or of a single point $u$ with $u_o \leq u \leq u_1$, (namely, when $u_o^r \leq \beta_r \leq u_1^r$); or it is the fixed pair $\{0, u_o\}$, (namely, when $\beta_r < u_o^r$). Here, $u_o$ corresponds to the point on $\Gamma$ for which a supporting tangent line passes through the beginning point $(0, 1)$ of $\Gamma$.

The reader will have no difficulty writing out the corresponding analytic formulae for the best upperbound in (6.39). For the case $s = 1$, $b = \infty$, these formulae were already given by Barlow and Marshall [2] p. 238, see also [1] p. 1236. See also (7.29) for the case $z_2 = b$.

As to the bounds (6.30), (6.31), (6.32), in the present case (where $n = 1$, $q = 2$, $q' = 1$) only the first bound happens to be sharp. Namely, it follows from the above geometrical considerations that presently every contact set $B(d)$ satisfies $[B(d)] \leq 2$ and $|B(d)| \leq 3$. However, there exist choices of the $z_j$ and $\sigma_j$ with $q = 2$ $(\sigma_o > \sigma_1 > \sigma_2)$ such that one can find a (single) contact set with $|B(d)| = 5$.

When $b = \infty$ and $s$ is large it is better to determine $L(\alpha_r)$ from the lowest point on the vertical through $(\alpha_r, 0)$ which is in the convex hull of the curve $\{((su)^r/\binom{s+r}{r}, h^*(su)): u \geq 0\}$ since this curve has an obvious limiting position as $s \to \infty$, compare (6.36).

(6.40) The remaining part of the present Section is concerned with the important application where it is given that $0 \leq Z \leq b$ has a distribution $\mu \in \mathcal{M}_s[0]$ and given power moments as in (6.19) and we ask for the best possible upperbound on $\Pr(Z \geq z)$, where $0 < z < b$ is a given number. This corresponds to the best lower bound in $Eh(Z) \geq L(c)$ where $h$ is defined by

$$\begin{aligned} h(x) &= 1 \quad \text{if} \quad x < z , \\ &= 0 \quad \text{if} \quad x \geq z . \end{aligned}$$

In this case,

$$\begin{aligned} 1 - h^*(u) &= 0 & \text{if} \quad 0 \leq u < z , \\ &= (1 - z/u)^s & \text{if} \quad u \geq z . \end{aligned}$$

Thus our problem largely reduces to a study of contact sets $B(d)$ corresponding to the inequality

$$\text{(6.41)} \qquad \begin{aligned} e_o + \sum_{j=1}^{n} e_j u^{r_j} &\geq 0 & \text{if} \quad 0 \leq u < z , \\ &\geq (1 - z/u)^s & \text{if} \quad z \leq u \leq b ; \end{aligned}$$

(compare (6.21); we are putting $e_o = 1 - d_o$ and $e_j = -d_j$ for $1 \leq j \leq n$). The corresponding contact set is the set of values $0 \leq u \leq b$ where (6.41) holds with the equality sign. The case $e_1 = e_2 = \dots = e_n = 0$ can be neglected since it only leads to the meagre contact set $\{b\}$. It follows from (6.30), (6.31) and (6.32) (where $q = 1$, $q' = 0$) that

$$|B(d)| \leq n + 2 \quad \text{and} \quad [B(d)] \leq 2 + n/2 .$$

(6.42) In the further discussion, we will restrict ourselves for convenience to the case $n = 2$, where two power moments $\alpha_{r_1}$ and $\alpha_{r_2}$ are given. The special case $s = 1$, $r_1 = 1$, $r_2 = 2$, $b = \infty$ has been completely determined by Royden [22] and Karlin and Studden [10] pp. 492-497.

In the present case $n = 2$ the sum $|B(d)|$ of the multiplicities for the contact set $B(d)$, relative to (6.41), satisfies $|B(d)| \leq 4$. Moreover, the multiplicity 1 can only occur at 0 or $b$, (not at $z$, even if $s = 1$). Consequently, all contact sets of interest must be of one of the following types:

$$B_o = \{0, u_o, b\}; B_1 = \{u_1', u_1''\}; \ B_2 = \{0, u_2\}; \ B_3 = \{u_3, b\} .$$

(6.43) The following reasoning is partly heuristic. Note that a contact set $\{u\}$ consisting of a single point $u$ is naturally quite possible. But altogether they can contribute only a set of (2-dimensional Lebesgue) measure 0 to the moment space $M$ of all possible points $c = (\alpha_{r_1}, \alpha_{r_2})$. Moreover, such contact sets $\{u\}$ are likely to be contained in a larger contact set $B_i$ and can thus be disregarded.

Recall further (see Lemma 4.12) that almost all moment points $c \in R^2$ arise from precisely one contact set. In other words, the different (maximal) contact sets make "largely disjoint" contributions to the moment space $M$.

Now the contribution of the 3-point set $B_o$ would have a positive (2-dimensional Lebesgue) measure, in fact, this contribution is a triangular region with its vertices corresponding to the measures $\mu_o$, $\mu_{u_o}$ and $\mu_b$. In view of this, one would expect $B_o$ to have 0 degrees of freedom; that is, the parameter $u_o$ in $B_o$ will have at most finitely many possible values.

Similarly, for each fixed $1 \leq i \leq 3$ , the contact sets of type $B_i$ will be of no great interest unless it has exactly one degree of freedom. Note that each fixed contact set $B_i$ would contribute a line segment to M and that M is of dimension 2 . Thus, one may expect that there is precisely one relation between $u_1'$ and $u_1''$ , while $u_2$ and $u_3$ are more or less free .

(6.44) Consider first the possible contact sets of the form $B_o = \{0, u_o, b\}$ with $0 < u_o < b$. One must be able to find a polynomial $e_o + e_1 u^{r_1} + e_2 u^{r_2}$ as in (6.41) attaining the equality sign in (6.41) at the points $0, u_o$ and $b$ ; (here, $0 < r_1 < r_2$ are preassigned). Since necessarily the corresponding multiplicity sum is $\leq 4$ , we must have that 0 and b are each of order 1 while $u_o$ is of order 2 . Clearly, $e_o = 0$ , hence, $z < u_o < b$ ; (for, $e_1 u^{r_1} + e_2 u^{r_2}$ cannot have a double zero at $0 < u_o < z$ without being identically zero) . Making use of the second remark (6.34), it is easily seen that the set $B_o = \{0, u_o, b\}$ with $z < u_o < b$ is a contact set <u>if and only if</u> the following system of equations has a solution with $e_1 > 0$ ; $e_2 < 0$ .

$$
\begin{aligned}
e_1 u_o^{r_1} + e_2 u_o^{r_2} &= (1 - z/u_o)^s , \\
r_1 e_1 u_o^{r_1 - 1} + r_2 e_2 u_o^{r_2 - 1} &= s(1 - z/u_o)^{s-1}(z/u_o^2) , \qquad (6.45) \\
e_1 b^{r_1} + e_2 b^{r_2} &= (1 - z/b)^s .
\end{aligned}
$$

Solving $e_1$ and $e_2$ from the first two equations, we see that the conditions $e_1 > 0$, $e_2 < 0$ are equivalent to the inequalities

$$(1 + s/r_2)z < u_o < (1 + s/r_1)z . \qquad (6.46)$$

In particular, the present case cannot occur unless $b > (1 + s/r_2)z$ . It remains to determine the numbers $u_o < b$ (if any) which satisfy (6.46) for which the system (6.45) has its determinant equal to 0 . In the special case that $0 < r_1 < r_2$ are integers this determinant would lead to an ordinary algebraic equation of degree $r_2 + s$ having $u_o = b$ as one of its roots.

(6.47) Next, consider the possible contact sets of type $B_1 = \{u', u''\}$ . Again we must be able to find constants $e_o$, $e_1$, $e_2$ satisfying (6.41) and with the equality sign at the points $u'$, $u''$, where $0 < u' < u'' < b$ . It is clearly impossible that $0 < u' < u'' < z$ . Let us restrict ourselves to the case $0 < u' < z < u'' < z$; (at least if $s = 1$ or $2$ , and possibly always, the situation $z < u' < u'' < b$ is impossible also).

Assuming further that $r_1 = 1$, $r_2 = 2$, the left hand side of (6.41) must be of the form $e_2(u - u')^2$ . At the point $u''$ the functions $e_2(u - u')^2$ and $(1 - z/u)^s$ have the same value and the same derivative, hence, the same logarithmic derivative. One finds that necessarily

$$u' = z - (\frac{2}{s}\frac{u''}{z} - 1)\ (u'' - z)\ .$$

In order that $0 < u' < z$ , one must have that $\max(1, \frac{s}{2}) < \frac{u''}{z} < \frac{s + 2}{2}$ . Also $u'' < b$ so that the present case can only occur when $b > (s/2)z$ . In any case, when $u''$ satisfies these inequalities the set $B_1 = \{u', u''\}$ will automatically be a contact set.

(6.48) Finally, one must investigate the possible contact sets $B_2 = \{0, u_2\}$ and $B_3 = \{u_3, b\}$ . For instance, suppose that $r_1 = 1$, $r_2 = 2$ . Then $\{0, u_2\}$ is a contact set if and only if $z < u_2 < b$ and, moreover, the curve $y = \frac{1}{u}(1 - \frac{z}{u})^s$ has at $u_2$ a positive slope and lies below its tangent line. This turns out to be the case precisely when $\frac{1}{2}(s + 1) \leq u_2 < (s + 1)z$ .

Therefore, if indeed $\frac{1}{2}(s + 1)z \leq u_2 \leq \min(b, (s + 1)z)$ (and only then) it will be true that the special random variable $Z'$, which has a positive mass at $0$ but otherwise is absolutely continuous having its density proportional to $(u_2 - x)_+^{s-1}$ , will have the largest value $\Pr(Z \geq z)$ among all random variables $0 \leq Z \leq b$ having the same first and second moments and having its distribution of class $\mathfrak{M}_s[0]$ .

This then describes (for the case $r_1 = 1$, $r_2 = 2$) the contribution made to the moment space $M$ by contact sets of the type $B_2 = \{0, u_2\}$ , in our computation of the best upper bound on $\Pr(Z \geq z)$ when regarded as a function on M (assuming again that $0 \leq Z \leq b$ has a distribution of class $\mathcal{M}_s[0]$). We shall not go into any further details.

## 7. Inequalities of the Camp-Meidell type

In the present Section, $Z$ will denote a nonnegative random variable with a single prescribed power moment

$$E(Z^r) = \alpha_r .$$

Here, $r$ is a fixed positive real number. Let further

$$F(x) = \Pr(Z \leq x); \qquad \bar{F}(x) = 1 - F(x) = \Pr(Z > x).$$

Already in 1821 Gauss proved that

$$\Pr(Z \geq z) \leq \frac{4}{9}\frac{\alpha_2}{z^2}, \tag{7.1}$$

given that $F$ is concave everywhere on $[0, \infty)$, see Cramèr [4], p. 183, 256 for a proof. There exists a large literature on (7.1) and its many generalizations, see Fréchet [8] pp. 156-168 for an extensive discussion of the older literature.

As to (7.1), the precise result is that

$$\begin{aligned} F(z - 0) &\geq (3\alpha_2)^{-1/2} z \quad \text{if } 0 < z \leq (\tfrac{4}{3}\alpha_2)^{1/2}, \\ &\geq 1 - (\tfrac{4}{9}\alpha_2) z^{-2} \quad \text{if } z \geq (\tfrac{4}{3}\alpha_2)^{1/2}, \end{aligned} \tag{7.2}$$

whenever $F$ is concave on $[0, \infty)$. The first bound is attained by the uniform distribution with range $[0, (3\alpha_2)^{1/2}]$. The second bound is attained by a distribution which has a mass at 0 but, otherwise, has a uniform distribution with range $[0, \frac{3}{2}z]$. The best lower bound on $F(z)$ would be 0 and is nearly attained by a large mass $1 - \epsilon$ at 0 together with a uniform distribution of mass $\epsilon$ over the interval $[0, (3\alpha_2/\epsilon)^{1/2}]$.

(7.3) In the sequel, $s$ will again denote a fixed positive integer and $x_0$ a fixed nonnegative real number. Further, $Z$ will denote a nonnegative random variable such that the distribution $\mu$ of $Z$ belongs to the class

$\mathcal{M}_s[x_o]$ . If $s = 1$ this means that $F$ is concave on $(x_o, \infty)$ . Finally, we impose the condition $Z \leq b$ where $b > x_o$ is prescribed, allowing $b = \infty$ . We also assume that we know the moment $E(Z^r) = \alpha_r$ where $r > 0$ is fixed.

We shall be interested in the quantity

$$(7.4) \qquad U = U(\alpha_r) = \sup_Z Eg(Z) ,$$

where $g$ is a given function while $Z$ ranges over all random variables which satisfy the above conditions. We shall only allow functions $g$ such that $g(x) \leq \gamma_o + \gamma_1 x^r$ for all $0 \leq x \leq b$ (with $\gamma_o$ and $\gamma_1$ as suitable positive constants) in which case $U(\alpha_r) < \infty$ .

The literature is largely concerned with the case $s = 1$ (often with $x_o = 0$) see the work of Camp [3], van Dantzig [5], Fréchet [8], Meidell [18], von Mises [19], Narumi [21], Royden [22], Smith [26] and Ulin [27]. Compare [28] pp. 94-97.

Most of our results below for prescribed power moments $EZ^r = \alpha_r$ easily extend to a given moment of the form $E\phi(Z) = c$ , provided the function $\phi$ on $[0, b]$ is non-decreasing and $\phi(0) = 0$ .

(7.5) We know from Theorem 5.28 that there is a 1:1 correspondence $\mu = T\nu$ between $\mathcal{M}_s[x_o]$ and the class $\mathcal{M}^*$ of all probability measures $\nu$ on $R$ . In order that $\mu$ is carried by $[0, b]$ it is necessary and sufficient that $\nu$ be carried by $[0, b]$ . If $\mu = T\nu$ then $\mu$ is absolutely continuous on $(x_o, b]$ and there its density $f$ is given by

$$(7.6) \qquad f(x) = \int_{x+0}^{\infty} \frac{s(u - x)^{s-1}}{(u - x_o)^s} \nu(du),$$

$$\text{for all } x_o < x \leq b ;$$

(if $s = 1$ then $f$ may jump and will be taken as the right hand derivative $f(x) = F'(x + 0)$ which is continuous to the right; the jump of $f$ at $x$ would equal $(u - x)^{-1} \nu(\{x\}))$ . Moreover,

$$(7.7) \qquad \mu(A) = \nu(A) \qquad \text{if } A \subset [0, x_o] .$$

Finally,

$$(7.8) \qquad E\phi(Z) = \int \phi(x)\ \mu(dx) = \int \phi^*(u)\ \nu(du) ,$$

(provided $\phi$ is Borel measurable and $E|\phi(Z)| < \infty$) . Here, $\phi^* = T\phi$ is defined by (5.39).

The important function $\psi^* = T\psi$ associated to $\psi(x) = x^r$ will be denoted as $\psi_r^*(u)$ Thus,

$$(7.9) \qquad \begin{aligned} \psi_r^*(u) &= \int_0^1 [(1-t)u + tx_0]^r\ st^{s-1}\ dt && \text{if } u > x_o , \\ &= u^r && \text{if } 0 \leq u \leq x_o . \end{aligned}$$

In particular,

$$(7.10) \qquad \psi_r^*(u) = \binom{r+s}{s}^{-1} u^r \qquad \text{if } x_o = 0 .$$

Note that the function $\psi_r^*(u)$ is continuous and strictly increasing in $u$ ; in the region $0 \leq x_o < u$ it is also strictly increasing in $x_o$ and strictly decreasing in $s$ .

In the special case that $r$ is a positive integer, (7.9) yields that

$$\psi_r^*(u) = \binom{r+s}{s}^{-1} \sum_{k=o}^{r} \binom{k+s-1}{s-1} x_o^k\ u^{r-k}, \qquad \text{for all } u > x_o .$$

If $s = 1$ we have from (7.9)

$$\psi_r^*(u) = \frac{1}{r+1}(u^{r+1} - x_o^{r+1})/(u - x_o), \qquad \text{for all } u > x_o ,$$

whether or not $r$ is an integer.

Note that $\psi_r^*(u)$ and $g^*(u)$ could also be defined as the moments $\int x^r \mu_u(dx)$ and $\int g(u)\mu_u(dx)$ of the special measure $\mu_u = T\epsilon_u$ , (which coincides with $\epsilon_u$ if $0 \leq u \leq x_o$ and, otherwise, is absolutely continuous with support $[x_o, u]$ and distribution $a(u - x)^{s-1}\ dx$) .

The following theorem can be interpreted as saying

that for the computation of $U(\alpha_r)$ as in (7.4) it suffices to consider measures $\mu \in \mathcal{M}_s[x_o]$ which are finite linear combinations of these special measures $\mu_u = T\epsilon_u$, (compare an analogous remark in (6.11)).

(7.11) THEOREM. *Consider the planar curve defined by*

(7.12) $$\Gamma(g) = \{\psi_r^*(u),\ g^*(u)) : 0 \leq u \leq b\} .$$

*Then*

(7.13) $$U(\alpha_r) = \sup\{w: (\alpha_r, w) \in \text{conv}\,\Gamma(g)\} .$$

*In particular, letting*

$$E_r = \sup_{0 < u \leq b} g^*(u)/\psi_r^*(u) ,$$

*we have* $U(\alpha_r) \leq E_r\,\alpha_r$ *(whatever* $\alpha_r$*) and here the constant* $E_r$ *cannot be replaced by any smaller constant.*

(7.14) *Proof.* Let

(7.15) $$M(g) = \{c = (EZ^r,\ Eg(Z))\}$$

denote the (convex) subset of $R^2$ consisting of all pairs of moments which can arise from a random variable $0 \leq Z \leq b$ having a distribution $\mu \in \mathcal{M}_s[x_o]$; ($\alpha_r$ is free now). Here, we only allow the case that $E|Z^r| < \infty$ and $E|g(Z)| < \infty$. Moreover, from (7.4),

(7.16) $$U(\alpha_r) = \sup\{w: (\alpha_r, w) \in M(g)\} .$$

Using (7.8), we can rewrite (7.15) as

$$M(g) = \{c = (\int \psi_r^*\,d\nu\ ,\ \int g^*\,d\nu)\} ,$$

where $\nu$ ranges through the collection of all probability measures on $[0, b]$ such that both $\psi_r^* = T(x^r) = T(|x^r|)$ and $T|g^*|$ are integrable relative to $\nu$, (compare (3.6)). Defined in the latter way, the set $M(g)$ would not be affected by adding the requirement that the probability measure $\nu$ on $[0, b]$ be of finite support. It follows that

(7.17) $M(g) = \text{conv}\,\Gamma(g)$ ,

where $\Gamma(g)$ is the set defined by (7.12)). Now (7.16) implies (7.13).

(7.18) As an important illustration of (7.12), let us now study the following generalization of (7.2). Namely, we want to know the best possible upperbound in

(7.19) $\Pr(Z \geq z) \leq U_s(\alpha_r, z)$ ,

where $0 < z < b$ is a given number. The corresponding optimal lower bound will be denoted by $L_s(\alpha_r, z)$ . As usual, we are assuming that $0 \leq Z \leq b$ has a distribution of class $\mathcal{M}_s[x_o]$ while $EZ^r = \alpha_r$ . Actually, the special case $x_o = 0$ of this problem is already covered by the treatment in (6.38), but there we did not write out the resulting explicit formulae.

Clearly, the best upperbound in (7.19) is precisely the quantity $U(\alpha_r)$ defined by (7.4), provided we replace there $g$ by the following function $g_z$ . Namely,

$$g_z(x) = 0 \quad \text{if} \quad 0 \leq x < z ,$$

$$= 1 \quad \text{if} \quad x \geq z .$$

It follows from (5.39) that $g_z^* = Tg_z$ is given by $g_z^* = g_z$ if $z \leq x_o$ , while in the case $z > x_o$ we have

$$g_z^*(u) = 0 \qquad \text{if} \quad u < z ,$$

$$= \left(\frac{u - z}{u - x_o}\right)^s \quad \text{if} \quad u \geq z .$$

In view of Theorem 7.11, we must determine the "upper boundary" of the convex hull of the curve $\Gamma(g_z)$ defined by (7.12) (with $g$ replaced by $g_z$ and, hence, $g^*$ by $g_z^*$) .

The case $z \leq x_o$ happens to be rather simple. Namely, in that case $g_z^* = g_z$ so that $\Gamma(g_z)$ consists of two horizontal line segments, namely, the segments $\{(v, 0): 0 \leq v < \psi_r^*(z) = z^r\}$ and $\{(v, 1): z^r \leq v \leq \psi_r^*(b)\}$ .

It follows from (7.13) that, for $z \leq x_o$ ,

$$U_s(\alpha_r, z) = \alpha_r z^{-r} \quad \text{if} \quad 0 \leq \alpha_r \leq z^r ,$$

(7.20)

$$= 1 \quad \text{if} \quad z^r \leq \alpha_r \leq \psi_r^*(b) .$$

Similarly, again for $z \leq x_o$ ,

$$L_s(\alpha_r, z) = 0 \quad \text{if} \quad 0 \leq \alpha_r \leq z^r ,$$

(7.21)

$$= (\alpha_r - z^r)/(\psi_r^*(b) - z^r)$$

$$\text{if} \quad z^r \leq \alpha_r \leq \psi_r^*(b) .$$

Naturally, one has in all cases that $\alpha_r \leq \psi_r^*(b)$ , (in order that $\alpha_r$ can be the r-th moment of $0 \leq Z \leq b$ with distribution in $\mathcal{M}_s[x_o]$) . If $x_o = 0$ this means that $\alpha_r \leq \binom{r+s}{s}^{-1} b^r$ .

(7.22) Let us now turn to the case $z > x_o$ . Then the curve $\Gamma(g_z)$ defined by (7.12) consists of the horizontal line segment

(7.23) $\quad \Gamma'_z = \{(v, 0): 0 \leq v \leq \psi_r^*(z)\}$ ,

together with the curve

(7.24) $\quad \Gamma''_z = \{(\psi_r^*(u) , (\frac{u - z}{u - x_o})^s): z \leq u \leq b\}$ .

It remains to determine the convex hull

(7.25) $\quad M_z = \text{conv}(\Gamma'_z \cup \Gamma''_z)$ .

After all, the upper boundary of $M_z$ determines $U_s(\alpha_r, z)$ and the lower boundary of $M_z$ determines $L_s(\alpha_r, z)$ . The possible values $\alpha_r$ are again given by $0 \leq \alpha_r \leq \psi_r^*(b)$.

(7.26) For a while (until (7.44)) we will assume that $x_o = 0$ . Then, by (7.10),

$$\psi_r^*(u) = (\lambda u)^r , \quad \text{where} \quad \lambda = \lambda_{r,s} = \binom{r+s}{s}^{-1/r} .$$

Thus

$$\Gamma''_z = \{(\lambda\ u)^r\ ,\ (1 - \frac{z}{u})^s\ :\ z \leq u \leq b\}$$

(7.27)

$$= \{(v,\ \chi_z(v)\ :\ (\lambda\ z)^r \leq v \leq (\lambda\ b)^r\}\ .$$

Here,

(7.28) $\chi_z(v) = (1 - \lambda\ zv^{-1/r})^s$ for $v \geq (\lambda\ z)^r$ .

In particular, $\alpha_r \leq (\lambda_{r,s} b)^r$ thus $b \geq \alpha_r^{1/r}/\lambda_{r,s}$ . Note that $\lambda_{r,s}^{-r} = (r +_r s) \sim s^r/r!$ thus $b = \infty$ if $\alpha_r > 0$ and $\mu \in \mathcal{M}_s[0]$ for all $s \geq 1$ , that is, $\mu \in \mathcal{M}_\infty[0]$; this would also follow from (5.7).

The function $\chi_z(v)$ is continuous and strictly increasing. When $s = 1$ it is concave everywhere while for $s \geq 2$ there is a unique point of inflection at the value $v_1 = [\frac{r+s}{r+1}z]^r$ . This corresponds to a u-value equal to $u_1 = \frac{r+s}{r+1}z$. In fact, the curve $\Gamma''_z$ is concave in the interval where $z \leq u \leq \min(\frac{r+s}{r+1}z\ ,\ b)$, while $\Gamma''_z$ is convex in the interval where $\frac{r+s}{r+1}z \leq u \leq b$ (that is, $(\frac{r+s}{r+1}\lambda z)^r \leq v \leq (\lambda b)^r$).

Another point on $\Gamma''_z$ of interest is the point for which the corresponding tangent line to the curve passes through the origin (which is the initial point of $\Gamma'_z$) . One finds that this happens if and only if $u = \frac{r+s}{r}z$ , thus, $v = (\frac{r+s}{r}\lambda z)^r$ . By the way, one may interpret $u = \frac{r+s}{r}z$ as the unique parameter of that special measure $\mu_u = T\epsilon_u^r$ for which $\int_z^\infty d\mu / \int x^r\, d\mu$ is maximal. Naturally, this value u is of no importance unless $b > \frac{r+s}{r}z$ . Also note that at this particular point $\chi_z(v) = (\frac{s}{r+s})^s$.

(7.29) The above remarks completely determine the upper boundary of the region (7.25) and one finds the following.

(i) If $\lambda z \leq \frac{r}{r+s}\alpha_r^{1/r}$ (which is equivalent to $\alpha_r \geq (\frac{r+s}{r}\lambda z)^r$ and implies that $z \leq \frac{r}{r+s}b$) then $U_s(\alpha_r, z)$ is given by

(7.30) $U_s(\alpha_r,\ z) = \chi(\alpha_r) = (1 - \lambda z\alpha_r^{-1/r})^s$ .

This upperbound is attained (only) by a special measure $\mu_u$ with parameter $\frac{r+s}{s}z \leq u \leq b$ having an r-th moment $\psi_r^*(u) = \alpha_r$. The corresponding cumulative distribution function is given by $F(x) = 1 - (1 - x/u)_+^s$ and is independent of $z$.

(ii) If $\frac{r}{r+s} \alpha_r^{1/r} \leq \lambda z \leq \frac{r}{r+s} \lambda b$ (that is, $\alpha_r \leq (\frac{r+s}{r} \lambda z)^r$ and $z \leq \frac{r}{r+s} b$) then $U_s(\alpha_r, z)$ is given by

$$(7.31) \qquad U_s(\alpha_r, z) = \frac{\alpha_r}{(\frac{r+s}{r}\lambda z)^r} \left(\frac{s}{r+s}\right)^s = \gamma_{r,s}\, \alpha_r z^{-r} ,$$

where

$$\gamma_{r,s} = \binom{r+s}{r} r^r s^s (r+s)^{-r-s} ,$$

($\gamma_{r,1} = (1 + r^{-1})^{-r}$). The upperbound (7.31) is attained (only) by a measure of the form $p\mu_o + (1-p)\mu_u$ where $\mu_o = \epsilon_o$ while $u = \frac{r+s}{s} z$. Further, $p$ is found from $(1-p)\psi_r^*(u) = \alpha_r$. The corresponding distribution function is given by

$$F(x) = 1 - (1-p)\left(1 - \frac{s}{r+s}\frac{x}{z}\right)_+^s .$$

(iii) If $\frac{r}{r+s} b \leq z \leq b$ then

$$(7.32) \qquad U_s(\alpha_r, z) = \alpha_r (\lambda b)^{-r} (1 - z/b)^r .$$

This case can occur only when $s \geq 2$. The upperbound (7.32) is attained (only) by a measure of the form $p\mu_o + (1-p)\mu_b$, thus, by a d.f. of the form $F(x) = 1 - (1-p)(1 - x/b)_+^s$.

REMARK. In the special case $s = 1$, we have from (7.30) and (7.31) that

$$U_1(\alpha_r, z) = 1 - z[(r+1)\alpha_r]^{-1/r}$$

if $z \leq \frac{r}{r+1} [(r+1)\alpha_r]^{1/r}$, while

$$U_1(\alpha_r, z) = \left(\frac{r}{r+1}\right)^r \alpha_r z^{-r}$$

if $z \geq \frac{r}{r+1} [(r+1)\alpha_r]^{1/r}$. These results are known and associated with the names of Gauss, Winkler, Camp [3], Meidell [18], Faber [6], von Mises [19] and Fréchet [8], see [8] p. 163 for a precise history. If one takes $s = 1$ and $r = 2$ one has the result (7.2).

Let us return to the case $s \geq 1$, still with $x_0 = 0$. Observe that $\Gamma''_z$ is always below the tangent line at the point with $v = (\frac{r+s}{r}\lambda z)^r$; moreover, this tangent line passes through the origin. Therefore, (7.31) must always be true at least with the inequality sign. That is, we always have

$$(7.33) \qquad U_s(\alpha_r, z) \leq \lambda_{r,s}\alpha_r z^{-r},$$

no matter what $\alpha_r$ and $z$ are. For fixed $\alpha_r$, the constant in (7.33) cannot be improved; for, then the domain of validity of (7.31) is always non-empty, since $\alpha_r \leq (\lambda b)^r$. For fixed $z$ with $z \leq \frac{r}{r+s} b$ the constant in (7.33) cannot be improved either; but it can be improved if $\frac{r}{r+s} b < z < b$, compare the second half of Theorem 7.11.

The universal bound (7.33) is due to van Dantzig [5] p. 39. He also noted that

$$\gamma_{r,\infty} = \lim_{s\to\infty} \gamma_{r,s} = e^{-r} r^r / r! .$$

In particular, $\gamma_{2,\infty} = 2e^{-2} = .271 < .444 = 4/9$.

(7.34) In the case $b = \infty$ one may ask for the best upper-bound $U_\infty(\alpha_r, z)$ for the quantity $\Pr(Z \geq z)$ given that $Z \geq 0$ has a distribution of class $\mathfrak{M}_\infty[0]$. This upper-bound is easily obtained by letting $s \to \infty$ in the formulae (7.30), (7.31) and one finds the following.

(i) If $z \leq r(\alpha_r/r!)^{1/r}$ then

$$(7.35) \qquad U_\infty(\alpha_r, z) = \exp[-z(r!/\alpha_r)^{1/r}] .$$

This bound is attained by the expontential distribution $e^{-ax}\, a\, dx$ on $[0, \infty)$ with $a = (r!/\alpha_r)^{1/r}$.

(ii) If $z \geq r(\alpha_r/r!)^{1/r}$ then

$$(7.36) \qquad U_\infty(\alpha_r, z) = (e^{-r} r^r / r!)\, \alpha_r z^{-r} = \gamma_{r,\infty}\, \alpha_r z^{-r} .$$

This bound is attained by a distribution $\epsilon(r/z)e^{-(r/z)x}\,dx$ on $[0, \infty)$, together with a mass $1 - \epsilon$ at $0$; here $\epsilon = (\alpha_r/r!)(r/z)^r \leq 1$.

(7.37) Still assuming $x_o = 0$, let us determine the best lower bound

$$(7.38) \qquad \Pr(Z \geq z) \geq L_s(\alpha_r, z),$$

assuming that $0 \leq Z \leq b$ has its distribution in $\mathcal{M}_s[0]$. It can be read off from the lower boundary of the region (7.25). Here, $\Gamma'_z = \{(v, 0): 0 \leq v \leq (\lambda z)^r\}$, while $\Gamma''_z$ is given by (7.27). Only the case $b < \infty$ is interesting, since, otherwise, one always has $L_s(\alpha_r, z) = 0$. One also has

$$(7.39) \qquad L_s(\alpha_r, z) = 0 \quad \text{whenever} \quad \alpha_r \leq (\lambda z)^r ,$$

that is, $z \geq \alpha_r^{1/r}/\lambda_{r,s}$. Thus, there only remains the case that $b < \infty$ and $\alpha_r > (\lambda z)^r$. If $s = 1$ then $\Gamma''_z$ is concave and we simply have

$$(7.40) \qquad L_1(\alpha_r, z) = \frac{\alpha_r - (\lambda z)^r}{(\lambda b)^r - (\lambda z)^r} \left(1 - \frac{z}{b}\right)$$

$$= \frac{(r + 1)\,\alpha_r - z^r}{b^r - z^r} \left(1 - \frac{z}{b}\right) ,$$

whenever $z^r \leq (r + 1)\,\alpha_r \leq b^r$; here $\lambda^r = \lambda^r_{r,1} = 1/(r + 1)$.

If $s \geq 2$ than $\Gamma''_z$ has a unique point of inflection at $u = \frac{r + s}{r + 1} z$. If $b = \frac{r + s}{r + 1} z$ then this point of inflection is not properly on $\Gamma''_z$ and $\Gamma''_z$ is in fact convex and part of the lower boundary of the region (7.25). Consequently,

$$(7.41) \qquad L_s(\alpha_r, z) \quad \chi_z(\alpha_r) \quad (1 - \lambda z \alpha_r^{-1/r})^s ,$$

whenever $\frac{r + 1}{r + s}\, b \leq z \leq \alpha_r^{1/r}/\lambda \leq b$.

It remains to consider the case that $s \geq 2$, $\alpha_r > (\lambda z)^r$

and $b > \frac{r+s}{r+1} z$ . Then there is a unique point on $\Gamma''_z$ with parameters $z < u_z < \frac{r+s}{r+1} z$ and $v_z = (\lambda u_z)^r$ at which the tangent line to $\Gamma''_z$ passes through the endpoint $(\beta, \chi_z(\beta)) = ((\lambda b)^r, (1 - z/b)^s)$ of $\Gamma''_z$ ; here, $\beta = (\lambda b)^r$ . This value $u_z$ can be obtained by solving the equation

$$(7.42) \qquad [\chi_z(\beta) - \chi_z(v_z)]/[\beta - v_z] = \chi'_z(v_z) \ ;$$

note that $v_z$ depends only on $b$ and $z$ , not on $\alpha_r$ ; in the special case that $r$ is a positive integer the equation (7.42) is equivalent to an algebraic equation of degree $r + s$ . After having found the quantity $v_z$ , one finds the value

$$(7.43) \qquad L_s(\alpha_r, z) = \chi_z(\beta) - \chi'_z(v_z)\,(\beta - \alpha_r) \ ,$$

provided $z \leq \min[\frac{r+1}{r+s} b, \alpha_r^{1/r}/\lambda]$ , where $\lambda = \lambda_{r,s}$ .

(7.44) We now turn to the only remaining case that both $x_o > 0$ and $x_o < z < b$ . For convenience, we shall restrict ourselves to the upperbound (7.19). In principle, it can be read off from the upper boundary of the region defined by (7.23), (7.24) and (7.25).

The first thing to do is to determine the best supporting line from above which passes through the origin. Its slope is given by

$$(7.45) \qquad E_r(z) = \max_{z \leq u \leq b} \left[ \left(\frac{u-z}{u-x_o}\right)^s / \psi_r^*(u) \right] .$$

Here, $E_r(z)$ is a strictly decreasing function of $z$ on $z > x_o$ ; (note that, for $u$ fixed, the function $(u - z)_+^s$ is non-increasing in $z$) . If $r$ is a positive integer then $\psi_r^*(u)$ is a polynomial of degree $r$ and $E_r(z)$ can be computed from an algebraic equation of degree $r + 1$ , namely, $s(z - x_o)\psi = (u - z)(u - x_o)\psi'$ , where $\psi = \psi_r^*$ .

Next, let $u(z)$ denote the largest value $z < u \leq b$ where the maximum in (7.45) is taken. Actually, we shall prove in (7.49) that the maximum in (7.45) is taken at a unique point $z \leq u \leq b$ . In any case, we have

$$U_s(\alpha_r, z) = \alpha_r E_r(z) \quad \text{if } \alpha_r \leq v(z) ,$$

(7.46)

$$< \alpha_r E_r(z) \quad \text{if } \alpha_r > v(z) .$$

Here, $v(z) = \psi_r^*(u(z))$ . If $\alpha_r \leq v(z)$ then $\mu = T\nu$ in $\mathfrak{M}_s[x_o]$ attains $U_s(\alpha_r, z)$ if and only if $\nu$ has the 2-point support $\{0, u(z)\}$ with a mass $\alpha_r/v(z)$; the corresponding measure $\mu$ has a mass at 0 and otherwise is absolutely continuous with support $[x_o, u(z)]$ and density proportional to $(u(z) - x)^{s-1}$.

We shall prove in (7.50) that the part of $\Gamma''_z$ with parameter $u > u(z)$ is concave. It follows that

(7.47) $$U_s(\alpha_r, z) = \left(\frac{u - z}{u - x_o}\right)^s , \quad \text{where } \psi_r^*(u) = \alpha_r ,$$

as soon as $v(z) < \alpha_r \leq \psi_r^*(b)$ . In this case, a measure $\mu = T\nu$ attains $U_s(\alpha_r, z)$ if and only if $\nu$ is supported by a single point $u(z) < u \leq b$ (satisfying $\psi_r^*(u) = \alpha_r$) so that $\mu = T\varepsilon_u$ is absolutely continuous with support $[x_o, u]$ and density proportional to $(u-x)^{s-1}$.

One could regard (7.47) as a parametric representation of the function $U_s(\alpha_r, z)$ in the range $v(z) < \alpha_r$ . Equivalently, $\alpha_r = \psi_r^*(x_o + (z-x_o)/(1-w))$, where $w = U_s(\alpha_r, z)^{1/s}$ .

(7.48) For the special case $s = 1$ , $b = \infty$ these formulae are actually known and due to von Mises [19], see [8] pp. 159-161 for a precise account. Since in this case

$$\psi_r^*(u) = (r + 1)^{-1}(u^{r+1} - x_o^{r+1})(u - x_o)^{-1}$$

(when $u > x_o$) we find that $u(z)$ is equal to the unique root $u > x_o$ of the equation

$$ru + x_o^{r+1}u^{-r} = (r + 1)z ,$$

(where the left hand side is strictly increasing for $u > x_o$) . However, $u(z) = b$ if $(r+1)z > rb + x_o^{r+1}b^{-r}$ .

(7.49) We must still prove an assertion preceding (7.46). Namely, that the tangent line $w = E_o v$ which support $\Gamma(g_z)$ from above has only one contact point (with parameter $u(z)$) . On the contrary, suppose there were two or more contact points. Then $\phi^*(u) = g_z^*(u) - E_o\psi_r^*(u)$ has at least 4 zeros in $[0,b]$ (counting $u = 0$ and multiplicities and allowing for the possibility that b is one of the zeros). It would follow from an obvious extension of Lemma 6.26 to the case $x_o > 0$ (using (5.43)) that $\phi(x) = g_z(x) - E_o x^r$ has at least 4 distinct zeros, provided a change of sign of $\phi$ at the discontinuity z would be counted as a zero. Since $g_z(x) = \text{sgn}(x - z) = 0$ or 1 , this is clearly impossible.

(7.50) We also must prove the assertion preceding (7.47). Namely, that the curve $\Gamma(g_z)$ is concave to the right of the above unique contact point with the supporting line $w = E_o v$ through the origin. If not there would be a straight line $w = a + cv$ supporting $\Gamma(g_z)$ from above with $a > 0$ and $c > 0$ , having two or more contact points with $\Gamma(g_z)$ . Then the function $\phi^*(u) = g_z^*(u) - a - c\psi_r^*(u)$ has at least 3 zeros in the interval $(u(z), b]$ (where $u(z) > z$) , counting multiplicities and allowing for the possibility that b is one of the zeros. Again by an extension of Lemma 6.26 to the case $x_o > 0$ , we would obtain that the function $\phi(x) = g_z(x) - a - cx^r$ has at least 3 distinct zeros counting a change of sign (if any) of $\phi$ at the unique discontinuity z of $g_z(x)$ . Since $a > 0$, $c > 0$, there can be no root at all with $x < z$ and one easily arrives at a contradiction.

(7.51) The results of the present Section illustrate the general method outlined in the Sections 3 and 4. In principle, the method would have worked just as well when the moments $EZ^r$ and $\Pr(Z \geq z)$ had been replaced by other moments.

I would like to end by indicating how the results of the present Section could have been obtained equally well by using the more elementary and intuitive method due to van Dantzig[5].

(7.52) An alternative method. Let $0 \leq Z \leq b$ have a distribution $\mu \in \mathcal{M}_S[x_o]$ with cumulative distribution function $F$ and density $f$, the latter only for $x > x_o$. In particular, $F$ is concave and thus continuous for $x > x_o$. Let further $x_o < z < b$ be fixed. We want to find the best possible upperbound on $Q_o = \Pr(Z \geq z)$ given the moment $\alpha_r = EZ^r$. This is equivalent to minimizing $\alpha_r$ given the number $\bar{F}(z) = \Pr(Z \geq z) = Q_o$.

Consider another distribution function $F_1$ and suppose that $F_1(x) \geq F(x)$ for all $x$, so that the measure $\mu_1$ corresponding to $F_1$ may be obtained from $\mu$ by "moving probability mass to the left". It follows that one has the lower bound

$$(7.53)\qquad \int \phi(x)\, dF(x) \geq \int \phi(x)\, dF_1(x) ,$$

holding for every non-decreasing function $\phi$ on $[0, b]$, in particular, when $\phi(x) = x^r$, $(r > 0)$.

Given $Q_o$ and given that $\mu \in \mathcal{M}_s[x_o]$ one now tries to construct $F_1$ in such a way that $\int x^r\, dF_1(x)$ is as large as possible; or at least, one would like to show that $F_1$ can in principle be chosen such that $\int x^r\, dF_1(x)$ exceeds a certain number which can be computed in terms of $s$, $x_o$ and $Q_o$.

The construction below of $F_1$ due to van Dantzig makes use of the value of the density $f(z) = F'(z + 0)$ at the point $z$ under consideration. This number cannot be arbitrarily large. For, by (7.6),

$$(7.54)\qquad f(z) = \int_{z+}^{b} \frac{s(u-z)^{s-1}}{(u - x_o)^s}\, \nu(du) ,$$

where $\nu$ is a probability measure on $[0, b]$. Determining the largest possible value of the integrand, one finds that

$$(7.55)\qquad \begin{aligned} f(z) &\leq \left(\frac{s-1}{s}\right)^{s-1} (z - x_o)^{-1} \quad \text{if } z \leq \frac{s-1}{s}x_o + \frac{1}{s}b ; \\ &\leq s(b - z)^{s-1}(b - x_o)^{-s} , \quad \text{otherwise.} \end{aligned}$$

Equality in (7.55) is obtained by concentrating the measure

$\nu$ entirely at the point $\min[b, sz - (s-1)x_o]$ ; if $b = \infty$ and $s$ is arbitrarily large (that is, $\mu \in \mathcal{M}_\infty[x_o]$) we obtain that $f(z) \leq e^{-1}(z - x_o)^{-1}$ which would also follow by minimizing the integrand in (5.7).

The bound (7.55) on $f(z)$ can be improved if we know the value

$$\bar{F}(z) = \int_z^b \left(\frac{u-z}{u-x_o}\right)^s \nu(du) , \tag{7.56}$$

compare (7.68). The main idea of van Dantzig [5] was to use the comparison (7.53) with $F_1$ of the special form

$$F_1(x) = 1 - (1-p)\left(\frac{\rho - x}{\rho - x_o}\right)_+^s , \quad \text{for } x \geq x_o , \tag{7.57}$$

where $0 \leq p < 1$ and $z < \rho \leq b$ . We put $F_1(x) = p$ for $0 \leq x \leq x_o$ . Thus the measure $\mu_1$ associated with $F_1$ has a mass $p$ at $0$ and otherwise is absolutely continuous with a support $[x_o, \rho]$ and density

$$f_1(x) = (1-p)s(\rho - x)_+^{s-1}(\rho - x_o)^{-s} \quad \text{if } x > x_o . \tag{7.58}$$

Clearly, $\mu_1 \in \mathcal{M}_s[x_o]$ . In fact, $\mu_1 = T\nu_1$ where $\nu_1$ has a 2-point support $\{0, \rho\}$ contained in $[0, b]$ with a mass $p$ at $0$ and a mass $1 - p$ at $\rho$ .

Van Dantzig choose $p$ and $\rho$ in such a way that $F$ and $F_1$ agree closely at $z$ in the sense that

$$F_1(z) = F(z) \quad \text{and} \quad f_1(z) = f(z) ; \tag{7.59}$$

(at least for $s = 1$ this is a very intuitive idea in view of the concavity of $F$). This leads to the pair of equations

$$\begin{aligned} (1-p)(\rho - z)^s(\rho - x_o)^{-s} &= \bar{F}(z) = Q_o ; \\ (1-p)s(\rho - z)^{s-1}(\rho - x_o)^{-s} &= f(z) . \end{aligned} \tag{7.60}$$

Thus, $\rho$ must be equal to

$$\rho = z + sQ_o/f(z) . \tag{7.61}$$

In order that $\rho \leq b$ we must have that

(7.62) $\overline{F}(z) \leq \frac{1}{s}(b - z)f(z)$ .

This happens to be true, see (7.67). The important thing is that it is possible to show that (7.57), (7.59) and $\rho \leq b$ together imply

(7.63) $F_1(x) \geq F(x)$ for all $x$ ,

see [5] p. 34. The fact that $p \geq 0$ is not used in the proof but follows from $p = F_1(x_o) \geq F(x_o) \geq 0$ . Another proof of (7.63) is given in (7.70).

Combining these results with (7.53), we now have the following result due to van Dantzig. When $s = 1$ the result is geometrically obvious and was already used by Narumi [21] and von Mises [19] , see [8] pp. 158-162.

(7.64) THEOREM <u>Let</u> $0 \leq x_o < z < b \leq \infty$ . <u>Let</u> $\mu$ <u>be a distribution with support</u> $[0, b]$ <u>and distribution function</u> $F$ <u>such that</u> $F(z) < 1$ . <u>Then the property</u> $\mu \in \mathcal{M}_s[x_o]$ <u>is sufficient for the existence of a distribution function</u> $F_1$ <u>of the special type</u> (7.57) <u>with</u> $z < \rho \leq b$ <u>and</u> $0 \leq p < 1$ <u>satisfying</u> (7.59) <u>and</u> (7.63). <u>Clearly,</u> $F_1$ <u>is unique. Moreover,</u> $F_1$ <u>attains the smallest value</u> $\int \phi(x)\, dF(x)$ <u>as in</u> (7.53) <u>relative to all</u> $F$ <u>satisfying</u> (7.63), <u>provided</u> $\phi$ <u>is a non-decreasing function</u>.

(7.65) Theorem 7.64 allows us to handle many of the moment problems considered in the present section (in particular the upperbounds on $\overline{F}(z)$ given $EZ^r = \alpha_r$ , that is, the lower bounds on $\alpha_r$ given $\overline{F}(z)$). It is not necessary to use any representation theorem as in (7.6).

Results analogous to Theorem 7.64 are implicit in (7.44). There, we proved (see (7.46) and (7.47)) that the upper boundary of the moment space corresponding to the pair $EZ^r$ , $\Pr(Z \geq z)$ is attained if and only if the corresponding measure $\nu$ (where $\mu = T\nu$) is supported by $\{0, u(z)\}$ or by a 1-point set $\{u\}$ with $u > u(z)$ . This would correspond to $F_1$ as in (7.57) with either $\rho = u(z)$ or $p = 0$ and $\rho > u(z)$ ; here, $u(z)$ depends also on

$r$, compare (7.45).

Since $\mathcal{M}_s[x_0] \subset \mathcal{M}_{s-1}[x_0]$ (see (5.3)) we have under the conditions of Theorem 7.64 that, for $j = 1, \ldots, s$, one can find a distribution function $F_{1,j}$ as in (7.57), (but with $s$, $\rho$ and $p$ replaced by $j$, $\rho_j$ and $p_j$, respectively) such that $F_{1,j}(z) = F(z)$; $F'_{1,j}(z+0) = f(z)$ while

$$F(x) \leq F_{1,s}(x) \leq F_{1,s-1}(x) \leq \cdots \leq F_{1,1}(x) \quad \text{for all } x.$$

Moreover, $p_s < p_{s-1} < \cdots < p_1$ and $z < \rho_1 < \cdots < \rho_{s-1} < \rho_s \leq b$.

(7.66) Finally, let us present some further comments on (7.62) and (7.63), as well as new proofs.

Let $0 < x_0 < z < b$ be fixed and consider the convex set $\Omega$ of all pairs of numbers $(f(z), \bar{F}(z))$ corresponding to the measures $\mu$ in $\mathcal{M}_s[x_0]$ which are supported by $[0, b]$, $(\mu = T\nu)$. From (7.54) and (7.56),

$$\Omega = \text{conv}[\{0\} \cup \Lambda).$$

Here, $0 = (0, 0)$ denotes the origin in the $(v, w)$-plane while

$$\Lambda = \{(v, w)\colon v = s(u-z)^{s-1}(u-x_0)^{-s}; w = (u-z)^s(u-x_0)^{-s};$$

$$z \leq u \leq b\}$$

If $s \geq 2$ then the origin is already on the curve $\Lambda$, namely, at the point corresponding to $u = z$. An equivalent equation for $\Lambda$ would be

$$\Lambda = \{(\tau(w), w) \quad 0 \leq w \leq (b-z)^s(b-x_0)^{-s}\},$$

where $\tau(w) = s(z-x_0)^{-1}(w^{1-1/s} - w)$; $\tau(1) = 0$. If $s = 1$ then $\tau(w) =$ is linear and $\tau(0) = s(z - x_0)^{-1}$. If $s \geq 2$ then the function $\tau(w)$ is concave and $\tau(0) = \tau(1) = 0$ while $\tau(w)$ has its largest value at $w = [1 - \frac{1}{s}]^s$. In both cases, we have that <u>in the</u> v-<u>direction</u> the lower boundary of $\Omega$ is a straight line segment $v = s(b - z)^{-1}w$, while the curve $\Lambda$ itself is the upper boundary of $\Omega$. It follows that the best lower bound on $f(z)$, given

$\overline{F}(z)$, is as follows:

(7.67) $$f(z) \geq s(b - z)^{-1}\overline{F}(z) = s(b - z)^{-1}Q_o .$$

This was already used in (7.61). The best upperbound on $f(z)$ in terms of $Q_o$ is given by

(7.68) $$f(z) \leq \tau(Q_o) = s(z - x_o)^{-1}Q_o(Q_o^{-1/s} - 1) .$$

It is uniquely attained by the measure $\mu_u = T\varepsilon_u$, where $u$ is such that $\mu_u([z, b]) = Q_o$. Note that this upperbound does not depend on $b$. If $s$ is large then $\tau(Q_o) \sim (z - x_o)^{-1}Q_o \log\frac{1}{Q_o}$.

(7.69) The above also throws more light on the choice of the measure $F_1$ as in (7.57) and (7.59). Namely, $F_1$ corresponds to a measure $\mu_1 = T\nu_1$ where $\nu_1$ is supported by $\{0, \rho\}$ with a mass $p$ at $0$ and a mass $1 - p$ at $\rho$. It follows that the point $S = (f(z), \overline{F}(z))$ in $\Omega$ must be equal to a simple convex linear combination of the special points $S_o$ and $S_\rho$ in $\Omega$ corresponding to $\mu_u = T\varepsilon_u$ with $u = 0$ and $u = \rho$, respectively. Here, $S_o$ is precisely the origin $0 = (0, 0)$ while the point $S_\rho$ is always on the curve $\Lambda$.

Consequently, $\rho$ is precisely the u-parameter of the projection $S_\rho$ of $S = (f(z), \overline{F}(z))$ onto $\Lambda$ by central projection from the origin. And it is now completely obvious that by means of this projection one finds the unique $F_1$ of the form (7.57) which satisfies (7.59).

(7.70) Finally, let us present another proof (besides the one in [5]) of the fact that this unique $F_1$ satisfies the surprisingly strong property (7.63). We shall do this by transforming the whole problem from the measure $\mu = T\nu \in \mathcal{M}_s[x_o]$ to a more or less arbitrary probability measure $\nu$, (see Section 5). We shall make use of the functions

(7.71) $$g_\zeta^*(u) = (u - \zeta)_+^s/(u - x_o)_+^s ,$$

one for each $x_o < \zeta \leq b$; ($g_\zeta^*$ is the T-transform of the function $g_\zeta(x) = \mathrm{sgn}(x - \zeta)$; note that $g_\zeta^*(u) = 0$ for $u \leq \zeta$). Let further

(7.72) $\quad h^*_\zeta(u) = s(u - \zeta)_+^{s-1}/(u - x_o)_+^s$ ,

again with $x_o < \zeta \le b$ ; we take $h^*_\zeta(\zeta) = 0$ even if $s = 1$ ; ($h^*_\zeta$ may be regarded as the T-transform of the $\delta$-function located at $\zeta$ , which in turn is the derivative of $g_\zeta$) .

Let $x_o < z \le b$ be fixed. In view of (7.54), (7.56), we are given two probability measures $\nu$ and $\nu_1$ on $[0,b]$ such that

(7.73) $\quad \nu(g^*_z) = \nu_1(g^*_z) > 0$ and $\nu(h^*_z) = \nu_1(h^*_z)$ ;

(here, $\nu(g) = \int g\, d\nu$) . It is also given that $\nu_1$ has a 2-point support $\{0, \rho\}$ . In view of (7.56) and (7.63), the above conditions apparently imply that

(7.74) $\quad \nu(g^*_\zeta) \ge \nu_1(g^*_\zeta)$ for all $x_o < \zeta \le b$ .

This would be equivalent to $\bar{F}(x) \ge \bar{F}_1(x)$ for all $x > x_o$ and implies in turn that, for $0 \le x \le x_o$, one has

$$\bar{F}_1(x) = \bar{F}_1(0) = 1 - p = \bar{F}_1(x_o) \le \bar{F}(x_o) \le \bar{F}(x) .$$

(7.75) Proof of (7.74). Put $\nu_1(\{0\}) = p$ thus $\nu_1(\{\rho\}) = 1 - p$ . Since $g^*_z(u) = 0$ for $u \le z$ and $\nu_1(g^*_z) > 0$ , we must have that $z < \rho \le b$ and $0 \le p < 1$. If $\zeta \ge \rho$ then $g^*_\zeta(0) = g^*_\zeta(\rho) = 0$ thus $\nu_1(g^*_\zeta) = 0$ and (7.74) is obvious. It remains to consider the case of a fixed value $\zeta$ with $x_o < \zeta < \rho$ .

Let $\alpha$ and $\beta$ be constants and define

$$\epsilon(u) = g^*_\zeta(u) - \alpha g^*_z(u) - \beta h^*_z(u) ,$$

thus $\epsilon(u) = 0$ for $u < \min(z, \zeta)$ , in particular, $\epsilon(0) = 0$ . It suffices to choose $\alpha$ and $\beta$ (depending on $z$, $\zeta$ and $\rho$) in such a way that

(7.76) $\quad \epsilon(u) \ge 0$ for all $u \ge 0$ ; $\epsilon(\rho) = 0$ .

For, then we have $\nu(\epsilon) \ge 0 = \nu_1(\epsilon)$ (since $\nu_1$ is supported by $\{0, \rho\}$) . Moreover, by (7.73), $(\nu - \nu_1)(g^*_z) = 0$ and $(\nu - \nu_1)(h^*_z) = 0$ , hence,

$$(\nu - \nu_1)(g^*_\zeta) = (\nu - \nu_1(\varepsilon) \geq 0 \ .$$

implying (7.74).

As to (7.76), note that $\varepsilon(u) \geq 0$ always for $u < z$ since there $g^*_\zeta(u) \geq 0$, $g^*_z(u) = h^*_z(u) = 0$. In order to achieve that $\varepsilon(u) \geq 0$ for $u < z$ and $\varepsilon(\rho) = 0$, we observe that the function

$$\theta(u) = \frac{g^*_\zeta(u)}{h^*_z(u)} = \frac{(u - \zeta)^s_+}{s(u - z)^{s-1}_+} = \frac{(u - z)_+}{s}[1 - \frac{\zeta - z}{u - z}]^s_+$$

is <u>convex</u> for $u > z$. Namely, in this region the function $\theta(u)$ is continuous. It is equal to 0 (and thus convex) when $z < u \leq \zeta$ (in case $z < \zeta$). Finally, observe that

$$\theta''(u) = (s-1)(\zeta - z)^2(u - z)^{-3}[1 - \frac{\zeta - z}{u - z}]^{s-2} > 0 \ ,$$

whenever $u > \max(z, \zeta)$. It follows that

$$\theta(u) \geq \theta(\rho) + \theta'(\rho)\ (u - \rho) \quad \text{for all } u > z \ ,$$

(recall that $\max(\zeta, z) < \rho \leq b)$. Letting $\alpha = s\theta'(\rho)$ and $\beta = \theta(\rho) + \theta'(\rho)(z - \rho)$, we have that

$$\delta(u) = \theta(u) - (\alpha/s)(u - z) - \beta$$

is nonnegative for all $u > z$, while $\delta(\rho) = 0$. Multiplying by $h^*_z(u)$, one obtains the desired result (7.76).

## References


1. R. E. Barlow and A. W. Marshall, "Bounds for distributions with monotone hazard rate, I", <u>Annals Math. Stat.</u> 35 (1964) 1234-1257.
2. R. E. Barlow and A. W. Marshall, "Bounds on interval probabilities for restricted families of distributions", <u>Proc. Fifth Berkeley Symp. Math. Statistics and Probability</u>, III pp. 229-257, Univ. of California Press, 1965.
3. B. H. Camp, "A new generalization of Tchebycheff's statistical inequality", <u>Bull. Amer. Math. Soc.</u> 28 (1922) 427-432.
4. H. Cramér, <u>Mathematical Methods of Statistics</u>, Princeton University Press, Princeton, 1946.

5. D. van Dantzig, "Une nouvelle généralisation de l'inégalité de Bienaymé", Ann. l'Inst. Henri Poicaré 12 (1951) 31-43.
6. A. Faber, Über nach Polynomen fortschreitende Reihen, Sitz. Ber. Math. Nat. Klasse Bayerischen Akad. Wiss. München (1922) 157-178.
7. W. Feller, An Introduction to Probability Theory and its Applications, II, Wiley and Sons, New York, 1966.
8. M. Fréchet, Recherches Théoriques Modernes sur le Calcul des Probabilités, I 2me ed., Gauthier Villars, Paris, 1950.
9. N. L. Johnson and C. A. Rogers, "The moment problem for unimodal distributions", Ann. Math. Stat. 22 (1951) 433-439.
10. S. Karlin and W. J. Studden, Tchebycheff Systems: with Applications in Analysis and Statistics, Interscience, New York, 1966.
11. J. H. B. Kemperman, "An analytical approach to the differential equations of the birth-and-death process", Michigan Math. J. 9 (1962) 321-361.
12. J. H. B. Kemperman, "On the sharpness of Tchebycheff type inequalities", Indag. Math. 27 (1965) 554-601.
13. J. H. B. Kemperman, "The general moment problem, a geometric approach", Ann. Math. Stat. 39 (1968) 93-122.
14. J. H. B. Kemperman, "On a class of moment problems", Proc. Sixth Berkeley Symp. Math. Statistics and Probability, to appear.
15. C. L. Mallows, "Note on the moment problem for unimodal distributions when one or both terminals are known", Biometrika 43 (1956) 224-227.
16. C. L. Mallows, "Generalizations of Tchebycheff's inequalities", J. Royal Stat. Soc. Ser. B 18 (1956) 139-176.
17. C. L. Mallows, "A generalization of the Chebyshev inequalities", Proc. London Math. Soc. 3d Ser. 13 (1963) 385-412.
18. B. Meidell, "Sur un problème du calcul des probabilités et les statistiques mathématiques", Comptes Rendus Acad. Sc. (Paris) 175 (1922) 806-808.
19. R. von Mises, "Sur une égalité pour les moments d'une distribution quasi-convexe", Bull. Sc. Math. (2) 62 (1938) 68-71.

20. H. P. Mulholland and C. A. Rogers, "Representation theorems for distribution functions," Proc. London Math. Soc. (3) 8 (1958) 179-223.
21. S. Narumi, "On further inequalities with possible applications to problems in the theory of probability," Biometrika 15 (1923) 245-253.
22. H. L. Royden, "Bounds on a distribution function when its first n moments are given," Ann. Math. Statist. 24 (1953) 361-376.
23. H. Richter, "Parameterfreie Abschätzung und Realisierung von Erwartungswerten," Blätter Deutschen Ges. für Versich. Math. 3 (1957) 147-161.
24. W. W. Rogosinsky, "Moments of non-negative mass," Proc. Roy. Soc. (London), Ser. A, 245 (1958) 1-27.
25. J. A. Shohat and J. D. Tamarkin, The Problem of Moments, Amer. Math. Soc., New York, 1943.
26. C. D. Smith, "On generalized Tchebycheff inequalities in mathematical statistics," Amer. J. Math. 52 (1930) 109-126.
27. B. Ulin, "An extremal problem in mathematical statistics," Skand. Aktuarietidskrift 36 (1953) 158-167.
28. J. E. Walsh, Handbook of Nonparametric Statistics, I, Van Nostrand, New York, 1962.

# VARIATIONAL METHODS IN ADAPTIVE FILTERING

Andrew P. Sage

Information and Control Sciences Center
SMU Institute of Technology
Dallas, Texas

Abstract

The practical implications of linear filter theory have long been recognized. The reduction of this theory to practice has generally been a long and difficult process. Much of the difficulty centers around problems of proper modeling of system parameters and the prior statistics of the plant and measurement noise and initial states. It is possible to pose various adaptive filter algorithms which, in effect, cause the filter to learn the system model as the system operates in real time. A very attractive feature of this approach is that it may be possible to reduce the time required to implement acceptable system design by reducing the time necessary for process modeling.

The plant and observation model are assumed to be given by

$$\dot{x} = f[x(t),t] + w(t) \tag{1}$$

$$z(t) = H(t)x(t) + v(t) \tag{2}$$

where the generalized state x(t) is composed of an original state y(t) and certain parameters associated with the system dynamics a(t) which one desires to identify. v(t) and w(t) are zero mean uncorrelated white Gaussian disturbances. A filter model of the form

$$\dot{\hat{x}} = f[\hat{x}(t),t] + K(t)[z(t) - H(t)\hat{x}(t)] \tag{3}$$

is postulated. The gain vector K(t) is determined such as

to maximize the likelihood function p[Z(t), K(t)]. It is shown that a least squares curve fit type cost function which is to be minimized, subject to the equality constraint of (3), results.

The use of the continuous time maximum principle of optimization theory leads to a two-point boundary value problem which is resolved by the method of invariant imbedding such as to obtain a sequential solution to the adaptive filter problem.

## I. Introduction

The Kalman-Bucy filter [1,2] provides a minimum variance unbiased estimate of the state of a linear dynamical system whose inputs and measurements of the states are corrupted by additive white Gaussian noise. If the noise is not Gaussian, the estimate is still the best obtainable from a linear filter. Also, various nonlinear extensions of the Kalman filter have been obtained [1-4]. Implementation of the Kalman-Bucy filter, however, presupposes exact knowledge of the parameters of the process as well as the process measurement noise moments to at least second order. Unfortunately, in a number of practical situations these moments are either unknown or known only approximately. The use of incorrect prior statistics in the implementation of a Kalman filter can lead to large estimation errors or even to filter divergence.

An optimal filtering scheme which yields estimates of the system states in the presence of unknown or inexactly known plant parameters and prior statistic information results in a nonlinear filtering scheme. The solution of such a problem invariably develops into a mechanization which requires extensive computations [5]. In order to decrease the computational load resulting from implementing the optimum filtering scheme, and to handle more general problems, suboptimal filtering schemes have been devised. These real-time schemes are adaptive in nature and result in recursive estimates for the noise covariance parameters and the system states.

A number of approaches to adaptive filtering can and have been taken. Many of them are summarized in recent survey papers [6,7] and a text concerning system identification [3].

This paper will discuss the use of techniques of variational calculus and optimization theory in order to solve the adaptive filtering problem. Section II will develop some models which may be used for adaptive filtering. The maximum a posteriori cost function for adaptive filtering is discussed in section III. Section IV discusses cost functions for maximum likelihood identification. In each of these sections it is shown that these stochastic adaptive filtering problems may be cast into a framework such that optimization theory or variational calculus may be used to obtain a two point boundary value problem, solution of which resolves the adaptive filtering problem.

For many engineering systems problems it is necessary to have an on line or sequential solution to the adaptive filtering problem. The method of invariant imbedding is presented in section V as a method which will resolve a two point boundary value problem in such a way as to obtain a sequential solution to the original optimization problem. The use of this technique to obtain nonlinear filtering and plant process parameter identification is presented in section VI. Section VII discusses adaptive filtering for problems with unknown noise covariances and the use of variational methods to resolve this problem.

For ease in presentation, this paper will develop algorithms for continuous time adaptive filtering although the discrete time algorithms have somewhat greater value when digital computer processing is contemplated. References [2], [3], [9] and [10] present considerable detail concerning the discrete approach.

## II. Models for Adaptive Filtering

In this section we will explore some of the models which may be used for adaptive filtering or system identification. For example, if the true value of a parameter to be identified were $\theta$ and we assumed a value $\hat{\theta}$, a suitable cost function might be $(\theta - \hat{\theta})^2$. Generally, the true value of the parameter, $\theta$, is not known with precision. Thus, some statistical measure of the deviation of $\hat{\theta}$ from $\theta$ is more meaningful. We may formulate this error in general terms for the case of a vector parameter $\theta$, where E denotes expectation, as

$$R = E\{C[\tilde{\theta}(Z)]|Z\} = \int_{-\infty}^{\infty} C[\tilde{\theta}(Z)]p(\theta|Z)\, d\theta =$$

$$\int_{-\infty}^{\infty}\int_{-\infty}^{\infty} \cdots \int_{-\infty}^{\infty} C[\tilde{\theta}(Z)]\; p(\theta|Z)\, d\theta_1\, d\theta_2 \ldots d\theta_N \tag{4}$$

Here, we define $C[\tilde{\theta}(Z)]$ as the cost of an error. The error is defined as

$$\tilde{\theta}(Z) = \theta - \hat{\theta}(Z) \tag{5}$$

where $\theta$ is the true value of the parameter and $\hat{\theta}(Z)$ is the estimate of the parameter based upon some observation Z. Equation (4) represents the conditioned expected value of the cost of an error in estimating the parameter and is obtained by direct application of the fundamental theorem of expectation. $p(\theta|Z)$ is the probability density of $\theta$ conditioned upon Z. Commonly used error costs are the squared error cost function

$$C[\tilde{\theta}(Z)] = \|\theta - \hat{\theta}(Z)\|_S^2 = [\theta - \hat{\theta}(Z)]^T\, S[\theta - \hat{\theta}(Z)] \tag{6}$$

where S is a positive definite symmetric matrix, and the uniform cost function

$$C[\tilde{\theta}(Z)] = \begin{cases} \dfrac{1}{\epsilon} & \text{if } \|\tilde{\theta}(Z)\| \geq \epsilon \\ 0 & \|\tilde{\theta}(Z)\| < \epsilon \end{cases} \tag{7}$$

Minimization of Eq. (4) with the squared error cost function of Eq. (6) by optimum choice of $\hat{\theta}(Z)$ easily leads to the conclusion that the best estimate of $\hat{\theta}(Z)$ is the conditional mean estimate

$$\hat{\theta}(Z) = \int_{-\infty}^{\infty} \theta p(\theta|Z)\, d\theta \tag{8}$$

The error criterion of Eq. (7) is usually considered for vanishingly small $\epsilon$ such that equivalent to Eq. (7) is the cost function

$$C[\tilde{\theta}(Z)] = -\prod_{i=1}^{N} \delta_D [\theta_i - \hat{\theta}_i(Z)] \tag{9}$$

By substitution of Eq. (7) into Eq. (4) and evaluation of the result as $\epsilon$ approaches zero, or by direct use of Eq. (9) in Eq. (4), there results the maximum a posteriori cost function

$$R = - p[\hat{\theta}(z) \mid Z] \tag{10}$$

where we desire to minimize R by choice of $\hat{\theta}(Z)$. This $\hat{\theta}(Z)$ is called the maximum a posterior (MAP) estimate since it is the estimate obtained by maximizing the conditional density function $p[\theta|Z]$ and normally obtained from solution of

$$\left. \frac{\partial p(\theta|Z)}{\partial \theta} \right|_{\theta = \hat{\theta}_{MAP}(Z)} = 0 \tag{11}$$

More classical than the maximum a posteriori cost function is the maximum likelihood (ML) cost function in which we desire to maximize the probability density of the observation conditioned upon the parameter $\theta$. This is obtained from

$$\left. \frac{\partial p(Z|\theta)}{\partial \theta} \right|_{\theta = \hat{\theta}_{ML}(Z)} = 0 \tag{12}$$

and is the estimate of $\theta$ which most likely caused Z to occur. The MAP and ML estimators are clearly related, since by Bayes rule

$$p(\theta|Z) = \frac{p(Z|\theta)p(\theta)}{p(Z)} \tag{13}$$

Thus, the MAP estimator is an ML estimator in which prior knowledge of the parameter, represented by the probability density $p(\theta)$, is considered in such a way as to improve the estimate. The ML estimator does not consider any prior knowledge concerning the value of $\theta$, the parameter to be estimated or identified.

The discrete time message and observation models which will be used here are given by

$$x(k+1) = \phi[x(k),k] + w(k) \tag{14}$$

$$z(k) = H(k)\, x(k) + v(k) \tag{15}$$

where

$x(k)$ = N dimensional state vector

$\phi[x(k),k]$ = N dimensional vector-valued function which includes any known inputs

$x(k)$ = M dimensional plant noise vector

$z(k)$ = R dimensional observation vector

$H(k)$ = R x N modulation matrix

$v(k)$ = R dimensional observation noise vector

$x(k)$ is used to represent the generalized state vector at the k-th sample time $x(t_k)$ or $x(kT_k)$. For the discrete estimation model, $w(k)$ and $v(k)$ are assumed to be independent, zero mean, Gauss-Markov white sequences such that

$$E\,\{w(k)w^T(j)\} = V_w(k)\delta_K(k-j) \tag{16}$$

$$E\,\{v(k)v^T(j)\} = V_v(k)\delta_K(k-j) \tag{17}$$

where $\delta_K(k-j)$ is the Kronecker delta funxtion, and $V_w(k)$ and $V_v(k)$ are symmetric non-negative definite M x M and R x R covariance matrices respectively.

A continuous estimation model may often be derived using a nonrigorous limiting procedure such that the continuous model follows directly from the discrete model as the samples become dense, that is as $t_{k+1} - t_k = T_k$ (the sample period) becomes zero $t_k \to t$. This continuous estimation model is defined by Eqs. (1) and (2) where $w(t)$ and $v(t)$ are assumed to be zero mean, white and independent, with Gaussian amplitude distributions such that

$$E\,\{x(t)w^T(\tau)\} = \Psi_w(t)\delta_D(t-\tau) \tag{18}$$

$$E\,\{v(t)v^T(\tau)\} = \psi_v(t)\delta_D(t-\tau) \tag{19}$$

The discrete and continuous models are related by the nonrigorous limits

$$f[x(t),t] \triangleq \lim_{\substack{T_k \to 0 \\ t_k \to t}} \frac{1}{T_k} \{\phi[x(k),k] - x(k)\} \tag{20}$$

$$\Psi_w(t) \triangleq \lim_{\substack{T_k \to 0 \\ t_k \to t}} T_k V_w(k) \tag{21}$$

$$\Psi_v(t) \triangleq \lim_{\substack{T_k \to 0 \\ t_k \to t}} T_k V_v(k) \tag{22}$$

We observe that the continuous nonlinear differential equation (1) could be written as the stochastic differential equation

$$dx(t) = f[x(t),t]dt + G[x(t),t]du(t) \tag{23}$$

where $du(t)$ is a Wiener process. In a similar manner, results of Eqs. (20) through (22) should be obtained rigorously from the stochastic calculus [2]. In this work emphasis will be on continuous time systems although all of the results to be obtained can also be obtained for discrete time systems.

The sequences $x(k_0),x(k_1), \ldots ,x(k_f)$ and $z(k_1),z(k_2), \ldots ,z(k_f)$ are denoted by $X(k_f)$ and $Z(k_f)$ respectively. In like manner, the continuous values of $x(t)$ and $z(t)$ in the interval $[t_0,t_f]$ are given by $X(t_f)$ and $Z(t_f)$.

$p[X(k_f)|Z(k_f)]$ and $p[X(t_f)|Z(t_f)]$ denote the conditional probability density function of X given measurements Z. We further assume that $p[x(k_0)]$ and $p[x(t_0)]$, if needed, are known and are normal with mean $\mu_{x0}$ and variance $V_{x0}$.

## III. Maximum A Posterior Cost Function For Adaptive Filtering

The best estimate of the generalized state vector x throughout an interval will, in general depend on the criteria used to determine the best estimate. Here, the term "best estimate" denotes that estimate derived from maximizing the conditional probability function $p[X|Z]$ as a function of x throughout the interval. The resulting estimator is known as the Bayesian maximum likelihood or maximum a posterior estimator [1,2].

Applying the Bayes rule to $p[X(k_f)|Z(k_f)]$ results in

$$p[X(k_f)|Z(k_f)] = \frac{p[Z(k_f)|X(k_f)]\, p[X(k_f)]}{p[Z(k_f)]} \tag{24}$$

From Eq. (14) it is clear that if x(k) is known, $p[z(k)|x(k)]$ is Gaussian since v(k) is Gaussian. If $X(k_f)$ is given, we have

$$p[Z(k_f)|X(k_f)] =$$

$$\prod_{k=k_0+1}^{k_f} \left[\frac{\exp\{-\frac{1}{2}\left(z(k)-H(k)x(k)\right)^T V_v^{-1}(k)\left(z(k)-H(k)x(k)\right)\}}{(2\pi)^{M/2}\det[V_v(k)]^{\frac{1}{2}}}\right] \tag{25}$$

Using the chain rule of probability

$$p[\alpha,\beta] = p[\alpha|\beta]\, p[\beta]$$

results in

$$p[X(k_f)] = p[x(k_f)|X(k_f-1)]\, p[x(k_f-1)|X(k_f-1)] \quad \ldots$$

$$\ldots\ p[x(k_1)|x(k_0)]\ p[x(k_0)]$$

Since w(k) is a white Gauss-Markov sequence, x(k) is Markov and

$$P[x(k_r)|X(k_r-1)] = p[x(k_r)|x(k_r-1)]$$

Thus, $p[X(k_f)]$ is composed of Gaussian terms and

$$p[X(k_f)] = p[x(k_0)] \prod_{k=k_0+1}^{k_f} p[x(k)|x(k-1)] \tag{26}$$

where $p[x(k)|x(k-1)]$ is Gaussian and from Eq. (14) has mean $\phi[x(k-1),k-1]$ and variance $V_w(k-1)$. $p[Z(k_f)]$ contains no terms in x(k), and $Z(k_f)$ is the known conditioning variable for the intended maximization. After a modest amount of manipulation, Eq. (24) may be written in terms of Eqs. (25) and (26) as

$$p[X(k_f)|Z(k_f)] =$$

$$A \exp\{-\frac{1}{2} \sum_{k=k_0+1}^{k_f} \| z(k) - H(k)x(k) \|^2_{V_v^{-1}(k)}$$

$$-\frac{1}{2} \sum_{k=k_0+1}^{k_f} \|x(k) - \phi[x(k-1),k-1]\|^2_{V_w^{-1}(k-1)}$$

$$-\frac{1}{2} \|x(k_0) - \mu_x(k_0)\|^2_{V_{x_0}^{-1}}\ \} \tag{27}$$

Where A is not a function of x(k). It is now clear that maximizing Eq. (27) with respect to a choice of $X(k_f)$ is equivalent to minimizing

$$J = \frac{1}{2} \|x(k_0) - \mu_x(k_0)\|^2_{V_{x_0}^{-1}} +$$

$$\frac{1}{2}\sum_{k=k_0}^{k_f-1} \|z(k+1) - H(k+1)x(k+1)\|^2_{V_v^{-1}(k+1)} + \frac{1}{2}\sum_{k=k_0}^{k_f-1} \|w(k)\|^2_{V_w^{-1}(k)} \tag{28}$$

subject to the equality constraint of Eq. (14).

Similarly, maximization of $p[X(t_f)|Z(t_f)]$ is equivalent to minimizing

$$J' = \frac{1}{2}\|x(t_0) - \mu_x(t_0)\|^2_{V_{x_0}^{-1}} +$$

$$\frac{1}{2}\int_{t_0}^{t_f} \{\|z(t) - H(t)x(t)\|^2_{\Psi_v^{-1}(t)} + \|w(t)\|^2_{\Psi_w^{-1}(t)}\}\, dt \tag{29}$$

subject to the differential equality constraint of Eq. (11). This relation, Eq. (29), is a least squares curve fit cost function and, if the prior variances are chosen correctly and appropriate Gaussian assumptions on $x(k_0)$, $v(k)$ and $w(k)$ hold, is equivalent to the maximum a posteriori cost function.

The two point boundary value problem for the continuous case may be obtained by using the continuous maximum principle [1] to minimize the cost function of Eq. (29) subject to the differential equality constraint of Eq. (1). If we use the latter approach, we define the Hamiltonian

$$\mathcal{H}[x(t),w(t),\lambda(t),t] = \frac{1}{2}\|z(t) - H(t)x(t)\|^2_{\Psi_v^{-1}(t)} +$$

$$\frac{1}{2}\| w(t) \|^2_{\Psi_w^{-1}(t)} + \lambda^T(t)\{f[x(t),t] + w(t)\} \tag{30}$$

and obtain the canonic equations from

$$\dot{\hat{x}} = \frac{\partial \mathcal{h}}{\partial \lambda}\ , \quad \lambda(t_0) = V_{x_0}^{-1}(t)\ [\hat{x}(t_0) - \mu_x(t_0)]$$

$$\dot{\lambda} = -\frac{\partial \mathcal{h}}{\partial \hat{x}}\ , \quad \lambda(t_f) = 0\ , \quad \frac{\partial \mathcal{h}}{\partial \hat{w}} = 0 \tag{31}$$

This results in the need to solve the two point boundary problem

$$\dot{\hat{x}} = f[\hat{x}(t),t] - \Psi_w(t)\ \lambda(t) \tag{32}$$

$$\dot{\lambda} = H^T(t)\Psi_v^{-1}(t)\ \{z(t) - H(t)\hat{x}(t)\} - \frac{\partial f^T[\hat{x}(t),t]}{\partial \hat{x}(t)}\lambda(t) \tag{33}$$

with the initial and terminal confitions

$$\lambda(t_0) = V_{x_0}^{-1}\ [\hat{x}(t_0) - \mu_x(t_0)],\quad \lambda(t_f) = 0 \tag{34}$$

The state variable can and should be written here as $\hat{x}(t|t_f)$ to indicate that if the two point boundary value problem is actually solved, the smoothing solution or estimate of x with observation through time $t_f$ is obtained. In this work, however, we will be interested in obtaining sequential solutions to the adaptive filtering problem. It is now of interest to relate the assumed message and observation models to the system identification problem. Consider the generalized estimation and identification problem in which the message model is

$$\dot{y} = f[y(t),a,t] + w(t) + b \tag{35}$$

The observation model is

$$z(t) = H(t)y(t) + c + v(t) \tag{36}$$

Here a, b, and c are constant parameters which are to be identified or adaptively estimated. Since they are constants, the differential relations

$$\dot{a} = 0, \qquad \dot{b} = 0, \qquad \dot{c} = 0 \tag{37}$$

are valid. This model is sufficiently general to represent a vest variety of identification or adaptive filtering situations. It is easily extended to random a, b, or c [3].

By defining the generalized state vector

$$x^T = [y^T \quad a^T \quad b^T \quad c^T] \tag{38}$$

it is clear that our identification problem has been cast into the framework of Eqs. (1) and (2). To obtain maximum a posteriori identification by minimization of Eq. (29), obtained by solution of the TPBVP of Eqs. (32) through (34), it is necessary that the random constant parameters a,b, and c have a Gaussian probability density function with known mean and variance. If these requirements are not satisfied, solution of the TPBVP still guarantees a least squares curve fit with the cost function of Eq. (27).

In many cases it is not desirable to assume that unknown parameters are constants. Time varying unknown parameter problems may be treated by regarding the unknown parameter as a stochastic process satisfying the Markov model

$$\dot{a} = A\, a(t) + B\, \eta(t) \tag{39}$$

where $\eta(t)$ is a Gaussian white noise with a known mean and variance coefficient. a(t) is the unknown time varying parameter. In order to obtain maximum a posteriori identification, it is necessary that the prior distribution of the parameter a(t) be Gaussian with known mean $\mu_a(t_0)$ and variance $V_a(t_0)$. Similar models for b and c may also be used.

## IV. Cost Functions for Maximum Likelihood Identification

In the previous development of cost functions for system identification, we assumed that we wished to adaptively estimate or identify system parameters and states. We now wish to relax the requirement that the system states formally be identified. Thus we wish to maximize $p[\theta|Z(k_f)]$ or $p[\theta|Z(t_f)]$ for maximum a posteriori identification where $\theta$ represents the constant unknown parameters (not including the system states) with which the maximization is to be conducted. For maximum likelihood identification, we maximize $p[Z(k_f)|\theta]$ or $p[Z(t_f)|\theta]$ with respect to $\theta$. If the distribution of the unknown parameter $\theta$ is uniform or if there is large uncertainty (large $V_\theta$) in the prior statistics, we have seen in the last section that maximum likelihood and maximum a posteriori identification of the system parameters are equivalent.

We consider the nonlinear message and observation models of Eqs. (14) and (15). We will let the symbol θ represent all of the unknown constants which are to be adaptively estimated or identified. We desire to maximize the expression $p[Z(k_f)|\theta]$ which may be rewritten, using the chain rule for probabilities, as

$$p[Z(k_f)|\theta] = \prod_{k=k_1}^{k_f} p[z(k)|Z(k-1),\theta] \quad (40)$$

Since there is assumed to be no observation at stage $k_0$

$$p[z(k_1)|z(k_0),\theta] = p[z(k_1)|\theta] \quad (41)$$

We define the conditional moments

$$\hat{x}(k|k-1,\theta) \triangleq E\{x(k)|Z(k-1),\theta\} \quad (42)$$

$$V_z(k|k-1,\theta) \triangleq \text{var}\{z(k)|Z(k-1),\theta\}$$

$$= V_{\hat{z}}(k|k-1,\theta) + V_v(k) \quad (43)$$

$$V_{\hat{z}}(k|k-1,\theta) \triangleq \text{var}\{H(k)x(k)|Z(k-1),\theta\} \quad (44)$$

The density function in Eq. (40) are in general non-Gaussian. However, it is possible to obtain a "pseudo Bayes" density function by assuming that the conditional densities in Eq. (40) are Gaussian such that

$$p[z(k)|Z(k-1),\theta] = \frac{1}{(2\pi)^{R/2}[\det V_z(k|k-1,\theta]^{\frac{1}{2}}}$$

$$\cdot \exp\{-0.5[z(k) - H(k)\hat{x}(k|k-1,\theta)]^T$$

$$\cdot V_z^{-1}(k|k-1,\theta)[z(k) - H(k)\hat{x}(k,|k-1,\theta]\} \quad (45)$$

With this pseudo Bayes assumption, the likelihood function

becomes

$$p[Z(k_f)|\theta] = \prod_{k=k_1}^{k_f} \frac{1}{(2\pi)^{R/2}\,[\det V_z(k|k-1,\theta)]^{\frac{1}{2}}} \cdot \exp\left\{-\frac{1}{2}\|\tilde{z}(k|k-1,\theta)\|^2_{V_z^{-1}(k|k-1,\theta)}\right\} \tag{46}$$

where $\tilde{z}(k|k-1,\theta)$ is the "innovations" process [2]

$$\begin{aligned}\tilde{z}(k|k-1,\theta) &\triangleq z(k) - H(k)\hat{x}(k|k-1,\theta)\\ &= z(k) - E\{z(k)|Z(k-1),\theta\}\end{aligned} \tag{47}$$

which represents the "new information" added by the observation z(k). It will often be more convenient to minimize the negative of the logarithm of Eq. (46) rather than to maximize Eq. (46). Thus maximum likelihood identification, for the discrete model considered, is obtained by minimizing the cost function

$$J = \frac{1}{2}\sum_{k=k_1}^{k_f} \ln\det\{V_z(k|k-1,\theta)\} + \|\tilde{z}(k|k-1,\theta)\|^2_{V_z^{-1}(k|k-1,\theta)} \tag{48}$$

with respect to θ. By using results from the Ito calculus, it is possible to show that as the samples become dense, minimization of Eq. (48) or maximization of Eq. (46) is not a meaningful operation. This is due to two factors. The variance of the plant and measurement noise is infinite in the continuous case. Also, the density function of Eq. (46) is infinite dimensional in the variable $Z(t_f)$ in the continuous case. If we disregard the possibility of identifying $V_v$ we may repose the identification problem as one of maximizing the likelihood ratio

$$L[Z(k_f)|\theta] = \frac{p[Z(k_f)|H_1,\theta]}{p[Z(k_f)|H_0]} \tag{49}$$

where $H_1$ represents the hypothesis that

$$z(k) = H(k)x(k) + v(k) \tag{50}$$

and $H_0$ represents the hypothesis that

$$z(k) = v(k) \ . \tag{51}$$

It is an easy matter to show that

$$p[z(k)|Z(k-1),H_0] = \frac{1}{(2\pi)^{R/2}\,[\det V_v(k)]^{\frac{1}{2}}} \cdot \exp\left\{-\frac{1}{2}\,\|z(k)\|^2_{V_v^{-1}(k)}\right\} \tag{52}$$

and that the likelihood ratio becomes

$$L[Z(k_f)|\theta] = \sum_{k=k_1}^{k_f} \frac{\det [V_v(k)]^{\frac{1}{2}}}{[\det V_{\hat{z}}(k|k-1,\theta) + V_v(k)]^{\frac{1}{2}}} \cdot \exp\{S[Z(k_f)|\theta]\} \tag{53}$$

where S is the sufficient statistic

$$S[Z(k_f)|\theta] = -\frac{1}{2}\sum_{k=k_1}^{k_f}\left\{\|\tilde{z}(k|k-1,\theta)\|^2_{V_z^{-1}(k|k-1,\theta)} - \|z(k)\|^2_{V_v^{-1}(k)}\right\} \tag{54}$$

If identification of $V_v$ is not involved, maximization of Eq. (53) with respect to θ is equivalent to maximization of Eq. (48) with respect to θ. We are now in a position to let the samples become dense.

It is possible to show [2] that, as the samples become

dense, maximization of Eq. (53) is equivalent to minimization of

$$J = \frac{1}{2} \int_{t_0}^{t_f} \tilde{z}^T(t|\theta)\Psi_v^{-1}(t)\tilde{z}(t|\theta)\, dt \tag{55}$$

where $\tilde{z}(t|\theta)$ is the innovation for the continuous time case

$$\tilde{z}(t|\theta) = z(t) - H(t)\hat{x}(t|\theta) \tag{56}$$

and $\hat{x}(t|\theta)$ is the optimum estimate defined by

$$\hat{x}(t|\theta) = E\,\{x(t)|Z(t),\theta\} \tag{57}$$

Although the estimate called for by Eq. (57) is the conditional mean estimate, it turns out that the approximate maximum a posteriori estimator as derived in this work is equivalent to the (first order) conditional mean estimate [1,2,3] defined by Eq. (8). Thus, if identification of $\Psi_v(t)$ is not attempted, then maximum likelihood estimation of $\theta$, such as the maximize $p[Z(t_f)|\theta]$, is equivalent to minimization of the cost function of Table 1 while satisfying the differential equation equality constraints of that table. Specification of the parameter $\theta$ to be identified, and any additional constraints associated with the parameter $\theta$, allows determination of a two point boundary value problem and solution of this two point boundary value problem leads to the optimum estimate of $\theta$.

## V. Invariant Imbedding

In essence, adaptive estimation of system parameters is nothing more or less than a problem of nonlinear filtering. Voluminous amounts of literature are available on linear and nonlinear filtering. [1-4, and their references]; the interested reader is directed to these sources for extensive discussions of nonlinear filtering. Only one approach to nonlinear filtering is developed here. The method of invariant imbedding is developed and applied to the two-point-boundary-value problems of system identification developed in the previous section. More extensive discussion of invariant imbedding may be found in references [1], [2], and [3] and the references associated with these texts.

In this section, we will consider the continuous formulation of the invariant imbedding algorithms although development of discrete time invariant imbedding algorithms has been accomplished [1,2,3]. In order to simplify the notation in the development and to maximize the generality of the results, we will base the derivation on a general two-point boundary value problem. We wish to find the solution of the TPBVP represented by

$$\dot{x}(t) = \gamma[x(t), \lambda(t), t] \tag{58}$$

$$\dot{\lambda}(t) = \beta[x(t), \lambda(t), t] \tag{59}$$

with the split boundary conditions

$$\lambda(t_0) = Ax(t_0) + b \quad , \quad \lambda(t_f) = 0 \tag{60}$$

This formulation encompasses all of the two-point boundary value problems developed in the previous two sections.

The basic concept of invariant imbedding is to change a specific problem into a more general problem. If one can solve the general problem then the specific problem is automatically solved. The amazing fact is that it is often easier to solve the general problem than the specific problem.

Here we will imbed the TPBVP of Eqs. (58),(59) and (60) by letting the terminal boundary condition on $\lambda(t)$ take a general value c rather than 0. In other words, the boundary condition $\lambda(t_f) = 0$ will be replaced by $\lambda(t_f) = c$. In addition, we will assume that both the boundary value c and the terminal time $t_f$ are variable; in particular, we wish to consider two neighboring trajectories, one which satisfies the boundary condition $\lambda(t_f) = c$ and one which satisfies $\lambda(t_f + \varepsilon) = c + \Delta c$.

For the trajectory which satisfies $\lambda(t_f) = c$, we assume that the terminal value for $x(t)$ is given by

$$x(t_f) = r(c, t_f) \tag{61}$$

In other words, the function $r(c, t_f)$ represents the relation between the boundary condition $\lambda(t_f) = c$ and the terminal value of $x(t)$. If this function were known, then the TPBVP could be trivially solved by integrating $x$ and $\lambda$ backward in time from the boundary conditions $\lambda(t_f) = c$ and $x(t_f) = r(c,t_f)$. But we do not know $r(c,t_f)$ and hence we wish to develop a technique for determining it.

Let us suppose that we have a trajectory which ends at $t_f$ with $\lambda(t_f) = c$ so that $x(t_f)$ is $r(c,t_f)$. Now let us extend the terminal time slightly so that it becomes $t_f + \epsilon$ and

$$\lambda(t_f + \epsilon) = c + \Delta c \tag{62}$$

and correspondingly

$$x(t_f + \epsilon) = x(t_f) + \Delta x = r(c,t_f) + \Delta x$$

where $\Delta c$ and $\Delta x$ are $0(\epsilon)$[1]. But $x(t_f + \epsilon)$ is also given by

$$x(t_f + \epsilon) = r(c + \Delta c, t_f + \epsilon).$$

Now let us equate the foregoing two equations to obtain

$$r(c,t_f) + \Delta x = r(c + \Delta c, t_f + \epsilon) \tag{63}$$

If we make a Taylor series expansion of the right hand side of Eq. (63) about $c$ and $t_f$, we obtain

$$\Delta x = \frac{\partial r(c,t_f)}{\partial c} \Delta c + \frac{\partial r(c,t_f)}{\partial t} \epsilon + 0(\epsilon^2) \tag{64}$$

By the use of Eqs. (58) and (59), we see that

$$\Delta x = \gamma[r(c,t_f),c,t_f]\epsilon + 0(\epsilon^2)$$

and

$$\Delta c = \beta[r(c,t_f),c,t_f]\epsilon + 0(\epsilon^2)$$

so that Eq. (64) becomes

$$\gamma[r(c,t_f),c,t_f]\epsilon = \frac{\partial r(c,t_f)}{\partial c} \beta[r(c,t_f),c,t_f]\epsilon$$

---

[1] The notation $0(\epsilon^j)$ is used to indicate that $\lim_{\epsilon \to \infty} 0(\epsilon^j)/\epsilon^{j-1} = 0$.

$$+ \quad \frac{\partial r(c,t_f)}{\partial t_f} \epsilon + O(\epsilon^2)$$

If we divide by $\epsilon$ and let $\epsilon$ approach zero, we obtain the following partial differential equation which $r(c,t_f)$ must satisfy

$$\gamma(r,c,t_f) = \frac{\partial r(c,t_f)}{\partial c} \beta(r,c,t_f) + \frac{\partial r(c,t_f)}{\partial t_f} \tag{65}$$

This equation is known as the invariant imbedding equation.

Unfortunately, this partial differential equation has no known general solution. However, we can often approximate the solution by a linear function given by

$$r(c,t_f) = \hat{x}(t_f) + P(t_f)c \tag{66}$$

Here $\hat{x}(t_f)$ is the solution if $c = 0$ which is the correct solution of the original TPBVP with $\lambda(t_f) = 0$. In other words, we are assuming that $r(c,t_f)$ is the optimal value for $x(t_f)$ if $c = 0$ (as yet unknown) plus a linear weighting of the deviation of $\lambda(t_f)$ from zero given by c. This should be a good approximation if $c \approx 0$; that is, if we are close to the true optimal solution. We will assume that c is small in the following development and ignore terms of $O(\|c\|^2)$ or higher.

If we substitute the assumed form of $r(c,t_f)$ as given by Eq. (66) into Eq. (65), we obtain

$$\gamma(\hat{x} + Pc, c, t_f) = P\beta(\hat{x} + Pc, c, t_f) + \dot{\hat{x}} + \dot{P}c \tag{67}$$

Next we expand $\gamma$ and $\beta$ in Taylor series about x, c, and $t_f$ retaining only first order terms in c. Eq. (67) becomes

$$\gamma(\hat{x},c,t_f) + \frac{\partial \gamma(\hat{x},c,t_f)}{\partial \hat{x}} Pc =$$

$$P\beta(\hat{x},c,t_f) + P \frac{\partial \beta(\hat{x},c,t_f)}{\partial \hat{x}} Pc + \dot{x} + \dot{P}c \tag{68}$$

In order to continue the development, it is necessary to substitute the actual functions for $\gamma$ and $\beta$ associated with the TPBVP of system identification.

## VI. Sequential Nonlinear Filtering and Plant Parameter Identification

In this section we will develop the invariant imbedding maximum a posteriori algorithms for filtering. We will use the TPBVP of Eqs. (32), (33) and (34). From Eqs. (32) and (33), the functions $\gamma$ and $\beta$ are given by

$$\gamma(\hat{x},c,t_f) = f(\hat{x},t_f) - \Psi_w(t_f)c \tag{69}$$

$$\beta(\hat{x},c,t_f) = H^T(t_f)\Psi_v^{-1}[z - H(t_f)\hat{x}(t_f)] - \frac{\partial f^T(\hat{x},t_f)}{\partial \hat{x}} c \tag{70}$$

Now if we substitute Eq. (69) and (70) into Eq. (68), we obtain the following expression where no terms which are of higher order than the first in c have been retained.

$$f(\hat{x},t_f) - \Psi_w(t_f)c + \frac{\partial f(\hat{x},t_f)}{\partial \hat{x}} Pc$$

$$= PH^T(t_f)\Psi_v^{-1}[z - H(t_f)\hat{x}(t_f)] - P\frac{\partial f^T(\hat{x},t_f)}{\partial \hat{x}} c$$

$$+ P\frac{\partial}{\partial \hat{x}}\{H^T(t_f)\Psi_v^{-1}[z - H(t)\hat{x}(t_f)]\} Pc + \dot{\hat{x}} + \dot{P}c$$

Since the expression is to be true for all c (which are sufficiently small), we may separately equate the coefficients of the zero and first power of c to obtain

$$\dot{\hat{x}} = f(\hat{x},t_f) - PH^T(t_f)\Psi_v^{-1}(t_f)\,[z - H(t)\hat{x}(t_f)] \tag{71}$$

$$\dot{P} = -\Psi_w(t_f) + P\frac{\partial f^T(\hat{x},t_f)}{\partial \hat{x}} + \frac{\partial f(\hat{x},t_f)}{\partial \hat{x}} P$$

$$+ PH^T(t_f)\Psi_v^{-1}(t_f)H(t_f)P \tag{72}$$

We may put this result into a slightly more familiar form by substituting P = -P. At the same time, we recognize that the terminal time is variable and may be written as just t. With these changes, Eqs.(71) and (72) become

$$\dot{\hat{x}}(t) = f[\hat{x}(t),t] + P(t)H(t)\Psi_v^{-1}(t)\,\{z(t) - H(t)\hat{x}(t)\} \tag{73}$$

$$\dot{P}(t) = \Psi_w(t) + P(t)\frac{\partial f^T[\hat{x}(t),t]}{\partial \hat{x}(t)} + \frac{\partial f[\hat{x}(t),t]}{\partial \hat{x}(t)}P(t) - P(t)H^T(t)\Psi_v^{-1}(t)H(t)P(t) \tag{74}$$

Initial conditions for Eq. (73) and (74) may be obtained from the boundary condition of Eq. (34) which in our current notation becomes

$$\lambda(t_0) = -V_{x0}^{-1}\,[x(t_0) - \mu_{x0}] \tag{75}$$

If we solve this expression for $x(t_0)$, we obtain

$$x(t_0) = \mu_{x0} - V_{x0}\lambda(t_0) \tag{76}$$

A simple comparison of Eqs. (66) and (76) reveals that we should use the initial conditions

$$\hat{x}(t_0) = \mu_{x0} \tag{77}$$

$$P(t_0) = V_{x0} \tag{78}$$

for Eqs. (73) and (74). The reader should remember that we made the substitution P = -P in writing Eqs. (73) and (74) from Eqs. (71) and (72) which explains the minus sign in Eq. (76).

Now we may integrate Eqs. (73) and (74) forward in time from the initial conditions of Eqs. (77) and (78) to obtain the sequential estimate of the "state" x(t) which often is augmented with the parameters that one desires to identify or adaptively estimate. The algorithm is summarized in Table 2 for easy reference.

Equation (74) looks very similar to the error variance algorithm for Kalman filtering. If f is a linear function of x it is identical to the error variance in filtering. It is not a difficult task to show that Eq. (74) is a first order approximation to the conditional error variance in estimation

$$V_{\tilde{x}}(t) = \text{var}\,\{x(t)|Z(t)\}$$

if the initial condition $x(t_0)$ and the noise imputs $w(t)$ and $v(t)$ are Gaussian, as assumed in the previous section. In a similar fashion the algorithms of this table are just those of the extended or first order Kalman filter for conditional mean nonlinear filtering.

## VII. Adaptive Estimation of Prior Statistics

In the previous section, we developed algorithms to adaptively estimate, in a sequential fashion, system states and parameters. The model considered there was of the form of Eq. (1). The variance of the plant and measurement noise was assumed known. Often this is not the case and the adaptive filter of Table 2 will not then be optimal. In this section we will develop a set of algorithms which resolve the maximum likelihood optimization problem of Table 1. The mean of the plant and observation noise will be assumed known although it is possible to develop algorithms by this approach to estimate unknown or random bias terms [8,9].

Instead of estimating or identifying the plant noise variance coefficient $\Psi_w(t)$, the gain vector $K(t)$ will be identified. This is advantageous in that the resulting algorithms are simpler than those which identify $\Psi_w(t)$, and the gain $K(t)$ is the only remaining unknown for the filter of Eq. (3). The development will be for the case of a scalar observation. Vector observations may be handled by processing the observations one at a time.

The continuous-time state and observation models are assumed to be described by Eqs. (1) and (2). In order to obtain a solution to the foregoing optimization problem two simplifying assumptions are made. They are:

i) The gain vector, $K(t)$, will be the parameter $\theta$, to be estimated.

ii) $K(t)$ will be assumed to evolve from the model

$$\dot{K}(t) = A(t)K(t) + \zeta(t) \qquad (79)$$

$\zeta(t)$ is assumed to be a zero-mean white Gaussian noise process uncorrelated with $v(\tau)$ and $w(\tau)$ for all $t, \tau$ and with covariance

$$\text{cov}\,\{\zeta(t),\zeta(\tau)\} = \xi(t)\delta_D(t-\tau)$$

Since $K(t)$ will be a time varying quantity in accordance with Eq. (79) the cost function to be chosen will be the unconditioned maximum likelihood [2] cost function

$$p[\theta(t_f),Z(t_f)] = p[Z(t_f)|\theta(t_f)]\; p[\theta(t_f)] \qquad (80)$$

Taking the natural logarithms of the foregoing results in

$$\ln p[\theta(t_f),Z(t_f)] = \ln p[Z(t_f)|\theta(t_f)] + \ln p[\theta(t_f)] \qquad (81)$$

The first term on the right in the foregoing is equivalent to the cost function of Table 1. If it is assumed that the $\theta(t_f)$ is equivalent to the gain $K(t)$, and that the probability density of this gain in Gaussian, then

$$\ln p[\theta(t_f)] = \text{constant} - \frac{1}{2}\left[\int_{t_0}^{t_f} \|\zeta(t)\|^2_{\xi^{-1}(t)}dt + \|K(t_0) - \mu_K(t_0)\|^2_{V_K^{-1}(t_0)}\right]$$

where $\mu_K(t_0)$ and $V_K(t_0)$ are the initial (prior) mean and variance of $K(t)$.

In view of the foregoing, the optimization problem is one of minimizing

$$J = \frac{1}{2}\left[\int_{t_0}^{t_f} \{[z(t)-H(t)\hat{x}(t)]^2\Psi_v^{-1}(t) + \|\zeta(t)\|^2_{\xi^{-1}(t)}\}\,dt\right.$$

$$+ \left\| K(t_0) - \mu_{K_0} \right\|^2_{V_K^{-1}} \Bigg] \tag{83}$$

subject to the constraints

$$\dot{\hat{x}}(t) = f[\hat{x}(t),t] + K(t)[z(t) - H(t)\hat{x}(t)] \tag{84}$$

$$\dot{K}(t) = A(t)K(t) + \zeta(t) \tag{85}$$

The optimization problem described by the foregoing three equations will be resolved by use of the maximum principle [1]. The Hamiltonian is defined as

$$\mathcal{h}[\hat{x}(t),K(t),\lambda_x(t),\lambda_K(t),t] = \tfrac{1}{2}[z(t) - H(t)\hat{x}(t)]^2 \Psi_v^{-1}(t)$$

$$+\tfrac{1}{2}\|\zeta(t)\|^2_{\xi^{-1}(t)} + \lambda_x^T(t)\{f[\hat{x}(t),t] + K(t)[z(t) - H(t)\hat{x}(t)]\}$$

$$+ \lambda_K^T(t)\ \{A(t)K(t) + \zeta(t)\} \tag{86}$$

and the canonic equations by

$$\frac{\partial \mathcal{h}}{\partial \zeta(t)} = 0, \quad \frac{\partial \mathcal{h}}{\partial \hat{x}(t)} = -\dot{\lambda}_x(t), \quad \frac{\partial \mathcal{h}}{\partial \hat{K}(t)} = -\dot{\lambda}_K(t),$$

$$\frac{\partial \mathcal{h}}{\partial \lambda_x(t)} = \dot{\hat{x}}(t), \quad \frac{\partial \mathcal{h}}{\partial \lambda_K} = \dot{\hat{K}}(t)$$

then the TPBVP becomes

$$\begin{aligned}
\hat{\zeta}(t) &= -\xi(t)\,\lambda_K(t) \\
\dot{\hat{K}}(t) &= A(t)\hat{K}(t) - \xi(t)\lambda_K(t) \\
\dot{\hat{x}}(t) &= f[x(t),t] + \hat{K}(t)\nu(t) \\
\dot{\lambda}_K(t) &= -\lambda_x(t)\nu(t) - A^T(t)\lambda_K(t) \\
\dot{\lambda}_x(t) &= -[F(t) - \hat{K}(t)H(t)]^T\lambda_x(t) + H^T(t)\Psi_v^{-1}(t)\nu(t)
\end{aligned} \tag{87}$$

where

$$F(t) \triangleq \frac{\partial f[\hat{x}(t),t]}{\partial \hat{x}(t)}$$

$$\nu(t) \triangleq z(t) - H(t)\hat{x}(t)$$

The two-point boundary conditions are

$$\hat{x}(t_0) = \mu_{x_0} \quad , \quad \lambda_x(t_f) = 0$$
$$\hat{K}(t_0) = \mu_{K_0} \quad , \quad \lambda_K(t_f) = 0$$

The optimization problem is now that of solving the TPBVP described by equations (87) above with the associated boundary conditions. Many methods are available for obtaining a solution; however, the method of invariant imbedding is particularly appropriate when one describes real-time filtering algorithms.

Thus, if there is defined

$$\lambda(t) \triangleq \begin{bmatrix} \lambda_x(t) \\ \lambda_K(t) \end{bmatrix}$$

$$y(t) \triangleq \begin{bmatrix} \hat{x}(t) \\ \hat{K}(t) \end{bmatrix}$$

$$\gamma[y,\lambda,t] \triangleq \begin{bmatrix} \dot{\lambda}_x(t) \\ \dot{\lambda}_K(t) \end{bmatrix} = \begin{bmatrix} \gamma_x[y,\lambda,t] \\ \gamma_K[y,\lambda,t] \end{bmatrix}$$

$$\beta[y,\lambda,t] = \begin{bmatrix} \dot{\hat{x}}(t) \\ \dot{\hat{K}}(t) \end{bmatrix} = \begin{bmatrix} \beta_x[y,\lambda,t] \\ \beta_K[y,\lambda,t] \end{bmatrix}$$

where $\lambda(t_f)$ 0, the problem takes the required for obtaining a solution by invariant imbedding. To this end, let

$$y(t_f) = r[C,t_f] \triangleq \begin{bmatrix} r_x(C,t_f) \\ r_K(C,t_f) \end{bmatrix}$$

where $\lambda(t_f)$ has been imbedded into a more general class of final conditions as

$$\lambda(t_f) = c = \begin{bmatrix} c_x \\ c_K \end{bmatrix}$$

Now, the invariant imbedding partial differential equation (65) with proper conditions on r, governs the dependence of the missing terminal conditions on y as a function of the process duration and terminal conditions on $\lambda$.

It is convenient to assume a solution of the form

$$r[C,t_f] = \hat{y}(t_f) - P(t_f)c \tag{88}$$

where

$$P(t_f) \triangleq \left[\begin{array}{c:c} P_x(t_f) & P_{xK}(t_f) \\ \hdashline P_{Kx}(t_f) & P_K(t_f) \end{array}\right]$$

On taking appropriate derivatives of r and substituting the results into (88), there results, where $t_f$ is replaced by the running time t,

$$\dot{\hat{y}}(t) - \dot{P}(t)c - P(t)\gamma[r,c,t] - \beta[r,c,t] = 0 \tag{89}$$

A solution of (89) is obtained by linearizing $\beta$ and $\gamma$ in a Taylor series about c = 0 and then equating those terms involving c and those not involving c. Thus one obtains

$$\dot{\hat{y}}(t) = \beta[\hat{y}(t),0,t] + P(t)\gamma[\hat{y}(t),0,t] \tag{90}$$

$$\dot{P}(t) = -\frac{\partial\beta}{\partial c}\bigg|_{c=0} - P(t)\frac{\partial\gamma}{\partial c}\bigg|_{c=0} \tag{91}$$

By substituing the definitions of $\beta$ and $\gamma$, and their derivatives into (90) and (91) the desired adaptive gain estimation algorithms result. These are displayed in Table 3. Since they are more general, maximum likelihood partial adaptive gain estimation algorithms are actually shown in this table. These result from letting

$$K(t) = T_1\beta(t) + T_2\alpha(t)$$

where $T_1$ and $T_2$ are known matrices, $\beta(t)$ is a known vector. $\alpha(t)$ is a Gauss-Markov process evolving from

$$\dot{\alpha} = A(t)\alpha(t) + \zeta(t) \tag{92}$$

Thus it is possible to incorporate known or assumed components into the final form of the filter and only attempt to identify certain portions of the gain K(t).

## VIII. Summary

Variational methods have been used to obtain algorithms for on line sequential adaptive filtering. Although no experimental results have been discussed in this paper, considerable success in applying the algorithms of the previous two sections to a variety of adaptive filtering problems has been achieved [1], [3], [8], [9], [10].

## IX. Acknowledgments

This work was supported by the Air Force Office of Scientific Research, under Contract F44620-68-C-0023.

## X. References

[1] Sage, A. P., Optimum Systems Control, Prentice-Hall, Inc., Englewood Cliffs, New Jersey, 1968.

[2] Sage, A. P. and Melsa, J. L., Estimation Theory with Applications to Communications and Control, McGraw-Hill, New York, 1971.

[3] Sage, A. P. and Melsa, J. L., System Identification, Academic Press, New York, 1971.

[4] Jazwinski, A. H., Stochastic process and Filter Theory, Academic Press, New York, 1970.

[5] Magill, D. T., "Optimal Adaptive Estimation of Sampled Stochastic Processes," IEEE Trans. on Automatic Control, Vol. AC-10, pp. 434-439, October, 1965.

[6] Sage, A. P. and Husa, G. W., "Algorithms for Sequential Adaptive Estimation of Prior Statistics," Proceedings of the 1969 IEEE Symposium on Adaptive Processes (8th), Penn. State University, November 17-19, 1969.

[7] Mehra, R. K., "Approaches to Adaptive Filtering,"

Proceedings 1970 IEEE Symposium on Adaptive Processes, December 1970.

[8] Sage, A. P. and Lin, J. L., "Maximum Likelihood Bias Estimation in Continuous Time Correlated Noise Filtering", Information Sciences, (to appear 1971).

[9] Lin, J. L. and Sage, A. P., "Maximum Likelihood Bias Estimation in Discrete Time Correlated Noise Filtering," IEEE Trans. on Systems, Man and Cybernetics, October 1971.

[10] Sage, A. P. and Wakefield, C. W., "On Invariant Imbedding and Maximum Likelihood Adaptive Estimation," Proceedings Joint IEEE-ORSA Conference on Major Systems, October 1971.

Table 1

Optimization Problem Equivalent to Maximum Likelihood Adaptive Estimation of θ

| | |
|---|---|
| Cost Function | $J = \frac{1}{2}\int_{t_0}^{t_f} \tilde{z}^T(t\|\theta)\Psi_v^{-1}(t)\tilde{z}(t\|\theta)\,dt$ |
| Innovation | $\tilde{z}(t\|\theta) = z(t) - H(t)\hat{x}(t\|\theta)$ |
| Filter | $\dot{\hat{x}}(t\|\theta) = f[\hat{x}(t\|\theta),t]+K(t\|\theta)[z(t)-H(t)\hat{x}(t\|\theta)]$ |
| Gain | $K(t\|\theta) = P(t\|\theta)H^T(t)\Psi_v^{-1}(t)$ |
| Error Variance In Estimation | $\dot{P}(t\|\theta) = F(t\|\theta)P(t\|\theta) + P(t\|\theta)F^T(t\|\theta) - P(t\|\theta)H^T(t)\Psi_v^{-1}(t)P(t\|\theta) + \Psi_w(t)$ |
| System Matrix | $F(t\|\theta) = \frac{\partial f[\hat{x}(t\|\theta),t]}{\partial \hat{x}(t\|\theta)}$ |
| Initial Conditions | $\hat{x}(t_0\|\theta) = E\{x(t_0)\} = \mu_x(t_0)$<br>$P(t_0\|\theta) = \mathrm{var}\,\{x(t_0)\} = P_x(t_0)$ |

Table 2

Continuous Invariant Imbedding Algorithm

| | |
|---|---|
| System Model | $x(t) = f[x(t),t] + w(t)$ |
| Observation Model | $z(t) = H(t)x(t) + v(t)$ |
| Statistical Parameters | $E\{x(t_0)\} = \mu_{x0}$ $\quad$ $\text{var}\{x(t_0)\} = V_{x0}$ <br> $E\{w(t)\} = E\{v(t)\} = 0$ <br> $\text{cov}\{w(t),w(\tau)\} = \Psi_w(t)\delta_D(t-\tau)$ <br> $\text{cov}\{v(t),v(\tau)\} = \Psi_v(t)\delta_D(t-\tau)$ |
| Filter Algorithm | $\dot{\hat{x}}(t) = f[\hat{x}(t),t] + P(t)H^T(t)\Psi_v^{-1}(t)\{z(t) - H(t)\hat{x}(t)\}$ |
| Error Variance Algorithm | $\dot{P}(t) = \Psi_w(t) + P(t)\dfrac{\partial f^T[\hat{x}(t),t]}{\partial \hat{x}(t)} + \dfrac{\partial f[\hat{x}(t),t]}{\partial \hat{x}(t)}P(t) - P(t)H^T(t)\Psi_v^{-1}(t) \cdot H(t)P(t)$ |
| Initial Conditions | $\hat{x}(t_0) = \mu_{x0}$, $\quad$ $P(t_0) = V_{x0}$ |

Table 3

Maximum Likelihood Adaptive Estimation Algorithms

| | |
|---|---|
| Message Model | $x(t) = f[x(t),t] + G(t)w(t)$ |
| Observation Model | $z(t) = H(t)x(t) + v(t)$ |
| Partial Gain Model | $\dot{\alpha}(t) = A(t)\alpha(t) + \zeta(t)$ |

Table 3 (cont.)

| | |
|---|---|
| Gain Model | $K(t) = T_1\beta(t) + T_2\alpha(t)$ |
| Prior Statistics | $\mathrm{Var}\,\{\gamma(t_0)\} = V_{\gamma_0}$<br>$\mathrm{Var}\,\{x(t_0)\} = V_{x_0}$<br>$E\,\{x(t_0)\} = \mu_{x_0}$; $E\,\{\gamma(t_0)\} = \mu_{\gamma_0}$<br>$\mathrm{cov}\,\{w(t),w(\tau)\} = \Psi_w(t)\delta_D(t-\tau)$ (unknown)<br>$\mathrm{cov}\,\{v(t),v(\tau)\} = \Psi_v(t)\delta_D(t-\tau)$<br>$\mathrm{cov}\,\{\zeta(t),\zeta(\tau)\} = \xi(t)\delta_D(t-\tau)$<br>$\mathrm{cov}\,\{v(t),w(\tau)\} = \mathrm{cov}\{v(t),\zeta(\tau)\} =$<br>$\mathrm{cov}\,\{\zeta(t),w(\tau)\} = \mathrm{cov}\,\{v(t),x(\tau)\} =$<br>$\mathrm{cov}\,\{\zeta(t),x(\tau)\} = 0$, for all $t,\tau$<br>$\mu_w(t) = \mu_v(t) = \mu_\zeta(t) = 0$ |
| Innovation | $\nu(t) = z(t) - H(t)\hat{x}(t)$ |
| Definition | $F(t) \triangleq \dfrac{\partial f[\hat{x}(t),t]}{\partial \hat{x}(t)}$ |
| System State Estimator | $\dot{\hat{x}}(t) = f[\hat{x}(t),t] + [T_1\beta(t) + T_2\hat{\alpha}(t)]\nu(t)$<br>$+P_x(t)H^T(t)\Psi_v^{-1}(t)\nu(t)$ |
| Partial Gain Estimator | $\dot{\hat{\alpha}}(t) = A(t)\hat{\alpha}(t) + P_{x\alpha}^T(t)H^T(t)\Psi_v^{-1}\nu(t)$ |
| Variance Algorithms | $\dot{P}_x(t) = P_x(t)\{F(t) - [T_1\beta(t) +$<br>$T_2\hat{\alpha}(t)]H(t)\}^T + \{F(t) - [T_1\beta(t) +$<br>$T_2\hat{\alpha}(t)]H(t)\}P_x(t) + \{T_2P_{x\alpha}^T(t) +$ |

Table 3 (cont.)

| | |
|---|---|
| Variance Algorithms (cont.) | $+ P_{x\alpha}(t)T_2^T\}\nu(t) - P_x(t)H^T(t)\Psi_v^{-1}(t)H(t)\cdot P_x(t)$<br>$\dot{P}_{x\alpha}(t) = P_{x\alpha}(t)A^T(t) + \{F(t) - [T_1\beta(t) + T_2\hat{\alpha}(t)]\}P_{x\alpha}(t) - P_x(t)H^T(t)\Psi_v^{-1}(t)\cdot H(t)P_{x\alpha}(t) + T_2P_\alpha(t)\nu(t)$<br>$\dot{P}_\alpha(t) = A(t)P_\alpha(t) + P_\alpha(t)A^T(t) - P_{x\alpha}^T(t)H^T(t)\Psi_v^{-1}(t)H(t)P_{x\alpha}(t)$ |
| Initial Conditions | $\hat{x}(0) = \mu_{x_0}$<br>$\hat{\alpha}(t_0) = \mu_{\alpha_0}$<br>$P_x(0) = 0$ if $V_{x_0}$ is given<br>$P_{x\alpha}(0) = 0$<br>$P_\alpha(0) = V_{\alpha_0}$ if known |

# NON LINEAR FILTERING*

G. Kallianpur

University of Minnesota

## 1. Introduction

In this paper I shall present, in brief outline, the approach to nonlinear filtering problems taken in a forthcoming paper by M. Fujisaki, H. Kunita and myself [6]. Sections 2 and 3 are thus expository in nature. The basic ideas will be described in the next section and one of the main results of [6] will be stated in Section 3. This theorem is an advance over previous work on the subject in the sense that it derives the stochastic differential equation for the nonlinear filtering problem for an important class of stochastically controlled signal processes. With the help of Theorem 1, I obtain, in Section 4, a "Kalman-Bucy" type result which is an extension both of a result due to R. S. Liptzer and A. N. Shiryaev [15] and a theorem of Wonham's [18]. Wonham's result was derived in connection with the existence of an optimal stochastic control in certain problems. It seems to me (although I have not worked out the details) that Theorem 2 enables one to extend Wonham's work to a more general situation.

## 2. A General Approach to Non-Linear Filtering Problems

Let me first set out the models relating the signal and the observation processes, indicate the differences with some of the earlier work, and then describe the approach adopted by in [6].

The signal process is denoted by $x_t(\omega)$, $t \in [0,T]$ and is assumed to take values in a complete metric space S.

*This work was supported by NSF Grant GP-1188-8.

The process $x_t(\omega)$ and the observation process $z_t(\omega)$ are supposed to be given on some probability space $(\Omega, A, P)$ and further related as follows:

$$z_t = \int_0^t h_u du + w_t \tag{2.1}$$

where

$$w_t \text{ is an N-vector standard Wiener process,} \tag{2.2}$$

and the N-vector process $h_t(\omega)$ satisfies the following conditions: $h_t(\omega)$ is $(t,\omega)$ measurable and

$$\int_0^T E\,|h_t|^2\,dt < \infty. \tag{2.3}$$

Here $|\cdot|$ denotes the norm of the N-vector. For each s $(0 \leq s < T)$

$$h_s(\omega) \text{ is } \sigma\{x_u, w_u, 0 \leq u \leq s\}\text{-measurable.} \tag{2.4}$$

Finally we need the assumption that the $\sigma$-fields

$$\sigma\{x_u, w_u, u \leq s\} \quad \text{and} \quad \sigma\{w_v - w_u, s \leq u < v \leq T\} \tag{2.5}$$

are independent. Let us introduce the notation

$$\mathfrak{F}_t = \sigma\{z_s, s \leq t\},\ G_t = \sigma\{x_s, w_s, s \leq t\}.$$

Since it is the assumptions (2.4) and (2.5) that introduce new difficulties and make possible a substantial generalization of earlier work some comments are in order. Condition (2.5) expresses the independence of the "past" of the signal process from the noise that is to follow. This is a more natural condition than the assumption of complete independence of $(x_t)$ and $(w_t)$. In fact, as we shall see below, many of the signal processes $(x_t)$ which are "controlled" by the present and past values of the observed process, $\{z_s, s \leq t\}$ satisfy (2.5). Condition (2.4) considerably widens the available class of observation processes as it is no longer necessary to restrict h to be a measurable function only of $x_t$.

The method used in proving the main results in [6] depends on three essential ideas:

(1) The "innovation process," which is a standard $\mathfrak{F}_t$-measurable Wiener process.

(2) A theorem of I. V. Girsanov.

(3) A stochastic integral representation of a square integrable $(\mathfrak{F}_t, P)$-martingale.

Let us write $\hat{h}_t$ for the conditional expectation $E(h_t | \mathfrak{F}_t)$. We note that the conditions we have imposed above ensure the existence of a modification of $E(h_t | \mathfrak{F}_t)$ which is jointly measurable and adapted to $(\mathfrak{F}_t)$. Define the innovation process

$$\nu_t = z_t - \int_0^t \hat{h}_s \, ds \, . \qquad (2.6)$$

The following result seems to have been known to several workers in the field, and its proof under varying hypotheses have appeared in earlier works (e.g., Kailath [10], [11], Frost [5], Liptzer and Shiryaev [15], Shiryaev [16]). Kailath, in particular, has stressed the usefulness of the innovation process approach ([10], [11]).

Lemma 1 [6].

Under assumptions (2.1) - (2.5), $(\nu_t, \mathfrak{F}_t, P)$ is an N-vector standard Wiener process. Furthermore, $\mathfrak{F}_s$ and $\sigma\{\nu_v - \nu_u, \ s \leq u < v \leq T\}$ are independent.

Remark. The lemma holds under a weaker condition than (2.3). (See [6]).

The importance of this result can be seen as follows. Obtaining a stochastic differential equation for $E[f(x_t) | \mathfrak{F}_t]$ where f is a suitable measurable, real valued function, is equivalent to getting a relation of the form

$$E[f(x_t) | \mathfrak{F}_t] = E[f(x_0) | \mathfrak{F}_0] + \int_0^t \alpha_1(u) \, du \ +$$

$$\sum_{i=1}^{N} \int_0^t \alpha_2^i(u)\, d\beta^i(u) \tag{2.7}$$

where $\alpha_1(t)$, $\alpha_2(t)$ are processes adapted to $(\mathfrak{F}_t)$, ($\alpha_2$ vector valued) and $(\beta(t), \mathfrak{F}_t, P)$ is a standard vector-valued Wiener process, the right hand side of (2.7) thus defining an Ito differential (see [17]).

$$[\alpha_2 = (\alpha_2^1, \ldots, \alpha_2^N) \text{ and } \beta = (\beta^1, \ldots, \beta^N)].$$

From (2.7) it is immediately seen that

$$Y_t \equiv E[f(x_t)|\mathfrak{F}_t] - E[f(x_0)|\mathfrak{F}_0] - \int_0^t \alpha_1(u)\, du \tag{2.8}$$

has to be a sample continuous $(\mathfrak{F}_t, P)$ martingale and square integrable if $f$ meets integrability requirements. Once the problem has been put in this perspective, the importance of Lemma 1 becomes immediately obvious since it suggests the choice $\beta = \nu$ in (2.7). It is also clear from (2.7) and (2.8) that the problem is solved if we can show that every square integrable $(\mathfrak{F}_t, P)$ martingale can be expressed as an Ito stochastic integral with respect to $\beta$. This, briefly, is the circle of ideas around which the proofs of the theorems in [6] are built. The approach just described yields the stochastic differential equation sought as an almost immediate consequence of the work of Cameron and Martin [1] and of Ito [9] if it can be established that for each t

$$\mathfrak{F}_t = \sigma\{\nu_s, 0 \leq s \leq t\}. \tag{2.9}$$

(2.9) is an appealing statement from a statistical point of view since it says that the data $\sigma$-fields can be generated by a Wiener process obtained by transforming the observed process suitably.

As can be easily verified (2.9) holds for the classical Kalman-Bucy (i.e., linear) problem, [14]. It has been conjectured that it is true in general, at least for the models considered here. Unfortunately no proof of (2.9) is available. Recently, J. M. C. Clark has proved (2.9) in the case when $(x_t)$ and $(w_t)$ are completely independent and $|h_t(\omega)| \leq N$ for a.e. $(t,\omega)$, N being a positive constant

[2]. Although (2.9) is of independent interest, in reality it is more than we need to derive (2.7). We bypass (2.9) and show by a different and independent argument based on Girsanov's theorem that every separable square integrable martingale $(Y_t, \mathfrak{F}_t, P)$ is sample continuous and has the representation

$$Y_t - E(Y_o) = \sum_{i=1}^{N} \int_0^t \Phi_s^i \, d\nu_s^i \tag{2.10}$$

where

$$\int_0^T E|\Phi_s|^2 \, ds < \infty \tag{2.11}$$

and $\Phi_s = (\Phi_s^1, \ldots, \Phi_s^N)$ is jointly measurable and adapted to $(\mathfrak{F}_s)$. Finding $\Phi_s$ is a separate problem. The details are given in [6] (Theorem 3.1). Girsanov's result which I shall state here for the reader's convenience is the following: Let $(\xi_t, \mathfrak{F}_t, P)$ be a Wiener process in $R^n$ and let $\varphi(t,\omega)$ be an n-vector jointly measurable process adapted to $(\mathfrak{F}_t)$ and such that

$$\int_0^T E|\varphi(t)|^2 \, dt$$

is finite. Then provided the measure

$$d\tilde{P} = \exp\left[\int_0^T \sum_{i=1}^{n} \varphi_i(u) \, d\xi_i(u) - \tfrac{1}{2}\int_0^T |\varphi(u)|^2 du\right] dP$$

is a probability measure, the process $\tilde{\xi}_t$ defined by

$$\tilde{\xi}_t = \xi_t - \int_0^t \varphi(s) \, ds$$

is a Wiener process with respect to $(\mathfrak{F}_t, \tilde{P})$. The precise manner in which this result is to be applied to yield (2.10) involves a somewhat delicate argument which would take me too far afield to reproduce here (see [6]).

In concluding this section the following brief

comment might be in order. For convenience consider only the scalar case (N = 1 in (2.7)). Suppose that $\alpha_1(t)$ is a process adapted to $(\mathfrak{F}_t)$ satisfying suitable conditions and such that $Y_t$ given by (2.8) is a square integrable $(\mathfrak{F}_t, P)$ martingale. We may then apply to $Y_t$ the decomposition theorem of H. Kunita and S. Watanabe (see ref. [12] in [6]) to get

$$Y_t = \sum_{n=1}^{M} \int_0^t g_n(u)\, d\zeta_n(u) \tag{2.12}$$

where the $\zeta_n$'s are mutually orthogonal $(\mathfrak{F}_t, P)$ martingales and the stochastic integrals are as given in the work of these authors. It is clear that (2.12) is of the form (2.7) except that M may be greater than one or even infinity (recall that N = 1). Furthermore from Lemma 1 we may take $\nu$ to be one of the $\zeta_n$'s say $\zeta_1$. Hence the approach in [6] may be regarded as a proof that M = 1 in (2.12) if the model is given by (2.1). The application of the above ideas to more general signal and observation processes awaits investigation.

## 3. Discussion of General Results

Let me state the result of [6] on the filtering problem involving stochastically controlled signal processes. Here I consider only one general model of such processes, viz. where the $x_t$ and the observation process $y_t$ are the solutions of a functional stochastic differential equation of the form

$$\begin{aligned} dx_t &= A(t, \pi_t x, \pi_t y)dt + B(t, \pi_t x, \pi_t y)dw_t \\ dy_t &= a(t, \pi_t x, \pi_t y)dt + b(t, \pi_t y)dw_t. \end{aligned} \tag{3.1}$$

Such equations have been specifically considered by Fleming and Nisio in studying the problem of existence of optimal stochastic controls [4].

In (3.1) $w_t$ is an M-vector standard Wiener process, a and A are N and M-N vector-valued. We denote by $C^k$ the space of all continuous mappings from [-T,0] to $R^k$ with the usual uniform topology. In (3.1) we make the following assumptions. a(t,f,g) and A(t,f,g) are N and (M-N)-vector functionals ($M > N$) respectively, defined on $[0,T] \times C^{M-N} \times C^N$. $b(t,f,g)$ and $B(t,f,g)$ are $N \times M$ and $(M - N) \times M$ matrix valued functionals on $[0.T] \times C^{M-N} \times C^N$. Furthermore all the functionals a, A, b and B are assumed Borel measurable on $[0,T] \times C^{M-N} \times C^N$. Also $\pi_t x \in C^{M-N}$ is defined for $(x_t)$ by

$$\begin{aligned}(\pi_t x)(u) &= x(t+u) && \text{if } -t \leq u \leq 0 \\ &= x(0) && \text{if } -T \leq u \leq -t.\end{aligned}$$

$\pi_t y$ is similarly defined. This notation means that in (3.1) the coefficients depend on the past values of the processes, e.g., a and A depend on t, $(x_s, 0 \leq s \leq t)$ and $(y_s, 0 \leq s \leq t)$. Further assume that the coefficients satisfy conditions that ensure the existence and uniqueness of a solution of the system (3.1). These conditions will be collectively referred to as (U). For (U) one may take the following: There exists a bounded measure $\Gamma$ on $[-T,0]$ and a positive constant K such that

$$|a(t,f,g) - a(t,\tilde{f},\tilde{g})|^2 + |b(t,f,g) - b(t,\tilde{f},\tilde{g})|^2 \leq K \int_{-T}^{0} (|f(s) - \tilde{f}(s)|^2 + |g(s) - \tilde{g}(s)|^2)\, d\Gamma(s). \tag{3.2}$$

$$|a(t,f,g)|^2 + |b(t,f,g)|^2 \leq K[1 + \int_{-T}^{0} |g(s)|^2 d\Gamma(s) + |L(t,f)|^2] \tag{3.3}$$

where $L(t,f)$ is a Borel measurable real valued functional on $[0,T] \times C^{M-N}$ such that

$$\int_0^T E|L(t,\pi_t x)|^2\, dt < \infty. \tag{3.4}$$

Similar Lipschitz conditions are also assumed for the coefficients A and B. For the initial random variables assume that $y_0 = 0$ and that

$$x_0 \quad 0 \text{ is independent of } \sigma\{w_s, 0 \leq s \leq T\}. \tag{3.5}$$

Then the equation (3.1) has a unique solution $(x_t, y_t)$ which is measurable with respect to $\sigma\{x_0, w_s, s \leq t\}$. In particular, for each t, $y_t$ is $\sigma\{x_u, w_u, 0 \leq u \leq t\}$ $(= G_t)$ measurable. Thus (3.1) furnishes us with a model in which $y_t$ is the observed process and the evolution of the system process $x_t$ depends on the values $(y_s, 0 \leq s \leq t)$. This, intuitively, is the notion of a "controlled" process.

For the stochastic differential equation for the non-linear filter we still need to put further conditions. For all $\xi$ in $R^N$

$$(c\xi, \xi) \geq K|\xi|^2 \quad \text{uniformly in } (t,g) \tag{3.6}$$

where $c = bb^t$ and $b^t$ is the transpose of b.

$$\int_0^T E|h'(t, \pi_t x, \pi_t y)|^2 \, dt < \infty \tag{3.7}$$

where

$$h'(t,f,g) = c^{-\frac{1}{2}}(t,g)\, a(t,f,g).$$

If we set

$$w_t' = \int_0^t c^{-\frac{1}{2}}(s, \pi_s y) b(s, \pi_s y)\, dw_s$$

and

$$z_t' = \int_0^t h'(s, \pi_s x, \pi_s y)\, ds + w_t' \tag{3.8}$$

then $w_t'$ is an N-vector standard Wiener process and we have

$$y_t = \int_0^t c_s^{\frac{1}{2}}\, dz_s'. \tag{3.9}$$

(3.8) and (3.9) show how the second equation in (3.1) is reduced to the "canonical" model introduced in (2.1). As in Section 2 (Lemma 1) if we define

$$\nu_t' = z_t' - \int_0^t \hat{h}_s'\, ds, \qquad \hat{h}_s' = E[h_s' | \mathcal{F}_s'] \tag{3.10}$$

and $\mathcal{F}'_t = \sigma\{z'_s, 0 \leq s \leq t\}$ then $\mathcal{F}'_t = \mathcal{F}_t \equiv \sigma\{y_s, s \leq t\}$ and $(\nu'_t, \mathcal{F}_t, P)$ is a standard Wiener process. We are now in a position to state the theorem which gives the stochastic differential equation for the optimal filter. For convenience we shall write $E^t[f(x_t)]$ for $E[f(x_t)|\mathcal{F}_t]$.

Theorem 1.

Let conditions (3.2) - (3.7) hold and let f be a real valued $C^2$-class function on $R^{M-N}$ satisfying

$$\int_0^T E|f(x_t)h'_t|^2 \, dt < \infty. \tag{3.11}$$

Then $E^t[f(x_t)]$ satisfies the stochastic differential equation

$$E^t[f(x_t)] = E[f(x_0)] + \int_0^t E^s[\tilde{A}_s f] \, ds$$
$$+ \int_0^t (c_s^{-\frac{1}{2}}[E[f(x_s)a_s] - E^s[f(x_s)] \, E^s[a_s] + b_s E^s[\tilde{D}_s f], d\nu'_s). \tag{3.12}$$

In (3.12) $\tilde{A}_t$ and $\tilde{D}_t$ are given by the following differential operators:

$$(\tilde{A}_t f)(\omega) = \sum_{i=1}^{M-N} A_i(t, \Pi_t x(\omega), \Pi_t y(\omega)) \, f'_{x_i}(x_t(\omega))$$
$$+ \tfrac{1}{2} \sum_{i,j=1}^{M-N} (BB^t)_{ij} \, (t, \Pi_t x(\omega), \Pi_t y(\omega)) \, f''_{x_i x_j}(x_t(\omega)). \tag{3.13}$$

$\tilde{D}_t = (\tilde{D}_t^1, \ldots, \tilde{D}_t^M)$ where

$$(\tilde{D}_t^i f)(\omega) = \sum_{j=1}^{M-N} B_{ji}(t, \Pi_t x(\omega), \Pi_t y(\omega) f'_{x_j}(x_t(\omega)). \tag{3.14}$$

The symbol
$$\int_0^t (f_s, d\nu'_s)$$

when $f_s = (f_s^1, \ldots, f_s^N)$ stands for the stochastic integral

$$\sum_{i=1}^{N} \int_0^t f_s^i \, d\nu_s'^{\,i}.$$

The reader who is interested in the proof of the above theorem is referred to [6] where it is deduced from more general theorems. Although the theory applies to a process $(x_t, y_t)$ which is not the solution even of a functional stochastic equation of type (3.1), the latter is interesting from the point of view of possible new applications and also because it contains much of the earlier work ([15], though not [12]) as a special case.

Remarks.

(a) If the coefficients $a, b, A, B$ depend only on $t$, $f(0)$ and $g(0)$ then it is well known that $((x_t, y_t), G_t, P)$ is Markov. In this case it is not hard to show (see [6]) that the operator $\tilde{A}_t$ in (3.13) is replaced by the generator of $(x_t, y_t)$.

(b) If $a, b, A, B$ depend on $t$, $f(0)$ and $g$ then $((x_t, y^t), G_t, P)$ is Markov where we write $y^t = (y_s, s \leq t)$. This is the case most often encountered in problems of stochastic control based on a partially observed process.

(c) It should be noted that the fact that $(x_t, y_t)$ is the solution of (3.1) does not necessarily imply that $(x_t, y_t)$ is a Markov process. However it is possible to introduce a related Markov process by enlarging the state space. Let $x^t = (x_s, s \leq t)$ and $y^t$ (as above) $= (y_s, s \leq t)$. Then it can be shown that $((x^t, y^t), G_t, P)$ is Markov. The identification of $\tilde{A}_t$ with the generator of the Markov processes in (b) and (c) can be made as in (a).

## 4. An Extension of Kalman and Bucy's Results

As an application of Theorem 1 of the previous section we shall derive from it a Kalman-Bucy type result for certain stochastically controlled processes $x_t$. Recently Wonham has obtained such an extension in connection with his 'separation theorem' of stochastic control. The result proved below holds for more general stochastic equations for $(x_t, y_t)$ than Wonham's [18]. It also contains as a particular case another extension of Kalman-Bucy's theorem due to Liptzer and Shiryaev ([15], Theorem 4). In order to keep the calculations less tedious we shall consider here only the scalar case, the general vector case being treated quite similarly. In other words, we take M = 2, N = 1 and consider the following special case of equation (3.1).

$$\begin{aligned} dx_t &= A_0(t,\pi_t y)\,dt + x_t A_1(t,\pi_t y)\,dt + \\ &\quad B_1(t,\pi_t y)\,dw_1(t) + B_2(t,\pi_t y)\,dw_2(t) \\ dy_t &= a_0(t,\pi_t y)\,dt + x_t a_1(t,\pi_t y)\,dt + \\ &\quad b_1(t,\pi_y y)\,dw_1(t) + b_2(t,\pi_t y)\,dw_2(t)\,. \end{aligned} \tag{4.1}$$

$w_1(t)$ and $w_2(t)$ are (one-dimensional) independent standard Wiener processes. In the notation of (3.1) we have

$$\begin{aligned} A(t,f,g) &= A_0(t,g) + f(0)A_1(t,g), \\ a(t,f,g) &= a_0(t,g) + f(0)a_1(t,g), \end{aligned} \tag{4.2}$$

$$w(t) = (w_1(t),w_2(t)),\ b = (b_1,b_2) \text{ and } B = (B_1,B_2).$$

For the Lipschitz condition (3.2) we assume the following: We assume all the conditions of Section 3 except that the global Lipschitz condition (3.2) does not seem appropriate and is replaced by similar conditions on $a_0$, $a_1$, $A_0$ and $A_1$:

$$\begin{aligned} &|\alpha_0(t,g) - \alpha_0(t,\tilde{g})|^2 + |\alpha_1(t,g) - \alpha_1(t,\tilde{g})|^2 \\ &\quad \le K \int_{-T}^{0} |g(s) - \tilde{g}(s)|^2\, d\Gamma(s), \end{aligned} \tag{4.3}$$

$$|\beta(t,g)| \le K \tag{4.4}$$

for all (t,g). Here $\alpha_0$ stands for $a_0$ or $a_1$ and $\alpha_1$ stands for $A_0$ and $A_1$ and $\beta$ stands for all the coefficients $a_0, a_1, A_0, A_1, b_1, b_2, B_1, B_2$. K is a positive constant and

$\Gamma$ is a bounded measure of $[-T,0]$ with a density $\gamma$ possessing a bounded derivative. We should remark that in imposing the above restrictions we have not tried to get the weakest possible conditions. It is easy now to verify that (4.3) and (4.4) imply the following conditions for a and A.

For every $N > 0$ there is a constant $L_N > 0$ such that if $|f(0)| \leq N$, $|\tilde{f}(0)| \leq N$,

$$|A(t,f,g) - A(t,\tilde{f},\tilde{g})|^2 \leq 2K^2|f(0) - \tilde{f}(0)|^2 + L_N \int_{-T}^{0} |g(s) - \tilde{g}(s)|^2 \, d\Gamma(s). \tag{4.5}$$

Finally the initial random variable $x_0$ is taken to be normally distributed with mean zero and variance $\sigma^2$ (say) and assumed independent of $\sigma\{w_s, 0 \leq s \leq T\}$. The existence of a unique solution $(x_t, y_t)$ of (4.1) is now proved essentially along the lines of the proof of Theorem 3 in Skorokhod's book ([17] pp. 51-55 (also see [7])). Our aim is to prove the following theorem under the assumptions (4.3) - (4.5).

Theorem 2.

Let $(x_t, y_t)$ be the solution of (4.1). Then for each t, the conditional distribution of $x_t$ given $(y_s, 0 \leq s \leq t)$ is Gaussian with mean $\hat{x}_t = E(x_t | \mathfrak{F}_t)$ and variance

$$R_t = E(x_t^2 | \mathfrak{F}_t) - (\hat{x}_t)^2$$

where $\hat{x}_t$ and $R_t$ satisfy the following differential equations.

$$d\hat{x}_t = (A_0 + A_1\hat{x}_t)dt + (b_1^2 + b_2^2)^{-\frac{1}{2}}[a_1R_t + b_1B_1 + b_2B_2]d\nu_t' \tag{4.6}$$

$$\frac{dR_t}{dt} = (B_1^2 + B_2^2) + 2A_1R_t - (b_1^2 + b_2^2)^{-\frac{1}{2}}[a_1R_t + (b_1B_1 + b_2B_2)]^2. \tag{4.7}$$

The coefficients in both equations are functions of t and $\Pi_t y$. In (4.6) $\nu_t'$ is the standard Wiener process on $(\mathfrak{F}_t, P)$ defined in (3.10).

Remark.

Given the data $(y_s, 0 \le s \le t)$ equation (4.7) is an ordinary (i.e., nonrandom) Riccati equation. The main part of the proof is to show that the conditional distribution of $x_t$ given $\mathfrak{F}_t$ is Gaussian. Equation (4.6) follows from (3.12) - (3.14) upon taking $f(x) = x$. To derive (4.7) use is made of the conditional Gaussian property through the relation

$$E^t(x_t^{\,3}) = (\hat{x}_t)^3 + 3\hat{x}_t R_t. \tag{4.8}$$

The calculations are entirely similar to those in [13] (p. 387). To prove that $x_t$ given $\mathfrak{F}_t$ is Gaussian we again make use of the filter equation (3.12) to derive stochastic differential equations for the cumulants (semi-invariants) $\lambda_k(t)$ $(k \ge 1)$ of the conditional distribution of $x_t$ given $\mathfrak{F}_t$. These equations and the initial condition that $x_0$ is a $N(0,\sigma^2)$ random variable yield a solution for which $\lambda_k(t) = 0$ for $k \ge 3$ for each t. The uniqueness of the solution of (4.1) then yields the desired conclusion. This is the method adopted by Liptzer and Shiryaev in the proof of their Theorems 4 and 5 of [15] and it works equally in our context with hardly any change. The only point-- an important one-- that is needed to ensure its validity is the existence of all the moments of $x_t$. This is achieved by means of the following lemmas. In fact we show the finiteness of $E[\exp(\delta x_t^2)]$ for some positive $\delta$.

Let us first observe that since $x_0$ is distributed $N(0,\sigma^2)$,

$$E(e^{\epsilon x_0^2}) < \infty \quad \text{for some } \epsilon > 0. \tag{4.9}$$

<u>Lemma 2.</u>

Suppose that $(x'_t)$ $(0 \le t \le T)$ is a sample continuous process satisfying

$$dx'_t = A_0(t,\Pi_t y)dt + x'_t A_1(t,\Pi_t y)dt + B_1(t,\Pi_t y)\, dw_1(t)$$

$$+ B_2(t,\Pi_t y)\, dw_2(t), \tag{4.10}$$

$$x'_0 = x_0$$

where $(y_t)$ is the second component of the solution process of (4.1). Assume

$$E(x_t'^2) < \infty \text{ for all } t. \tag{4.11}$$

Then for each t

$$E[x_t' - x_t]^2 = 0, \tag{4.12}$$

and in fact it then follows that $x_t' = x_t$ for all t with probability one.

Proof:

The proof is almost immediate. Noting that $(x_t, y_t)$ satisfies (4.1) and using (4.10), (4.11) and the conditions on the coefficients, we get

$$E[x'(t) - x(t)]^2 \leq (K^2T) \int_0^t E[x'(s) - x(s)]^2 \, ds. \tag{4.13}$$

(4.13) easily yields the inequality

$$E[x'(t) - x(t)]^2 \leq \frac{(K^2T)^{n+1} \quad T^n}{n!} \int_0^T E[x'(s) - x(s)]^2 ds. \tag{4.14}$$

Since n is an arbitrary integer (4.12) follows.

Lemma 3.

For some positive $\lambda$

$$E \exp[\lambda \left( \int_0^t B_i(s, \pi_s y) \, dw_i(s) \right)^2 ] < \infty \tag{4.15}$$

and

$$E \exp[\lambda \int_0^T \left( \int_0^t B_i(s, \pi_s y) \, dw_i(s) \right)^2 dt] \infty, \ (i=1,2). \tag{4.16}$$

Proof:

Note that the lemma will be proved under the assumption (4.4) of boundedness of the

$B_i$ ($|B_i| \leq K$). Using it we obtain the bound

$$E\left[\int_0^t B_i(s)dw_i(s)\right]^{2n} \leq \frac{(2tK^2)^n \,\Gamma(n+\frac{1}{2})}{\Gamma(\frac{1}{2})} \tag{4.17}$$

for n = 0,1,2, ... where we have written $B_i(s)$ for $B_i(s,\Pi_s y)$. The proof of (4.17) is rather tedious and will not be given here. I believe it is a known result although I have been unable to locate a reference.

For $\lambda > 0$ the series

$$\sum_{n=0}^{\infty} \frac{E\left[\int_0^t B_i(s)\, dw_i(s)\right]^{2n} \lambda^n}{n!} \leq \sum_{n=0}^{\infty} \frac{(2TK^2\lambda)^n \,\Gamma(n+\frac{1}{2})}{\Gamma(\frac{1}{2})n!} \tag{4.18}$$

and converges for $2TK^2\lambda < 1$. This proves (4.15) for $\lambda$ satisfying

$$0 < \lambda < \frac{1}{2TK^2} . \tag{4.19}$$

Now from the Jensen convexity inequality applied to the integration in t we have a.s.,

$$\begin{aligned} &\exp\{\lambda \int_0^T \left(\int_0^t B_i(s)\, dw_i(s)\right)^2 dt\} = \\ &\exp\{\frac{1}{T}\int_0^T \lambda T \left(\int_0^t B_i(s)\, dw_i(s)\right)^2 dt\} \\ &\leq \frac{1}{T}\int_0^T \exp[\lambda T \left(\int_0^t B_i(s)\, dw_i(s)\right)^2] \,dt \end{aligned} \tag{4.20}$$

so that taking expectations

$$E\exp\{\lambda \int_0^T \left(\int_0^t B_i(s)\, dw_i(s)\right)^2 dt\} \leq$$

$$\frac{1}{T}\int_0^T E\left\{\exp\left[\lambda T\left(\int_0^t B_i(s)\,dw_i(s)\right)^2\right]\right\}dt$$

$$\leq \sum_{n=0}^{\infty} \frac{(2T^2K^2\lambda)^n\,\Gamma(n+\frac{1}{2})}{\Gamma(\frac{1}{2})n!} \tag{4.21}$$

where we have used the bound given on the right hand side of (4.18) with $\lambda$ replaced by $\lambda T$. The above series converges if

$$0 < \lambda < \frac{1}{2T^2K^2} \quad . \tag{4.22}$$

The lemma is proved if $\lambda$ is chosen to satisfy (4.19) and (4.22).

Let us now set

$$\eta(t) = \int_0^t B_1(s)\,dw_1(s) + \int_0^t B_2(s)\,dw_2(s). \tag{4.23}$$

From Lemma 3, the Schwarz inequality and the elementary inequality

$$|a+b|^n \leq 2^{n-1}(|a|^n + |b|^n)$$

(n,a positive integer) it immediately follows that for some $\epsilon_1 > 0$

$$E\exp[\epsilon_1\eta^2(t)] < \infty \text{ for each } t, \tag{4.24}$$

and

$$E\exp[\epsilon_1\int_0^T \eta^2(t)\,dt] < \infty. \tag{4.25}$$

<u>Lemma 4.</u>

Assume the conditions imposed at the beginning of this section. Then for each t in [0,T] there is a positive $\delta$ (which is, in fact, chosen independently of t) such that

$$E\exp(\delta x_t^2) < \infty. \tag{4.26}$$

Proof:

For brevity write $A_0(s,\omega)$ for $A_0(s,\Pi_s y(\omega))$ etc. Let $x_0(t,\omega) \equiv 0$ for all $\omega$ and all $t$ in $[0, T]$ and for $n > 1$ define $x_n(t,\omega)$ recursively by

$$x_n(t,\omega) = x(0,\omega) + \int_0^t [A_0(s,\omega) + x_{n-1}(s,\omega)\, A_1(s,\omega)]\, ds + \eta(t,\omega). \tag{4.27}$$

Setting

$$\Delta_n x(t,\omega) = x_n(t,\omega) - x_{n-1}(t,\omega) \quad (\Delta_1 x(t,\omega) = x_1(t,\omega))$$

and noticing that

$$[\Delta_1 x(t,\omega)]^2 \leq 2x^2(0,\omega) + 4K^2 t + 4\eta^2(t,\omega)$$

it is easy to show that for almost all $\omega$ the following inequality holds for every $t$ in $[0,T]$ and $n = 1,2, \ldots$

$$[\Delta_n x(t,\omega)]^2 \leq \frac{4M\, M_1^{n-1}}{(n-1)!} + \frac{4M\, M_1^{n-2}}{(n-2)!} \int_0^T \eta^2(s,\omega)ds + \frac{2M_1^{n-1}}{(n-1)!} x^2(0,\omega), \tag{4.28}$$

where $M = TK^2$ and $M_1 = MT$.

Let $n > m$ and proceed as in Doob ([3], p.281) (dropping $\omega$) to get

$$[x_n(t) - x_m(t)]^2 \leq \left(\sum_{j=m+1}^{n} 2^{-j}\right) \left(\sum_{j=m+1}^{n} 2^j [\Delta_j x(t)]^2\right), \tag{4.29}$$

and

$$\sum_{j=m+1}^{n} 2^j [\Delta_j x(t)]^2 \leq \sum_{j=m+1}^{n} 2^j \left[\frac{4M\, M_1^{j-1}}{(j-1)!} + \right.$$

$$\frac{4M\, M_1^{j-2}}{(j-2)!} \int_0^T \eta^2(s)ds + \frac{2\, M_1^{j-1}}{(j-1)!} \; x^2(0) \Big] \qquad (4.30)$$

From (4.29) and (4.30) we find that a.s.

$$x_n(t) \to \text{ a limit } x'(t) \qquad (4.31)$$

uniformly in $t$ and it is easy to see that $x'(t)$ is a.s. continuous and satisfies (4.10). From (4.29) taking $m = 1$ and using the inequality for $x_1^2(t)$ (prior to (4.28)) we obtain

$$x_n^2(t) \leq c_1 + c_2 x^2(0) + c_3 \int_0^T \eta^2(s)\, ds + 8\eta^2(t) \qquad (4.32)$$

where $c_1$, $c_2$, $c_3$ are constants independent of $t$ and $\omega$. Hence

$$E(\exp[\delta x_n^2(t)]) \leq \exp(\delta c_1)\; E \exp[\delta c_2 x^2(0) + \qquad (4.33)$$

$$\delta c_3 \int_0^T \eta^2(s)\, ds + 8\delta\eta^2(t)].$$

Using (4.24) and (4.25) it follows by two applications of the Schwarz inequality that the right hand side expectation is finite for some $\delta > 0$. Making $n \to \infty$ (4.31) and Fatou's lemma give

$$E \exp[\delta x'^2(t)] < \infty. \qquad (4.34)$$

From Lemma 2 we immediately have (4.26).

Proof of Theorem 2:

As remarked above we proceed as in Liptzer and Shiryaev [15] to derive stochastic diferential equations for the semi-invariants $\lambda_k(t)$ of the conditional distribution of $x_t$. However we shall work with the conditional moment generating function instead of the conditional characteristic function of $x_t$. It will be shown below that this is permissible. It makes for simplicity because taking logarithms will not introduce any difficulties.

Fix $v_0$ real and positive and let $f(x) = e^{v_0 x}$. From (3.6) and the boundedness condition (4.4) on $a_0$ and $a_1$ and observing that

$$h'_t = (b_1^2 + b_2^2)^{-\frac{1}{2}} (a_0 + x_t a_1)$$

we have

$$E[f(x_t)h'_t]^2 = E[e^{2vx_t} (b_1^2 + b_2^2)^{-1}(a_0 + x_t a_1)^2$$

$$\leq 2K \, E[(1 + x_t^2) \, e^{2vx_t}].$$

Hence it follows from (4.26) of Lemma 4 that if $v_0$ is taken (and fixed) sufficiently small and independent of $t$ then

$$\int_0^T E[f(x_t)h'_t]^2 \, dt < \infty$$

thus verifying (3.11) of Theorem 1. For any $C^2$-class function $f$ satisfying (3.11) we obtain from (3.12), (3.13) and (3.14) the following equation

$$E^t (f(x_t)) = E (f(x_0)) + \int_0^t [A_0 \, E^s(f'(x_s))$$

$$+ A_1 E^s(x_s f'(x_s)) + \tfrac{1}{2}(B_1^2 + B_2^2)E^s f''(x_s))]ds$$

$$+ \int_0^t (b_1^2 + b_2^2)^{-\frac{1}{2}} \{a_1[E^s(x_s f(x_s)) - E^s(x_s)E^s(f(x_s))] \tag{4.35}$$

$$+ (b_1 B_1 + b_2 B_2)E^s(f'(x_s))\}d\nu'_s \, .$$

We may now take $f(x) = e^{vx}$ in (4.35) where $|v| \leq v_0$. It follows that if we define

$$\Phi_t(v) = E^t[e^{vx_t}] \tag{4.36}$$

then $\Phi_t(v)$ is finite for all $t$ in $[0,T]$ with probability one. A similar reasoning shows that the conditional moments $E^t(|x_t|^k)$ $(k = 1,2, \ldots)$ are finite for all $t$ with probability one. Hence with probability one, for each t,

$$\frac{\partial^k \Phi_t(v)}{\partial v^k} \quad \text{and hence} \quad \frac{\partial^k \Psi_t(v)}{\partial v^k}$$

where $\Psi_t(v) = \log \Phi_t(v)$ exist at $v = 0$ for $k \geq 1$. By definition, the k-th semi-invariant

$$\lambda_k(t) = \left( \frac{\partial^k \Psi_t(v)}{\partial v^k} \right)_{v = 0} .$$

Hence for all t with probability one the $\lambda_k(t)$ exist $(k \geq 1)$. Now substituting $f(x) = e^{vx}$ (with $|v| \leq v_0$) in (4.35) and simplifying we obtain

$$\Phi_t(v) = \Phi_0(v) + \int_0^t \{v[A_0\Phi_s(v) + A_1 \frac{\partial \Phi_s}{\partial v}] + \frac{v^2}{2}(B_1^2 + B_2^2)\, \Phi_s(v)\}\, ds + \int_0^t (b_1^2 + b_2^2)^{-\frac{1}{2}}[a_1 \left( \frac{\partial \Phi_s}{\partial v} - \lambda_1(s)\, \Phi_s(v) \right) + v(b_1B_1 + b_2B_2)\, \Phi_s(v)]\, d\nu'_s. \tag{4.37}$$

Using (4.37) and applying Ito's lemma, we get

$$d\Psi_t(v) = [v(A_0 + A_1 \frac{\partial \Psi_t}{\partial v}) + \frac{v^2}{2}(B_1^2 + B_2^2) - \tfrac{1}{2}(b_1^2 + b_2^2)^{-1} \{a_1 \left( \frac{\partial \Psi_t}{\partial v} - \lambda_1(t) \right) + v(b_1B_1 + b_2B_2)\}^2]dt + (b_1^2 + b_2^2)^{-\frac{1}{2}} [a_1 \left( \frac{\partial \Psi_t}{\partial v} - \lambda_1(t) \right) + v(b_1B_1 + b_2B_2)]\, d\nu'_s . \tag{4.38}$$

Since

$$\lambda_k(t) = \left( \frac{\partial^k \Psi_t}{\partial v^k} \right)_{v=0}$$

we obtain from (4.38) stochastic equations for $\lambda_k(t)$ in a similar manner. For $k \geq 3$ we have (see [15])

$$d\lambda_k(t) = \{k\lambda_k(t)A_1 - (b_1^2 + b_2^2)^{-1} \sum_{j=0}^{k-1} C_{k-1}^j [(b_1B_1 + b_2B_2)a_1(\lambda_{j+2}(t)\delta_{1,k-j-1} + \lambda_{k-j}(t)\delta_{0,j}) + a_1^2 \lambda_{k-j}(t)\lambda_{j+2}(t) - a_1^2 \lambda_1(t) \lambda_{j+2}(t) \delta_{0,k-j-1}]\}dt + \frac{a_1\lambda_{k+1}(t)}{b_1^2 + b_2^2} d\nu_t' . \quad (4.39)$$

Since $x(0)$ is $N(0,\sigma^2)$, $\lambda_k(0) = 0$ for $k \geq 3$. From this fact, (4.39) and the remarks preceding the lemmas it follows that for all $t$ in $[0,T]$ with probability one

$$\lambda_k(t) = 0 \quad (k \geq 3). \quad (4.40)$$

Familiar arguments now yield the fact that a.s. for all $t$, $\Phi_t(v)$ is the characteristic function of a normal distribution. The proof of the theorem is complete.

References


[1] Cameron, R. H. and Martin, W. T. (1947). The orthogonal development of non linear functionals in series of Fourier-Hermite functionals. Ann. Math. 48.

[2] Clark, J. M. C. (1969). Conditions for one-to-one correspondence between an observation process and its innovation. Control Systems Research Report, Imperial College.

[3] Doob, J. L. (1952). Stochastic Processes. Wiley, New York.

[4] Fleming, W. H. and Nisio, M. (1966). On the existence of optimal stochastic controls. J. of Math. and Mech. 15, 777-794.

[5] Frost, P. A. (1968). Estimation of continuous-time

nonlinear systems. Ph.D. dissertation, Department of Electrical Engineering, Stanford University.

[6] Fujisaki, M., Kallianpur, G. and Kunita, H. (1971). Stochastic differential equations for the non linear filtering problem. To appear in Osaka J. Of Math.

[7] Gikhman, I. I. and Skorokhod, A. V. (1968). Stochastic Differential Equations (in Russian). Kiev.

[8] Girsanov, I. V. (1960). On transforming a certain class of stochastic processes by absolutely continuous substitution of measures. Theory of Prob. and its Appl. 5 285-301.

[9] Ito, K. (1951). Multiple Wiener integral. J. Math. Soc. Japan 3 158-169.

[10] Kailath, T. (1968). An innovations approach to least squares estimation, Part I: Linear filtering with additive white noise. IEEE Transactions on Automatic Control, AC-13 6 646-655.

[11] Kailath, T. (1970). A further note on a general likelihood formula for random signals in Gaussian noise. IEEE Transactions on Information Theory IT-16.

[12] Kallianpur, G. and Striebel, C. (1969). Stochastic differential equations occurring in the estimation of continuous parameter stochastic processes. Theory of Prob. and its Appl. 4 597-622.

[13] Kallianpur, G. and Striebel, C. (1969). Stochastic differential equations in stochastic estimation problems. Multivariate Analysis 2. Academic Press.

[14] Kalman, R. E. and Bucy, R. S. (1961). New results in linear filtering and prediction theory. J. Basic. Engrg. 83.

[15] Liptzer, R. S. and Shiryaev, A. N. (1968). Non linear filtering of Markov diffusion processes. Trudy MIAN 135-180.

[16] Shiryaev, A. N. (1966). Stochastic equations of non linear filtering of Markovian jump processes. Problemi Peredachi Informatsii 3 3-22.

[17] Skorokhod, A. V. (1965). Studies in the theory of random processes (English translation). Addison-Wesley.

[18] Wonham, W. M. (1968). On the separation theorem of stochastic control. SIAM J. Control 6(2) 312-326.

# A CONVERGENCE THEOREM FOR NON NEGATIVE ALMOST SUPERMARTINGALES AND SOME APPLICATIONS*

H. Robbins and D. Siegmund

Columbia University

## 1. Introduction and Summary

The purpose of this paper is to give a unified treatment of a number of almost sure convergence theorems by exploiting the fact that the processes involved possess a common "almost supermartingale" properties To be precise, let $(\Omega,F,P)$ be a probability space and $F_1 \subset F_2 \subset \ldots$ a sequence of sub-$\sigma$-algebras of $F$. For each $n = 1,2, \ldots$ let $z_n$, $\beta_n$, $\xi_n$, and $\zeta_n$ be non-negative $F_n$-measurable random variables such that

$$E(z_{n+1}|F_n) \leq z_n(1 + \beta_n) + \xi_n - \zeta_n. \qquad (1)$$

In Section 2 below we prove

Theorem 1. $\lim_{n \to \infty} z_n$ exists and is finite and $\sum_1^\infty \zeta < \infty$ a.s. on

$$\{ \sum_1^\infty \beta_n < \infty, \quad \sum_1^\infty \xi_n < \infty\}.$$

In Sections 3-6 we apply this result (and the closely related Theorem 2) to obtain a number of convergence theorems, including those of Kolmogorov (strong law of large numbers for independent, non-identically distributed random variables), Blackwell (minimax theorem for vector- valued payoffs in game theory), MacQueen (cluster analysis), and some results in stochastic approximation.

*Research supported by NIH Grant 5-R01-GM-16895-03 and ONR Grant N00014-67-A-0108-0018.

The approach we use is similar to those in [7] and [8]. (We thank V. Fabian for pointing out the reference [7] to us.)

## 2. Proof of Theorem 1, Theorem 2, and Inequalities

To prove Theorem 1, let

$$A = \{\sum_1^\infty \beta_n < \infty, \sum_1^\infty \xi_n < \infty\}$$

and define

$$z_n' = z_n \prod_1^{n-1} (1+\beta_k)^{-1}, \quad \xi_n' = \xi_n \prod_1^{n} (1+\beta_k)^{-1},$$

and

$$\zeta_n' = \zeta_n \prod_1^{n} (1+\beta_k)^{-1}.$$

Then from (1)

$$E(z_{n+1}' | F_n) \leq z_n' + \xi_n' - \zeta_n' \tag{2}$$

and

$$\Sigma \xi_n' \leq \Sigma \xi_n < \infty \text{ on } A. \tag{3}$$

Let

$$u_n = z_n' - \sum_1^{n-1} (\xi_k' - \zeta_k') \tag{4}$$

and for $a > 0$

$$t = \inf \{n: \sum_1^n \xi_k' > a\} . \tag{5}$$

On $\{t > n\}$, we have by (2)

$$E(u_{n+1} | F_n) = E(z_{n+1}' - \sum_1^n (\xi_k' - \zeta_k') | F_n) \leq u_n,$$

and hence

$$E(u_{t\wedge(n+1)} | F_n) = u_t I_{\{t \leq n\}} + E(u_{n+1} | F_n) I_{\{t > n\}} \leq u_{t\wedge n}$$

so $\{u_{t\wedge(n\ 1)}, F_n, 1 \leq n < \infty\}$ is a supermartingale. Since

$$u_{t\wedge n} \geq -\sum^{t\wedge n-1} \xi_k \geq -a$$

for all n it follows from the martingale convergence theorem that

$$\lim_{n\to\infty} u_{t\wedge n} \text{ exists}$$

and is finite a.s., i.e., $\lim_{n\to\infty} u_n$ exists and is finite on

$$\{t = \infty\} = \{\sum_1^\infty \xi_n' \leq a\}.$$

Since a is arbitrary, we see that $\lim_{n\to\infty} u_n$ exists and is finite a.s. on $\{\Sigma\, \xi_n' < \infty\}$. Hence by (4) $\lim_{n\to\infty} z_n'$ exists and is finite and $\Sigma\, \zeta_n' < \infty$ on $\{\Sigma\, \xi_n' < \infty\}$, and it follows from (3),

$$z_n = z_n' \prod_1^{n-1} (1+\beta_k),$$

and the inequality

$$\zeta_n \leq \zeta_n' \prod_1^\infty (1+\beta_k)$$

that $\lim_{n\to\infty} z_n$ exists and is finite and $\sum_1^\infty \zeta_n < \infty$ a.s. on A.

<u>Theorem 2.</u> Suppose (1) is satisfied, and for $b > 0$ let

$$\beta_n(b) = \beta_n I\,\{z_n \leq b\},\quad \xi_n(b) = \xi_n I\,\{z_n \leq b\},$$

(6)

$$\zeta_n(b) = \zeta_n I\,\{z_n \leq b\}. \tag{6}$$

Then $\lim_{n\to\infty} z_n$ exists a.s. on $B = \bigcap_b \{\Sigma\beta_n(b) < \infty, \Sigma\xi_n(b) < \infty\}$, and $\sum_1^\infty \zeta_n < \infty$ a.s. on $B \cap \{\lim_n z_n < \infty\}$.

Proof: By (1)

$$E(z_{n+1} \wedge b | F_n) \leq [(1 + \beta_n)z_n \quad \xi_n - \zeta_n] \wedge b. \tag{7}$$

Hence on $\{z_n \leq b\}$

$$E(z_{n+1} \wedge b | F_n) \leq (1 + \beta_n)z_n + \xi_n - \zeta_n =$$

$$(1 + \beta_n(b))\ (z_n \wedge b) + \xi_n(b) - \zeta_n(b)$$

by (6), while on $\{z_n > b\}$

$$E(z_{n+1} \wedge b | F_n) \leq b = (1+ \beta_n(b))\ (z_n \wedge b) + \xi_n(b) - \zeta_n(b),$$

so that (1) holds for $z_n \wedge b$. It follows from Theorem 1 that $\lim_{n \to \infty} z_n \wedge b$ exists and

$$\sum_1^\infty \zeta_n(b) < \infty$$

a.s. on $\{\sum \beta_n(b) < \infty, \sum \xi_n(b) < \infty\}$. Since b is arbitrary, the first assertion of the theorem follows at once. If $\omega \in B$ and $\lim_{n \to \infty} z_n(\omega) < \infty$, then for some b, $\lim_{n \to \infty} z_n(\omega) < b$ and hence $\zeta_n(b)\ (\omega) = \zeta_n(\omega)$ for all $n = 1,2,\ldots$ from which the second assertion of the theorem follows.

A. Dvoretzky, who read the original manuscript of this paper, is responsible for shortening the proof of Theorem 1 and improving the form of Theorem 2. He has pointed out to us that the present method enables one to prove the following generalization of Theorem 1. Theorem 2 may be similarly generalized, and even more general versions of both theorems are possible.

Theorem 1'. For each $n = 1,2, \ldots$, let $\theta_n > 0$ and $\eta_n$ be $F_n$-measurable random variables, and suppose that

$$E(z_{n+1}|F_n) \leq \theta_n z_n + \eta_n. \qquad (8)$$

Then $\inf \sum_{k=1}^{n} \eta_k \prod_{i=1}^{k} \theta_i^{-1} > -\infty$ a.s. on

$$\{\sup_n \sum_{k=1}^{n} \eta_k \prod_{i=1}^{k} \theta_i^{-1} < \infty\},$$

and $\lim_{n\to\infty} z_n$ exists and is finite and $\lim_{n\to\infty} \sum_{k=1}^{n} \eta_k \prod_{i=k+1}^{n} \theta_i^{-1}$ exists and is finite a.s. on $\{\lim_{n\to\infty} \prod_1^n \theta_k$ exists and is finite,

and $\lim_{n\to\infty} \sum_{k=1}^{n} \eta_k \prod_{i=1}^{k} \theta_i^{-1}$ exists and is not $+\infty\}$.

The following inequalities are simple and useful generalizations of well-known results in martingale theory. Some applications are given in Sections 5 and 7. Moreover, Proposition 2 below yields an upcrossing inequality from which Theorem 1 follows easily (cf. problems 10 and 11 on p. 40 of [4]). Doob's submartingale convergence theorem may then be deduced directly from Theorem 1. (Suppose that $\{x_n, F_n, 1 \leq n < \infty\}$ is a submartingale and that

$$\sup_n Ex_n^+ < \infty.$$

To prove that $\lim_{n\to\infty} x_n$ exists it suffices to show that

$$\lim_{n\to\infty} \max(x_n, a)$$

exists for every $-\infty < a < \infty$, and since $\{\max(x_n, a) - a, F_n, 1 \leq n < \infty\}$ is a nonnegative submartingale, we may assume without loss of generality that $x_n \geq 0$ and $\sup_n Ex_n < \infty$. Now $E(x_{n+1}|F_n) = x_n + \xi_n$, where

$$\xi_n = E(x_{n+1}|F_n) - x_n \geq 0$$

and

$E\left(\sum_1^\infty \xi_n\right) \leq \sup_n E(x_n) < \infty$, so by Theorem 1, $\lim_n x_n$ exists a.s.)

Proposition 1. If (1) holds, then for any $m = 1,2, \ldots$ and finite stopping time $t \geq m$,

$$E(z_t|F_m) \leq z_m + E\left[\sum_{k=m}^{t-1} (\beta_k z_k + \xi_k)|F_m\right] \tag{9}$$

Proof Let $a > 0$, let $s$ = first $n \geq m$ such that $z_n \geq a$,

$$\sum_m^n \beta_k \geq a, \quad \text{or} \quad \sum_m^n \xi_k \geq a; \; s = \infty$$

if no such n exists, and let

$$u_n^{(m)} = z_n - \sum_{k=m}^{n-1} (\beta_k z_k + \xi_k). \tag{10}$$

As in the proof of Theorem 1, it may be verified that

$$\{u^{(m)}_{t\wedge s\wedge n}, F_n, m \leq n < \infty\}$$

is a supermartingale. Hence, for any $n \geq m$,

$$E(u^{(m)}_{t\wedge s\wedge n}|F_m) \leq u_m^{(m)},$$

or equivalently by (10),

$$E(z_{t\wedge s\wedge n}|F_m) \leq z_m + E\left[\sum_m^{t\wedge s\wedge n-1} (\beta_k z_k + \xi_k|F_m)\right]. \tag{11}$$

Letting $a \to \infty$, $n \to \infty$, we obtain (9) from (11) and the monotone convergence theorem and Fatou's lemma for conditional expectations.

Proposition 2.

If (1) holds, for any $a > 0$, $m = 1,2, \ldots$, and $n \geq m$

$$P\{\max_{m \le k \le n} z_k \ge a | F_m\} \le a^{-1} [z_m + E\left(\sum_m^{n-1} \xi_k | F_m\right)] + E\left(\sum_m^{n-1} \beta_k \, | F_m\right). \tag{12}$$

Proof. Let $t = \inf \{k: k \ge m, z_k \ge a\}$. Then $aP\{z_{t\wedge n} \ge a | F_m\} \le E(z_{t\wedge n} | F_m)$, which together with (9) and the definition of t proves (12).

## 3. The Strong Law of Large Numbers

The line of reasoning in the following application of Theorem 1 to the strong law of large numbers is typical of most of our applications. Application 1 has been obtained by Neveu [9], p. 148. Corollary 1 is Kolmogorov's classical result. Corollary 2, which is stated here for use in Section 6, was obtained independently of Neveu by Dubins and Freedman [5].

Application 1. Let $y_1, y_2, \ldots$ be a sequence of random variables such that $E|y_k| < \infty$ and $E(y_{n+1} | F_n) = 0$ $(n = 1, 2, \ldots)$. Assume that for each $n = 2, 3, \ldots$, $c_n$ is a positive $F_{n-1}$-measurable random variable with $c_1 \le c_2 \le \ldots$. Then

$$\lim_{n \to \infty} c_n^{-1} \left| \sum_1^n y_k \right|$$

exists and is finite a.s. on $\{\Sigma c_k^{-2} E(y_k^2 | F_{k-1}) < \infty\}$; it equals 0 a.s. on

$$\{c_n \uparrow \infty, \Sigma c_k^{-2} E(y_k^2 | F_{k-1}) < \infty\}.$$

Proof. Let

$$z_n = c_n^{-2} \left(\sum_1^n y_k\right)^2,$$

$$\zeta_n = \frac{c_{n+1}^2 - c_n^2}{c_{n+1}^2} z_n,$$

and

$$\xi_n = c_{n+1}^{-2} E(y_{n+1}^2 | F_n).$$

It is easily verified that (1) is satisfied with $\beta_n = 0$ $(n \geq 1)$, and hence $\lim_{n \to \infty} z_n$ exists and is finite and

$$\sum_1^\infty \zeta_n < \infty \quad \text{a.s on}$$

$$\{\Sigma \xi_n < \infty\} = \{\Sigma c_k^{-2} E(y_k^2 | F_{k-1}) < \infty\}.$$

By Lemma 1 below

$$\sum_1^\infty \left\{ \frac{c_{n+1}^2 - c_n^2}{c_{n+1}^2} \right\} = \infty \text{ on } \{c_n \uparrow \infty\},$$

and hence the convergence of $z_n$ together with that of $\Sigma \zeta_n$ implies that $\lim_{n \to \infty} z_n = 0$ a.s. on

$$\{c_n \uparrow \infty, \Sigma c_k^{-2} E(y_k^2 | F_{k-1}) < \infty\}.$$

<u>Corollary 1.</u> Let $y_1, y_2, \ldots$ be independent random variables with $Ey_n = 0$ $(n = 1,2, \ldots)$. If

$$\sum_1^\infty n^{-2} Ey_n^2 < \infty,$$

then

$$\lim_{n \to \infty} n^{-1} \sum_1^n y_k = 0 \quad \text{a.s.}$$

<u>Corollary 2.</u> Let $r_1, r_2, \ldots$ be random variables assuming the values of 0 or 1, and let $p_k = P\{r_k = 1 | r_1, \ldots, r_{k-1}\}$. Assume $p_1 = P\{r_1 = 1\} > 0$. Then

$$\lim_{n \to \infty} \frac{r_1 + \cdots + r_n}{p_1 + \cdots + p_n}$$

exists and is finite a.s. It is equal to 1 on $\{\Sigma p_k = \infty\}$.

Proof. We employ Application 1 with $y_n = r_n - p_n$, $F_n = B(r_1, \ldots, r_n)$, $c_n = p_1 + \cdots + p_n$ and hence

$$E(y_{n+1}^2 \mid F_n) = p_n(1 - p_n) \le p_n.$$

Obviously

$$\Sigma \frac{p_n(1 - p_n)}{(p_1 + \cdots + p_n)^2} \le \Sigma \frac{p_n}{(p_1 + \cdots + p_n)^2} < \infty$$

on $\{\Sigma p_n < \infty\}$, and by Lemma 1 below this series also converges on $\{\Sigma p_n = \infty\}$. Hence

$$\lim_{n \to \infty} \left( \frac{r_1 + \cdots + r_n}{p_1 + \cdots + p_n} - 1 \right) = \lim_{n \to \infty} \frac{\sum_1^n (r_k - p_k)}{\sum_1^n p_k}$$

exists and is finite a.s. and equals 0 a.s. on $\{\Sigma p_n = \infty\}$.

Lemma 1. Let $0 < a_n \uparrow \infty$. Then

$$\sum_1^\infty \frac{a_{n+1} - a_n}{a_{n+1}^2} < \infty \tag{15}$$

and

$$\sum_1^\infty \frac{a_{n+1} - a_n}{a_{n+1}} = \infty \quad . \tag{16}$$

Proof. To prove (15), note that

$$\sum_1^\infty \frac{a_{n+1} - a_n}{a_{n+1}^2} \le \sum_1^\infty \int_{a_n}^{a_{n+1}} x^{-2} dx = \int_{a_1}^\infty x^{-2} dx < \infty.$$

To prove (16), first suppose $\sup_n a_{n+1}/a_n = \infty$. Then infinitely many of the terms

$(a_{n+1} - a_n)/a_{n+1}$ exceed ½ and hence (16) holds. On the other hand if $\sup_n a_{n+1}/a_n = a < \infty$, then

$$\sum_1^\infty \frac{a_{n+1} - a_n}{a_{n+1}} \geq \sum_1^\infty \frac{a_n}{a_{n+1}} \int_{a_n}^{a_{n+1}} x^{-1} dx \geq a^{-1} \int_{a_1}^{\infty} x^{-1} dx = \infty.$$

## 4. Stochastic Approximation

Let $\{y(x), -\infty < x < \infty\}$ be a family of random variables with $Ey(x) = M(x)$, Var $y(x) = \sigma^2(x) < \infty$ $(-\infty < x < \infty)$. The Robbins-Monro process [10] is designed to find, under certain assumptions, the location of the root $M(x) = 0$ of the regression fuction M. Starting with an arbitrary random variable $x_1$, we define successively $x_2, x_3, \ldots$, by

$$x_{n+1} = x_n - a_n y_n, \tag{17}$$

where $a_n$ is a non-negative (measurable) function of $x_1, y_1, \ldots, y_{n-1}$, and conditional on $x_1, y_1, \ldots, y_{n-1}$ the random variable $y_n$ has the distribution of $y(x_n)$. The following application of Theorem 1 generalizes results of Blum [2] and Gladyshev [7].

<u>Application 2.</u> If $\sigma$ and M are measurable,

$$\sigma(x) + |M(x)| \leq a + b|x| \quad \text{for some } a, b > 0, \tag{18}$$

there exists a real number $\theta$ such that for each $\epsilon > 0$

$$\inf_{\epsilon < x-\theta < \epsilon^{-1}} M(x) > 0, \qquad \sup_{\epsilon < x-\theta < \epsilon^{-1}} M(x) < 0, \tag{19}$$

$$\sum a_n^2 < \infty \text{ for every sequence } x_1, y_1, \ldots, \tag{20}$$

and

$$\sum a_n = \infty \text{ for every sequence } x_1, y_1, \ldots \text{ such that } \sup |x_n| < \infty, \tag{21}$$

then $\lim_{n\to\infty} x_n = \theta$ with probability one.

Proof. Let $F_n = B(x_1, y_1, \ldots, y_{n-1})$, $z_n = (x_n - \theta)^2$. By (17), (18), (19) and the inequality

$$(a + b)^2 \leq 2(a^2 + b^2)$$

$$E(z_{n+1}|F_n) = z_n - 2a_n(x_n - \theta)E(y_n|F_n) + a_n^2 E(y_n^2|F_n)$$

$$z_n - 2a_n|x_n - \theta||M(x_n)| + a_n^2(\sigma^2(x_n) + M^2(x_n))$$

$$\leq z_n(1 + 4b^2a_n^2) - 2a_n|x_n - \theta||M(x_n)| + a_n^2(2a^2 + 4b^2\theta^2).$$

Putting

$$\beta_n = 4b^2a_n^2, \quad \xi_n = 2a_n^2(a^2 + 2b^2\theta^2),$$

and

$$\zeta_n = 2a_n|x_n - \theta||M(x_n)|,$$

we see from (20) and Theorem 1 that $\lim_{n \to \infty} z_n$ exists and is finite and

$$\sum_1^\infty \zeta_n = \sum_1^\infty a_n |x_n - \theta||M(x_n| < \infty$$

a.s. Hence by (19) and (21) $\lim_{n \to \infty} |x_n - \theta| = 0$ with probability one.

Dvoretzky [6] has given an example which shows that the growth condition on M given by (18) cannot be dropped completely. If $M(x) = x|x|$, $\sigma^2 = 0$, $a_n = 1/n$, and $x_1 = 6$, then it is easy to see that $\sup x_n = +\infty$, $\liminf x_n = -\infty$, and the sequence $(x_n)$ has no finite point of accumulation. It is also easy to give examples which show that the growth condition on $\sigma$ cannot be dropped completely.

The following simple modification of the Robbins-Monro process permits removal of the growth condition on M. (See also Remarks (i) and (ii) below.) Let $\underline{\theta} = \bar{\theta}$ and define

$$\tilde{x}_{n+1} = \begin{cases} (\tilde{x}_n - a_n y_n) I_{\{\tilde{x}_n - a_n y_n \geq \underline{\theta}\}} + \underline{\theta}\, I_{\{\tilde{x}_n - a_n y_n < \underline{\theta}\}} & \text{if } \tilde{x}_n > \bar{\theta} \\ \tilde{x}_n - a_n y_n & \text{if } \underline{\theta} \leq \tilde{x}_n \leq \bar{\theta} \\ (\tilde{x}_n - a_n y_n) I_{\{\tilde{x}_n - a_n y_n \leq \bar{\theta}\}} + \bar{\theta}\, I_{\{\tilde{x}_n - a_n y_n > \bar{\theta}\}} & \text{if } \tilde{x}_n \leq \underline{\theta}, \end{cases}$$

where $\tilde{x}_1$ is arbitrary, and as before $a_n$ is a non-negative function of $\tilde{x}_1, y_1, \ldots, y_{n-1}$ and conditional on $\tilde{x}_1, y_1, \ldots, y_{n-1}$ the random variable $y_n$ has the distribution of $y(\tilde{x}_n)$.

Application 3. If $\sigma$ and $M$ are measurable, $M$ is locally bounded,

$$\sigma^2(x) \leq a + bx^2 \quad \text{for some } a, b > 0 \tag{23}$$

(19), (20), and (21) hold (with $\tilde{x}$'s instead of $x$'s), then $\lim_{n \to \infty} \tilde{x}_n = \theta$ with probability one.

Proof. Without loss of generality we may assume that $\theta = 0$. By (22) and Chebyshev's inequality, on $\{\tilde{x}_n > \bar{\theta}\}$ we have

$$E(\tilde{x}_{n+1}^2 | F_n) \leq E((\tilde{x}_n - a_n y_n)^2 | F_n) + \underline{\theta}^2 P\{\tilde{x}_n - a_n y_n < \underline{\theta} | F_n\}$$

$$\leq \tilde{x}_n^2 - 2a_n \tilde{x}_n M(\tilde{x}_n) + a_n^2 E(y_n^2 | F_n) + \underline{\theta}^2 P\{a_n y_n > \bar{\theta} - \underline{\theta} | F_n\}$$

$$\leq \tilde{x}_n^2 - 2a_n \tilde{x}_n M(\tilde{x}_n) + a_n^2 (\sigma^2(\tilde{x}_n) + M^2(\tilde{x}_n)) \left( 1 + \frac{\underline{\theta}^2}{(\bar{\theta} - \underline{\theta})^2} \right).$$

Similar computations on $\{\underline{\theta} \leq \tilde{x}_n \leq \bar{\theta}\}$ and $\{\tilde{x}_n < \underline{\theta}\}$ show that in general

$$E(\tilde{x}_{n+1}^2 | F_n) \leq \tilde{x}_n^2 - 2a_n \tilde{x}_n M(\tilde{x}_n)$$

$$+ a_n^2 (\sigma^2(\tilde{x}_n) + M^2(\tilde{x}_n)) \left( 1 + \frac{\bar{\theta}^2 + \underline{\theta}^2}{(\bar{\theta} - \underline{\theta})^2} \right).$$

Hence by Theorem 2 and the line of reasoning used in the

proofs of Applications 1 and 2,

$$\lim_{n \to \infty} \tilde{x}_n = 0$$

a.s. on $\{\liminf |\tilde{x}_n| < \infty\}$. To complete the proof it suffices to show that $P\{\liminf |\tilde{x}_n| = \infty\} = 0$. Let

$$\tilde{z}_n = [(\tilde{x}_n - \theta \vee \bar{\theta})^+]^2.$$

Using (19) and (23) it may be shown that

$$\begin{aligned} E(\tilde{z}_{n+1}|F_n) &\leq \tilde{z}_n(1+2ba_n^2) + a_n^2(a+2b(\theta\vee\bar{\theta})^2) && \text{if } \tilde{z}_n > 0 \\ &\leq Ca_n^2 && \text{if } \tilde{z}_n = 0, \end{aligned}$$

where C depends on $\underline{\theta}$, $\bar{\theta}$, $\theta$,

$$\sup_{\underline{\theta} \leq x \leq \theta\vee\bar{\theta}} |M(x)|, \qquad \text{and} \qquad \sup_{\underline{\theta} \leq x \leq \theta\vee\bar{\theta}} \sigma(x).$$

Hence in general

$$E(\tilde{z}_{n+1}|F_n) \leq \tilde{z}_n(1 + 2ba_n^2)$$

$$+ a_n^2(a + 2b(\theta\vee\bar{\theta})^2) + Ca_n^2 I_{\{\tilde{z}_n = 0\}},$$

and by Theorem 1, $P\{\lim \tilde{z}_n = \infty\} = P\{\lim \tilde{x}_n = \infty\} = 0$. Similarly, $P\{\lim \tilde{x}_n = -\infty\} = 0$. From (22) if $\limsup \tilde{x}_n = +\infty$ and $\liminf \tilde{x}_n = -\infty$, then $\tilde{x}_n \in [\underline{\theta},\bar{\theta}]$ infinitely often and hence $\liminf |\tilde{x}_n| < \infty$. Thus $P\{\liminf |\tilde{x}_n| = \infty\} = 0$, which completes the proof.

Remarks. (i) The conditions (18) and (23) may be dropped if (20) is appropriately strengthened. For example, for the modified Robbins-Monro process defined by (22), if the there exists a known measurable function $\tilde{\sigma}$ (which may be assumed to vanish on an arbitrary bounded interval) such that

$$\sigma(x) = O(\tilde{\sigma}(x)) \qquad \text{as} \qquad |x| \to \infty,$$

then

$$\Sigma a_n^2 \; (1 + \tilde{\sigma}^2(\tilde{x}_n)) < \infty \qquad \text{for every sequence } \tilde{x}_1, y_1, \ldots \tag{24}$$

together with (19), (21), and the local boundedness of M and $\sigma$ implies the convergence of $\tilde{x}_n$ to $\theta$ with probability one.

(ii) The preceding modification of the Robbins-Monro process has practical as well as theoretical interest. The experimenter who has some prior knowledge of the value of $\theta$ may wish to use (22) with the hope that the speed of convergence will be increased if in fact $\theta \in [\underline{\theta},\bar{\theta}]$, while being protected against the possibility that $\theta \notin [\underline{\theta},\bar{\theta}]$ in the sense that convergence to the true value of $\theta$ still occurs with probability one.

(iii) Schmetterer [11] has shown that Dvoretzky's example for the unmodified Robbins-Monro process with $\sigma^2$ bounded depicts a general phenomenon, i.e., with probability one that either $\lim x_n = \theta$ or that $\limsup x_n = +\infty$, $\liminf x_n = -\infty$, and $(x_n)$ has no finite point of accumulation. This result follows easily from the method of proof of Aplication 3.

In [6] Dvoretzky proved a general convergence theorem, which includes Blum's result for the Robbins-Monro process and the corresponding result for the Kiefer-Wolfowitz method for estimating the maximum of a regression function as special cases. The probability one part of Dvoretzky's theorem (he also proved mean square convergence) follows from Theorem 1, as the following generalization of his result shows.

Application 4. Let X be a Hilbert space. For each $n = 1, 2, \ldots$ let $X_n = X \times X \times \ldots \times X$ (n factors) and suppose that $T_n : X_n \to X$ is a Borel-measurable transformation satisfying for some $\theta \in X$

$$||T_n(x_1, \ldots, x_n) - \theta|| \quad \max(a_n, (1+b_n)||x_n - \theta|| - c_n) \tag{25}$$

where $a_n$, $b_n$, and $c_n$ are non-negative Borel measurable functions defined on $X_n$ such that

$$\lim_{n \to \infty} a_n = 0 \qquad \text{for all } x_1, x_2, \ldots \tag{26}$$

$$\Sigma\, b_n < \infty \quad \text{for all} \quad x_1, x_2, \ldots \tag{27}$$

and

$$\Sigma\, c_n = \infty \text{ for all } x_1, x_2, \ldots \text{ such that } \sup \|x_n\| < \infty. \tag{28}$$

Let $x_1, y_1, y_2, \ldots$ be random variables taking values in X and define

$$x_{n+1} = T_n(x_1, \ldots, x_n) + y_n \qquad (n = 1,2, \ldots). \tag{29}$$

If

$$E(y_n | x_1, y_1, \ldots, y_{n-1}) = 0 \quad \text{a.s. } (n = 1,2, \ldots) \tag{30}$$

and

$$\Sigma\, E(\|y_n\|^2 | x_1, y_1, \ldots, y_{n-1}) < \infty \quad \text{a.s.} \tag{31}$$

then $\lim_{n \to \infty} \|x_n - \theta\| = 0$ with probability one.

Proof. The proof is a modification of the first part of Dvoretzky's argument. It may be assumed that $\theta = 0$. Let

$$F_n = B(x_1, y_1, \ldots, y_{n-1}), \quad z_n = [(\|x_n\| - a_{n-1})^+]^2.$$

For any $F_n$-measurable random variable $u_n$ such that $\|u_n\| \leq a_n$ it follows from the triangle inequality that

$$z_{n+1} \leq \|x_{n+1} - u_n\|^2$$

and hence by (29) and (30)

$$E(z_{n+1} | F_n) \leq \|T_n - u_n\|^2 + E(\|y_n\|^2 | F_n) . \tag{32}$$

Putting

$$u_n = T_n I_{\{\|T_n\| \leq a_n\}} + a_n \frac{T_n}{\|T_n\|} I_{\{\|T_n\| > a_n\}},$$

we obtain from (32)

$$E(z_{n+1} | F_n) \leq [(\|T_n\| - a_n)^+]^2 + E(\|y_n\|^2 | F_n). \tag{33}$$

Now by (25), if $\|x_n\| \leq a_n$

$$(\|T_n\| - a_n)^+ \leq (b_n a_n - c_n)^+$$

while if $\|x_n\| > a_n$

$$(\|T_n\| - a_n)^+ \leq [(1 + b_n)(\|x_n\| - a_n) + b_n a_n - c_n]^+,$$

so in either case

$$[(\|T_n\| - a_n)^+]^2 \leq [(1 + b_n)(\|x_n\| - a_n)^+ - c_n + a_n b_n]^2.$$

Using the inequality $(a + b)^2 \leq (1 + b)a^2 + b(1 + b)$ which is valid for all $b \geq 0$ and all real a, we obtain

$$[(\|T_n\| - a_n)^+]^2 \leq (1 + a_n b_n)(1 + b_n)^2 z_n - c_n z_n^{\frac{1}{2}}$$

$$+ \ (1 + a_n b_n)c_n^2 + a_n b_n(1 + a_n b_n).$$

Hence by (33), (1) holds with

$$\beta_n = b_n(1 + a_n b_n)(2 + b_n), \ \zeta_n = c_n z_n^{\frac{1}{2}}$$

and

$$\xi_n = (1 + a_n b_n)c_n^2 + a_n b_n(1 + a_n b_n) + E(\|y_n\|^2 | F_n). \tag{34}$$

Now if (25) holds for a given sequence $(c_n)$, it holds a fortiori for any smaller sequence. Hence, replacing $c_n$ by $\min(1,c_n)$, we may assume that $c_n \leq 1$, since the series

$$\sum_1^\infty \min(1,c_n)$$

diverges if $\sum_1^\infty$ does. Moreover, defining

$$c'_n = \begin{cases} 0 & \text{if } c_1 + \dots + c_n = 0 \\ c_n/(c_1 + \dots + c_n) & \text{if } c_1 + \dots + c_n > 0, \end{cases}$$

we see that (25) continues to hold with $c_n'$ in place of $c_n$, and by Lemma 1

$$\Sigma c'_n = \infty \text{ for all } x_1, x_2, \ldots \text{ such that } \sup \|x_n\| < \infty$$

$$\Sigma (c'_n)^2 \leq \Sigma \frac{c_n}{(c_1 + \cdots + c_n)^2} < \infty \text{ for all } x_1, x_2, \ldots$$

Hence it may be assumed that

$$c_n \leq 1, \quad \Sigma c_n^2 < \infty \quad \text{for all } x_1, x_2, \ldots \quad . \tag{35}$$

Then by (26), (27), (34) and (35) $P\{\Sigma \beta_n < \infty, \Sigma \xi_n < \infty\} = 1$, and it follows from Theorem 1 that

$$\lim_{n \to \infty} z_n$$

exists and is finite and $\Sigma \zeta_n = \Sigma c_n z_n^{\frac{1}{2}} < \infty$ a.s. By (28) $\lim_{n \to \infty} z_n = 0$ a.s. which by (26) implies that

$$\lim_{n \to \infty} \|x_n\| = 0$$

a.s.

## 5. Minimax Theorem for Vector Payoffs

In [1] Blackwell considers two person games in which each player has a finite number of possible actions, and the return to Player I if he takes action i ($1 \leq i \leq r$) and Player II takes action j ($1 \leq j \leq s$) is a random variable distributed according to m(i,j), a probability distribution over a closed, bounded, convex set X (not depending on (i,j)) in N space. The question Blackwell posed is whether in a sequence of repeated plays of such a game Player I can force the center of gravity $\bar{x}_n$ of his rewards $x_1, \ldots, x_n$ in the first n plays to converge to a given set S no matter what Player II does. Letting $\delta_n$ denote the squared distance from $\bar{x}_n$ to S, Blackwell shows that under certain hypotheses on S and the matrix (m(i,j)) there exists a particular sequence of strategies for Player I such that no matter what Player II does

$$E(\delta_n | \delta_1, \ldots, \delta_{n-1}) \leq (1 - 2/n)\delta_{n-1} + c/n^2 \text{ if } \delta_{n-1} > 0, \tag{36}$$

$$0 \leq \delta_n \leq a, \tag{37}$$

and $$|\delta_n - \delta_{n-1}| \leq b/n, \tag{38}$$

where a,b, and c depend only on the size of the bounded set X. He then applies the following lemma, which we state as

Application 5. A sequence of random variables $\delta_1, \delta_2, \ldots$ satisfying (36), (37), and (38) converges to 0 at a rate which depends only on a, b, and c; i.e., for every $\epsilon > 0$ there exists an $n_0$ depending only on $\epsilon$,a,b, and c such that

$$P\{\sup_{n \geq n_0} \delta_n \geq \epsilon\} \leq \epsilon$$

uniformly for all sequences $(\delta_n)$ satisfying (36), (37) and (38).

Proof. Let $z_n = \delta_n^2$, $F_n = B(\delta_1, \ldots, \delta_n)$ $(F_0 = \{\emptyset, \Omega\})$. Then by (36), (37), and (38), on

$$\{\delta_{n-1} > 0\}$$

$$E(z_n | F_{n-1}) = E((\delta_n - \delta_{n-1} + \delta_{n-1})^2 | F_{n-1})$$

$$= E((\delta_n - \delta_{n-1})^2 | F_{n-1}) + 2\delta_{n-1} E(\delta_n - \delta_{n-1} | F_{n-1}) + \delta_{n-1}^2$$

$$\leq \left(\frac{b}{n}\right)^2 + 2\delta_{n-1}(-2/n\ \delta_{n-1} + c/n^2) + \delta_{n-1}^2$$

$$\leq z_{n-1}(1 - 4/n) + \frac{2ac + b^2}{n^2} .$$

Hence by (38), with probability one

$$E(z_n | F_{n-1}) \leq (1 - 4/n) z_{n-1} + \frac{2ac + b^2}{n^2} \tag{39}$$

$(n \geq 4)$. It follows at once from Theorem 1 with

$$\xi_n = \frac{2ac + b^2}{(n+1)^2}$$

and $\zeta_n = (4/(n+1)) z_n$ that

$\lim_{n \to \infty} z_n$ exists and is finite and

$$\sum_1^\infty \zeta_n = 4 \sum_1^\infty z_n / n+1 < \infty \quad \text{a.s.};$$

hence $\lim_{n \to \infty} z_n = 0$ a.s. To go farther and prove the required uniformity, note that from (37) and (39) we obtain by iteration for $n_0 \geq m \geq 4$

$$Ez_{n_0} \leq a^2 \prod_{k=m+1}^{n_0} (1 - 4/k) + (2ac + b^2) \sum_{k=m+1}^{n_0} k^{-2} \quad . \tag{40}$$

Now $\prod (1 - 4/k)$ diverges and $\sum k^{-2}$ converges. Hence for any $\epsilon > 0$ there exists $m = m(a,b,c,\epsilon)$ so large that

$$(2ac + b^2) \sum_{m+1}^{\infty} k^{-2} < \epsilon^2/4$$

and $n_0 = n_0(a,b,c,\epsilon) \geq m$ such that

$$a^2 \prod_{m+1}^{n_0} (1 - 4/k) < \epsilon^2/4.$$

Then by (12) and (40)

$$P\{\sup_{n \geq n_0} z_n \geq \epsilon\} \leq \epsilon^{-1}[Ez_{n_0} + (2ac + b^2) \sum_{k=n_0+1}^{\infty} k^{-2}]$$

$$\leq \epsilon^{-1}[\epsilon^2/2 + \epsilon^2/4] < \epsilon.$$

## 6. Cluster Analysis

In [8] MacQueen describes a method for partitioning an N-dimensional population into k sets on the basis of a sample. Let $y_1, y_2, \ldots$ be independent random vectors in N-dimensional Euclidean space each having a continuous distribution p. For a given k-tuple $(x_1, \ldots, x_k)$, $x_i \in E_n$, $i = 1,2, \ldots, k$, we define a <u>minimum distance partition</u> $S(x) = (S_1(x), \ldots, S_k(x))$ of $E_N$ by

$$S_1(x) = T_1(x),\ S_2(x) = T_2(x) \cap S_1^c(x),\ \ldots,$$
$$S_k(x) = T_k(x) \cap S_1^c(x) \cap \ldots \cap S_{k-1}^c(x),$$

where

$$T_i(x) = \{\xi:\ \xi \in E_N,\ \|\xi - x_i\| \leq \|\xi - x_j\|,\ j = 1,2,\ \ldots,k\}.$$

( $\|\ \|$ denotes the usual Euclidean distance.) Sample <u>k-means</u>

$$x^n = (x_1^n, \ldots x_k^n)$$

with associated integer weights $(w_1^n, \ldots, w_k^n)$ are defined by

$$x_i^1 = y_i,\quad w_i^1 = 1 \quad (i = 1,2, \ldots, k)$$

and for $n = 1,2, \ldots$, if $y_{k+n} \in S_i(x^n)$

$$x_i^{n+1} = \frac{w_i^n x_i^n + y_{k+n}}{w_i^n + 1},\qquad x_j^{n+1} = x_j^n \quad \text{for } j \neq i$$
$$w_1^{n+1} = w_i^n + 1,\qquad w_j^{n+1} = w_j^n \quad \text{for } j \neq i. \tag{41}$$

For each k-tuple $x = (x_1, \ldots, x_k)$ define

$$W(x) = \sum_1^k \int_{S_i(x)} \|y - x_i\|^2 \, dp(y)$$

and

$$V(x) = \sum_1^k \int_{S_i(x)} \|y - u_i(x)\|^2 \, dp(y),$$

where

$$u_i(x) = \int_{S_i(x)} y\,dp(y)/p(S_i(x)) \quad \text{if } p(S_i(x)) > 0$$

and $= x_i$ if $p(S_i(x)) = 0$. For ease of notation let

$S_i^n = S_i(x^n)$, $p_i^n = p(S_i^n)$, and $u_i^n = u_i(x^n)$. MacQueen's main result is

Application 6. If there exists a closed, bounded convex set $R \subset E_N$ such that $p(R) = 1$, if $p(A) > 0$ for every open subset $A \subset R$, and if $p$ is absolutely continuous, then

$$\lim_{n \to \infty} W(x^n) \text{ exists with probability one} \tag{42}$$

and equals $V(x)$ for some $x$ in the class of points $x^*$ having the property that $x_i^* \neq x_j^*$ $(i \neq j)$ and $x_i^* = u_i(x^*)$ $(i = 1,2, \ldots ,k)$. Moreover with probability one

$$\lim_{m \to \infty} \sum_{n=1}^{m} \left( \sum_{i=1}^{k} p_i^n \|x_i^n - u_i^n\| \right) /m = 0. \tag{43}$$

Using only the hypothesis that $R$ is bounded, we shall prove (42) and (43). For the remainder of the proof of Application 6, which is straightforward, see [8].

Let $F_n = B(z_1, \ldots ,z_{k+n-1})$. From

$$W(x^{n+1}) = \sum_{i=1}^{k} \int_{S_i^{n+1}} \|y - x_i^{n+1}\|^2 \, dp(y) \leq$$

$$\sum_{i=1}^{k} \int_{S_i^n} \|y - x_i^{n+1}\|^2 \, dp(y), \ (41),$$

and the relation

$$\int_A \|y - x\|^2 \, dp(y) = \int_A \|y - u\|^2 \, dp(y) + p(A)\|x - u\|^2,$$

if $\int_A (y - u) \, dp(y) = 0$, it follows by straightforward computation that

$$E[W(x^{n+1})|F_n] \leq W(x^n) + \sum_{i=1}^{k} \sigma_{ni}^2 \, (p_i^n)^2/(w_i^n + 1)^2 \tag{44}$$

$$-\sum_{i=1}^{k} \|x_i^n - u_i^n\|^2 (p_i^n)^2 (2w_i^n + 1)/(w_i^n + 1)^2,$$

where $\sigma_{ni}^2 = \int_{S_i^n} \|y - u_i^n\|^2 \, dp(y)/p_i^n$ is uniformly bounded.

Let $z_n = W(x^n)$,

$$\xi_n = \sum_{i=1}^{k} \sigma_{ni}^2 (p_i^n)^2/(w_i^n + 1)^2,$$

and

$$\zeta_n = \sum_{i=1}^{k} \|x_i^n - u_i^n\|^2 (p_i^n)^2 (2w_i^n + 1)/(w_i^n + 1)^2.$$

Now

$$w_i^{n+1} = 1 + \sum_{j=1}^{n} I_i^j \quad , \text{ where } I_i^j = I_{\{y_{k+j} \in S_i^j\}} \text{ and hence}$$

$E(I_i^j | F_j) = p_i^j$. Hence by Corollary 2 and Lemma 1 of Section 3

$$\sum_{n=1}^{\infty} \frac{(p_i^n)^2}{(w_i^n + 1)^2} \leq \sum_{n=1}^{\infty} \frac{p_i^n}{(p_i^1 + \cdots + p_i^n)^2} \quad \cdot$$

$$\left( \frac{p_i^1 + \cdots + p_i^n}{w_i^{n+1}} \right)^2 < \infty \text{ a.s.}$$

$(i = 1,2, \ldots,k)$. It follows that $\sum \xi_n < \infty$ a.s. and hence by Theorem 1 (42) holds and $\sum \zeta_n < \infty$ a.s. Since $w_i^n \leq n$, we have

$$\sum_{n=1}^{m} (p_i^n)^2 \|x_i^n - u_i^n\|^2/m \to 0 \quad \text{a.s.} \quad (i = 1,2, \ldots ,k).$$

Hence by the Schwarz inequality

$$\frac{1}{m}\sum_{n=1}^{m} p_i^n \|x_i^n - u_i^n\| \leq \left(\frac{1}{m}\sum_{n=1}^{m} (p_i^n)^2 \|x_i^n - u_i^n\|^2\right)^{\frac{1}{2}} \to 0$$

a.s. $(i = 1,2, \ldots ,k)$ which proves (43).

## 7. Hájek-Rényi-Chow Inequality

In the preceding sections we have emphasized the use of Theorems 1 and 2 in proving the almost sure convergence of a number of different processes. In this section we give two simple illustrative applications of the fundamental inequalities (9) and (12). (See also Section 5.)

Let $\{y_n, F_n, 1 \leq n \leq n_0\}$ be a non-negative submartingale and $c_n$ $(1 \leq n \leq n_0)$ positive $F_{n-1}$-measurable random variables $(F_0 = \{\emptyset, \Omega\})$ such that $c_1 \geq c_2 \geq \ldots$ . Chow's generalization [3] of the Hájek-Rényi inequality states that

$$P\{\max_{1\leq n\leq n_0} c_n y_n \geq 1\} \leq E(c_1 y_1) + \sum_{k=2}^{n_0} E[c_k(y_k|F_{k-1}) - y_{k-1})]. \tag{45}$$

To prove (45), let $z_n = c_n y_n$. Then

$$E(z_{n+1}|F_n) = c_{n+1}(y_n + E(y_{n+1}|F_n) - y_n)$$

$$\leq z_n + c_{n+1}(E(y_{n+1}|F_n) - y_n)$$

so that (1) is satisfied with $\beta_n = 0$, $\zeta_n = 0$, and

$$\xi_n = c_{n+1}(E(y_{n+1}|F_n) - y_n) \geq 0.$$

Hence by (12) with $m = 1$

$$P\{\max_{1\leq k\leq n_0} c_k y_k \geq 1\} \leq E(z_1) + E\left(\sum_{k=1}^{n_0-1} \xi_k\right)$$

$$= E(c_1 y_1) + \sum_{k=2}^{n_0} E[c_k(E(y_k|F_{k-1}) - y_{k-1})].$$

In the special case that

$$y_n = (y + x_1 + \dots + x_n)^2, \quad c_n = (v+b_n)^{-2},$$

where $E(x_n|F_{n-1}) = 0$,

$$E(x_n^2|F_{n-1}) = \sigma_n^2, \quad b_n = \sigma_1^2 + \dots + \sigma_n^2, \text{ and}$$

$v > 0$, the inequality (9) becomes (for $m = 0$)

$$E\left(\frac{y + x_1 + \dots + x_t}{v + b_t}\right)^2 \leq (y/v)^2 + E\left(\sum_{n=0}^{t-1} \frac{\sigma_{n+1}^2}{(v + b_{n+1})^2}\right)$$

$$\leq (y/v)^2 + E\left(\sum_0^\infty \int_{b_n}^{b_{n+1}} \frac{du}{(v+u)^2}\right) \leq (y/v)^2 + 1/v, \tag{46}$$

an inequality of Dubins and Freedman [5]. The special case $y = 0$ has proved useful in problems of optimal stopping (cf. [4], p. 88).

## References

[1] Blackwell, D. , An analogue of the minimax theorem for vector payoffs, Pac. Jour. Math., 6, (1956) 1-8.

[2] Blum, J., Approximation methods which converge with probability one, Ann. Math. Statist., 25, (1954) 382-386.

[3] Chow, Y. S., A martingale inequality and the law of large numbers, Proc. Amer. Math. Soc. 11 (1960) 107-111.

[4] Chow, Y. S., Robbins, H., and Siegmund, D. Great Expectations: The Theory of Optimal Stopping, Houghton Mifflin, Boston, 1971.

[5] Dubins, L. and Freedman, D., A sharper form of the Borel-Cantelli lemma and the strong law, Ann. Math. Statist. 36, (1965) 800-807.

[6] Dvoretzky, A., On stochastic approximation, Proc. Third Berk. Symp. Math. Statist. and Probab. Vol. I, University of California Press, Berkeley, 1956, pp. 39-55.

[7] Gladyshev, E. G., On stochastic approximation, Theory

of Prob. and Its Applications, 10 (1965) 275-278.

[8] MacQueen, J., Some methods for classification and analysis of multivariate observations, Proc. Fifth Berk. Symp. Math. Statist. and Probab., Vol. I, Univ. Calif. Press, Berkeley, 1967 , pp. 281-297.

[9] Neveu, J., Mathematical Foundations of the Calculus of Probabilities, Holden-Day, San Francisco, 1965, p. 148.

[10] Robbins, H. and Munro, S., A stochastic approximation method, Ann. Math. Statist., 22, (1951), 400-407.

[11] Schmetterer, L., Stochastic approximation, Proc. Fourth Berk. Symp. Math. Statist. and Probab., Vol. I, Univ. of Calif., Berkeley, 1961, pp. 587-609.

# ON RELATIONSHIPS BETWEEN THE NEYMAN-PEARSON PROBLEM AND LINEAR PROGRAMMING

Richard L. Francis

The Ohio State University

## Introduction

The purpose of this paper is to develop and explore the parallel structure between duality relationships for the linear functional version of the generalized Neyman-Pearson problem [19], and the duality relationships of linear programming as they apply to the bounded-variable linear programming problem. The duality relationships are developed by establishing the equivalence of the dual problem considered in this paper to a "Lagrangian" dual problem which is related to the generalized Neyman-Pearson problem in a known manner. In the discussion to follow it will be assumed that the reader may not be acquainted with linear programming.

It is first convenient to state the problems of interest. The version of the generalized Neyman-Pearson problem to be considered, which for convenience will be referred to as P-1, will be stated as follows:

$$\text{maximize } g(y) = \int_{E_n} p(z)\, y(z)\, dz$$

subject to

$$\int_{E_n} q_i(z)\, y(z)\, dz \leq a_i, \quad i = 1, \ldots, m \tag{1}$$

$$b(z) \leq y(z) \leq c(z) \quad \text{for all } z \in E_n. \tag{2}$$

In the above statement of P-1, all integrals are Lebesgue

integrals, as are all integrals to follow, $E_n$ denotes Euclidean n-space, and the functions p, $q_i$, ... ,$q_m$, b, and c are given, bounded, Lebesgue-measurable functions with domain $E_n$, while $a_1$, ... ,$a_m$ are given real numbers. It is assumed, of course, that $b(z) < c(z)$ for all $z \in E_n$. Any measurable function y satisfying (1) and (2) will be called a <u>feasible solution to P-1</u>, and any feasible solution which maximizes g will be called a <u>solution to P-1</u>. The problem P-1 may be considered to be the primal problem.

Corresponding to P-1 is a second problem, which may be considered to be the dual problem. and which will be referred to as D-1. Several notational conventions will be useful in stating D-1. Denote the nonnegative orthant of $E_m$ by $E_m^+$; for every vector $u = (u_1, \ldots, u_m) \in E_m^+$, the sum

$$\sum_{i=1}^{m} a_i u_i$$

will be represented by a u, while for every $z \in E_n$, the sum

$$\sum_{i=1}^{m} u_i q_i(z)$$

will be represented by uQ(z). In the statement of the problem D-1 below, the functions v and w are assumed to be bounded and measurable, with domain $E_n$. The dual problem may now be stated as follows:

$$\text{minimize } h(u,v,w) = a\,u - \int_{E_n} b(z)v(z)dz + \int_{E_n} c(z)w(z)dz$$

subject to

$$uQ(z) - v(z) + w(z) = p(z) \quad \text{for all } z \in E_n, \tag{3}$$

$$u \in E_m^+, \tag{4}$$

$$v(z) \geq 0 \quad \text{for all } z \in E_n, \tag{5}$$

$$w(z) \geq 0 \quad \text{for all } z \in E_n. \tag{6}$$

A <u>feasible solution to D-1</u> will be denoted by (u,v,w), and will be defined to be a vector u in $E_m$ satisfying (4), together with bounded, measurable functions v and w satisfying (3), (5), and (6). A solution to D-1 will be defined to be a feasible solution to D-1 which minimizes h.

In order to establish relationships between the primal and dual problems, use will be made of a third problem. In order to state the third problem, it is convenient to define the following sets for every vector $u \in E_m^+$:

$$S_1(u) = \{z \in E_n: p(z) - uQ(z) \leq 0\},$$

$$S_2(u) = \{z \in E_n: p(z) - uQ(z) > 0\} = E_n - S_1(u).$$

The third problem, which will be referred to as D-0, may be now stated as follows:

$$\text{minimize} \quad f(u) = a\,u + \int_{S_1(u)} b(z)\,[p(z) - uQ(z)]\,dz$$

$$+ \int_{S_2(u)} c(z)\,[p(z) - uQ(z)]\,dz\ , \quad \text{subject to } u \in E_m^+.$$

The problem D-0 may be thought of as a Langrangian problem, for if the Lagrangian L(y,u) is defined by

$$L(y,u) = \int_{E_n} p(z)\,y(z)\,dz - \sum_{i=1}^{m} u_i \left[\int_{E_n} q_i(z)y(z)dz - a_i\right]$$

$$= a\,u + \int_{E_n} [p(z) - uQ(z)]\,y(z)\,dz,$$

then it can be shown, for any $u \in E_m^+$, that

$$f(u) = \sup_{y \in C} L(y,u)\ ,$$

where C is the class of all measurable functions y satisfying (2).

The analogy between P-1 and D-1, and the bounded

variable linear programming problem and its dual is readily seen. If the integrals in P-1 are replaced by sums over a set, say $J = \{1, \ldots, t\}$ of t positive integers, $E_n$ is replaced everywhere by J, and the functions in the statement of P-1 are considered to have a domain of J, then the bounded variable linear programming problem is obtained. Likewise, the dual of the bounded variable programming problem is obtained by replacing all integrals in D-1 by sums over J, replacing $E_n$ everywhere by J, and considering the functions in the statement of D-1 to have a domain of J.

The original development of duality in linear programming is due to Von Neumann, and his work was subsequently extended by Gale, Kuhn, and Tucker [13]. An excellent discussion of the development of duality theory is given by Dantzig [7]; in addition to Dantzig's book, the ones by Charnes and Cooper [4], and by Gale [14], are recommended. The literature on the subject of linear programming duality theory is extensive, and a discussion of it is beyond the scope of this paper. For a recent list of references, as well as for new duality results for linear programming, see the paper by Ben-Israel [1].

The bounded variable linear programming problem has received considerable attention, most of which has involved capitalizing upon the special structure of the problem to develop efficient algorithms for solving the problem. After the initial study of the problem by Charnes and Lemke [2], and by Dantzig [6], alternative approaches to solving the problem were developed by Eisemann [11], and H. Wagner [29]. Subsequently, Dantzig and Van Slyke developed solution algorithms for a generalized version of the problem [8], [9].

For references and a discussion of various versions of the generalized Neyman-Pearson problem as well as of applications of the problem, see the papers by Francis and Wright [12], and by D. Wagner [28]. Wagner shows that the basic ideas embodied in the Neyman-Pearson Lemma may be extended to give sufficient conditions for solutions to certain nonlinear problems. Recently some duality results for nonlinear versions of the generalized Neyman-Pearson problem, which extend those of Francis and Wright, have been obtained by Meeks and Francis [21]. For results on abstract duality, as well as an extensive reference list and discussion of references, the paper by Van Slyke and Wets is recommended [27]. Luenberger [20] also provides a treatment of Lagrangian duality; his treatment permits the development of duality for many nonlinear problems.

It is interesting to note, although the original interest of Dantzig in the generalized Neyman-Pearson problem provided much of the impetus for the subsequent development of linear programming [7], and although, as Dantzig points out, his work with Wald[10] provides an early proof of one form of the important duality theorem of linear programming, that the notion of duality appears to be a relatively unexplored one in Statistics, at least in the English literature. Particularly in Economics, duality theory has been found to be of substantial direct interest [14]; whether or not this will be the case in Statistics seems still, to some extent, to be in question. In this paper the problems D-0, D-1, and P-1 will be treated only as problems in optimization, and no attempt will be made to obtain any direct statistical interpretations. However, there are at least some indications that the study of duality in Statistics may be of direct interest, as can be seen by noting the resemblance of the sets $S_2(u)$ and $S_1(u)$ to those found in establishing critical regions and their complements when using the likelihood ratio approach to hypothesis testing. Also the closed form expressions for the partial derivatives of the function f given in [12] suggest relationships between likelihood ratio tests and duality.

Since the dual of the bounded variable linear programming problem is also a linear programming problem, it can be studied in a relatively direct manner. However, the problem D-1 appears rather awkward to study directly, since a feasible solution to D-1 consists of a vector as well as two functions, and the establishment of duality relationships between D-1 and P-1 by establishing the equivalence of D-0 and D-1 appears to be a simpler approach than that of working with D-1 and P-1 directly; the results establishing the equivalence of D-0 and D-1 are believed to be substantially new. Previously established duality relationships between the problems D-0 and P-1 due to Francis and Wright [12] provide the basis for the establishment of duality relationships between the problems P-1 and D-1. It may be worth noting that the work in [12] is primarily motivated by the study by Kuhn of a location problem which he terms the General Fermat Problem [18]. A version of the General Fermat Problem was first formulated by Fermat early in the seventeenth century, and there exists a duality result for a version of the problem, due to Fasbender, published in 1846.

This paper is intended to be complementary to the one by Krafft [17]; Krafft discusses duality results in Statistics in the German literature for problems quite closely related to the generalized Neyman-Pearson problem, including an interpretation of the dual problem. The results discussed by Krafft are developed in a quite general, measure theoretic context. The approach to follow, on the other hand, while less general, capitalizes upon the equivalence of the problems D-0 and D-1 to develop duality results in a manner requiring substantially less machinery. In order to avoid unnecessary duplication, the reader is referred to Krafft's paper for his discussion, as well as relevant references in the German literature not included in the list of references for this paper.

At this point it is useful to state several relationships between the problems P-1 and D-0 which will be used subsequently. In order to state the first relationship, the notion of a constraint qualification is required. The problem P-1 will be said to satisfy Slater's constraint qualification [26] if there exists at least one feasible solution to the problem P-1 for which none of the inequalities (1) hold as equalities. (We note that an additional constraint qualification required in the version of Lemma 1 given in [12] has since been found to be unnecessary). The relationships of interest may now be stated.

Lemma 1: Suppose P-1 satisfies the constraint qualification. If y is a solution to P-1, then there exists a solution u to D-0, and f(u) = g(y).

Lemma 2: Suppose f is differentiable at the point $u \in E_m^+$. If u is a solution to D-0, then there exists a solution y to P-1, and f(u) = g(y).

Karlin [15] points out that if there exists at least one feasible solution to P-1 then there exists a solution to P-1. Thus when P-1 satisfies the constraint qualification, the existence of solutions to both the problems P-1 and D-0 is assured by Lemma 1.

## Relationships Between the Problems D-0 and D-1

In this section it is established that the problems D-0 and D-1 are equivalent in the sense that a solution to one problem provides a solution to the other, and the values of the solutions are equal. The equivalence of the two problems is established in Properties 1 and 2. In order to establish Property 1 a number of facts must be established first.

Proposition 1: For any $u \in E_m^+$, there exist functions v and w such that (u,v,w) is a feasible solution to D-1, and

$$f(u) = h(u,v,w) .$$

Proof: Let $u \in E_m^+$ and define the functions v and w as follows:

$$v(z) = uQ(z) - p(z) \quad \text{for all } z \in S_1(u),$$

$$v(z) = 0 \quad \text{for all } z \in S_2(u),$$

$$w(z) = 0 \quad \text{for all } z \in S_1(u),$$

$$w(z) = -uQ(z) + p(z) \quad \text{for all } z \in S_2(u).$$

Then clearly the functions v and w are always nonnegative, and v and w satisfy (3). The definition of v and w implies that v and w are bounded and measurable, and so it follows that (u,v,w) is a feasible solution to D-1. Further, from the definition of v and w

$$f(u) = a\,u - \int_{S_1(u)} b(z)\, v(z)\, dz + \int_{S_2(u)} c(z)\, w(z)\, dz$$

$$- \int_{S_2(u)} b(z)\, v(z)\, dz + \int_{S_1(u)} c(z)\, w(z)\, dz$$

$$= a\,u - \int_{E_n} b(z)\, v(z)\, dz + \int_{E_n} c(z)\, w(z)\, dz$$

$$= h(u,v,w).$$

Note that the approach used in constructing a feasible solution to D-1 in the proof of Proposition 1 implies there always exist feasible solutions to D-1.

A result similar to the following Proposition 2, for the dual of the bounded variable linear programming problem, has been established by Charnes, Cooper, and Mellon [3], [4]. Their result, in a linear programming context, is also true for solutions; however, a different approach, used in Proposition 4, is needed in the present case for solutions, as Charnes, Cooper, and Mellon make use of the notions of basic feasible solutions and linear independence of vectors, and there appears to be no simple way to use these notions in the case in question.

Proposition 2: If $(u,v,w)$ is a feasible solution to D-1 such that

$$v(z)\, w(z) = 0 \text{ a.e. on } E_n,$$

then

$$f(u) = h(u,v,w).$$

Proof: Given any feasible solution $(u,v,w)$ satisfying the hypotheses, let

$$S_1'(u) = \{z \in E_n : uQ(z) - p(z) > 0\},$$

$$S_1''(u) = \{z \in E_n : uQ(z) - p(z) = 0\} = S_1(u) - S_1'(u).$$

Then it follows readily from the hypotheses that

$$v(z) = uQ(z) - p(z) \quad \text{a.e. on } S_1'(u),$$

$$w(z) = 0 \quad \text{a.e. on } S_1'(u),$$

$$v(z) = 0 \quad \text{a.e. on } S_2(u),$$

$$w(z) = -uQ(z) + p(z) \quad \text{a.e. on } S_2(u).$$

Thus

$$h(u,v,w) = au - \int_{S_1'(u)} b(z)\,[uQ(z) - p(z)]\,dz \; -$$

$$\int_{S_1''(u)} b(z)\,v(z)\,dz + \int_{S_2(u)} c(z)\,[p(z) - uQ(z)]\,dz$$

$$+ \int_{S_1''(u)} c(z)\,w(z)\,dz.$$

Now it follows again from the hypotheses that

$$v(z) = 0 \qquad \text{a.e. on } S_1''(u),$$

$$w(z) = 0 \qquad \text{a.e. on } S_1''(u)$$

so that

$$\int_{S_1''(u)} c(z)\,w(z)\,dz = 0$$

and

$$- \int_{S_1''(u)} b(z)v(z)dz = 0 = - \int_{S_1''(u)} b(z)[uQ(z) - p(z)]\,dz.$$

It now readily follows that

$$h(u,v,w) = a\,u + \int_{S_1(u)} b(z)\,[p(z) - uQ(z)]\,dz$$

$$+ \int_{S_2(u)} c(z)\,[p(z) - uQ(z)]\,dz = f(u).$$

The following proposition is needed in order to establish Proposition 4, which is the key proposition needed to establish Property 1.

Proposition 3: Let $(u,v,w)$ be any feasible solution to D-1 and define the function $m(z)$ on $E_n$ by

$$m(z) = \min\,[v(z),w(z)].$$

If T is any measurable subset of $E_n$ and functions $\hat{v}$ and $\hat{w}$ are defined on $E_n$ by

$$\begin{aligned}
\hat{v}(z) &= v(z) && \text{for all } z \in E_n - T,\\
\hat{v}(z) &= v(z) - m(z) && \text{for all } z \in T,\\
\hat{w}(z) &= w(z) && \text{for all } z \in E_n - T,\\
\hat{w}(z) &= w(z) - m(z) && \text{for all } z \in T,
\end{aligned}$$

then $(u,\hat{v},\hat{w})$ is a feasible solution to D-1 and

$$h(u,v,w) - h(u,\hat{v},\hat{w}) = \int_T [c(z) - b(z)]m(z)\, dz.$$

Proof: It follows immediately from the definitions of $\hat{v}$ and $\hat{w}$ that $(u,\hat{v},\hat{w})$ is a feasible solution to D-1. Further,

$$h(u,v,w) - h(u,\hat{v},\hat{w}) = -\int_T b(z)\, v(z)\, dz + \int_T c(z)w(z)\, dz$$
$$+ \int_T b(z)\, \hat{v}(z)\, dz - \int_T c(z)\, \hat{w}(z)\, dz = -\int_T b(z)\, m(z)\, dz$$
$$+ \int_T c(z)\, m(z)\, dz = \int_T [c(z) - b(z)]\, m(z)\, dz.$$

Proposition 4: If there exists a solution to D-1, then there exists a solution $(u,v,w)$ to D-1 such that

$$v(z)\, w(z) = 0 \quad \text{a.e. on } E_n. \tag{7}$$

Proof: Let $(\hat{u},\hat{v},\hat{w})$ be a solution to D-1 and define the following sets:

$$\begin{aligned}
T_1 &= \{z \in E_n : \hat{v}(z)\, \hat{w}(z) > 0,\ c(z) - b(z) > 0\},\\
T_2 &= \{z \in E_n : \hat{v}(z)\, \hat{w}(z) > 0,\ c(z) - b(z) = 0\},\\
T_3 &= \{z \in E_n : \hat{v}(z)\, \hat{w}(z) = 0,\ c(z) - b(z) \geq 0\}.
\end{aligned}$$

Note that the sets $T_1, T_2, T_3$ are disjoint, measurable, and that their union is $E_n$. It will be first shown that the measure of $T_1$ is zero. Suppose that the measure of $T_1$ is positive, and define functions v* and w* as follows:

$$v^*(z) = \hat{v}(z) \qquad \text{for all } z \in E_n - T_1,$$

$$v^*(z) = \hat{v}(z) - m(z) \qquad \text{for all } z \in T_1,$$

$$w^*(z) = \hat{w}(z) \qquad \text{for all } z \in E_n - T_1,$$

$$w^*(z) = \hat{w}(z) - m(z) \qquad \text{for all } z \in T_1,$$

where

$$m(z) = \min\,[\hat{v}(z), \hat{w}(z)] \text{ for all } z \in E_n.$$

Then, by Proposition 3, $(\hat{u}, v^*, w^*)$ is a feasible solution to D-1 and

$$h(\hat{u},\hat{v},\hat{w}) - h(\hat{u},v^*,w^*) = \int_{T_1} [c(z) - b(z)]\, m(z)\, dz > 0$$

so that $h(\hat{u},\hat{v},\hat{w}) > h(\hat{u},v^*,w^*)$, which contradicts the fact that $(\hat{u},\hat{v},\hat{w})$ is a solution to D-1. Thus the measure of $T_1$ is zero. Next, define functions v and w by

$$v(z) = \hat{v}(z) \qquad \text{for all } z \in E_n - T_2,$$

$$v(z) = \hat{v}(z) - m(z) \qquad \text{for all } z \in T_2,$$

$$w(z) = \hat{w}(z) \qquad \text{for all } z \in E_n - T_2,$$

$$w(z) = \hat{w}(z) - m(z) \qquad \text{for all } z \in T_2.$$

Again by Proposition 3, $(\hat{u}, v, w)$ is a feasible solution to D-1, and, in this case,

$$h(\hat{u},\hat{v},\hat{w}) - h(\hat{u},v,w) = \int_{T_2} [c(z) - b(z)]\, m(z)\, dz = 0$$

by the definition of $T_2$, so that $(\hat{u}, v, w)$ is also a solution to D-1. By definition of v and w,

$$v(z)\, w(z) = 0 \quad \text{for all } z \in T_2.$$

Also, the defintion of v, w, and $T_3$ implies

$$v(z)\, w(z) = 0 \quad \text{for all } z \in T_3.$$

Since the measure of $T_1$ is zero, it follows that v and w as defined satisfy (7). The conclusion now follows on taking u to be $\hat{u}$.

It is now easy to establish the first property relating the problems D-0 and D-1.

Property 1: If (u,v,w) is a solution to D-1, then u is a solution to D-0, and

$$f(u) = h(u,v,w).$$

Proof: Let (u,v,w) be a solution to D-1. Then it follows from Propositions 2 and 4 that

$$f(u) = h(u,v,w).$$

Let $\hat{u}$ be any feasible solution to D-0. From Proposition 1, there exist functions $\hat{v}$ and $\hat{w}$ such that $(\hat{u},\hat{v},\hat{w})$ is a feasible solution to D-1 and such that

$$f(\hat{u}) = h(\hat{u},\hat{v},\hat{w}).$$

Since $$h(u,v,w) \leq h(\hat{u},\hat{v},\hat{w}),$$

it follows that $$f(u) \leq f(\hat{u}),$$

so that u is a solution to D-0, and the conclusion now follows.

Given Property 1, one might suspect that there should be a converse of Property 1, if properly worded; such a converse is given in Property 2. The proposition below is needed in order to establish Property 2.

Proposition 5: If (u,v,w) is any feasible solution to D-1, then

$$f(u) \leq h(u,v,w).$$

Proof: For any feasible solution (u,v,w) to D-1, the following identity is readily established:

$$h(u,v,w) - f(u) = \int_{S_1(u)} b(z)[uQ(z) - v(z) + w(z) - p(z)]dz$$

$$+ \int_{S_2(u)} c(z)\,[uQ(z) - v(z) + w(z) - p(z)]\,dz$$

$$+ \int_{S_1(u)} [c(z) - b(z)]\,w(z)\,dz + \int_{S_2(u)} [c(z)-b(z)]v(z)\,dz.$$

In the above identity the first two integrals on the right side are clearly zero and the last two integrals are clearly nonnegative, so that the conclusion now follows.

<u>Property 2:</u> If u is a solution to D-0, then there exist functions v and w such that (u,v,w) is a solution to D-1 and

$$f(u) = h(u,v,w).$$

<u>Proof:</u> Let $(\hat{u},\hat{v},\hat{w})$ be any feasible solution to D-1. Then, by Proposition 5,

$$f(\hat{u}) \leq h(\hat{u},\hat{v},\hat{w}),$$

and so certainly $\qquad f(u) \leq h(\hat{u},\hat{v},\hat{w}).$

But by Proposition 1 there exist functions v and w such that (u,v,w) is a feasible solution to D-1 and

$$f(u) = h(u,v,w). \tag{8}$$

Thus

$$h(u,v,w) \leq h(\hat{u},\hat{v},\hat{w}), \tag{9}$$

and the conclusion now follows from (8) and (9).

## Relationships Between the Problems D-1 and P-1

Properties 1 and 2, together with Lemmas 1 and 2, can now be used to establish readily a number of relationships between the problems D-1 and P-1. The properties of this section are new in the sense that they rely upon Lemmas 1 and 2 and Properites 1 and 2, but are primarily expository in nature, as they all have their counterparts in the duality theory of linear programming, and are what might be expected, given a linear programming point of view.

The use of Lemma 1 and Property 2 gives the following property immediately.

Property 3: Suppose P-1 satisfies the constraint qualification. If y is a solution to P-1, then there exists a solution (u,v,w) to D-1, and g(y) = h(u,v,w).

An immediate consequence of Lemma 2 and Property 1 is the following property.

Property 4: Suppose f is differentiable at the point $u \in E_m^+$. If (u,v,w) is a solution to D-1, then there exists a solution y to P-1, and g(y) = h(u,v,w).

At this point it is useful to introduce the notion of complementary slackness, which is of fundamental importance in linear programming duality theory. Given any feasible solution y to P-1 and any feasible solution (u,v,w) to D-1, the conditions (10), (11), and (12) below will be called the complementary slackness conditions:

$$u_i[a_i - \int_{E_n} q_i(z)\, y(z)\, dz] = 0, \quad i = 1, \ldots, m, \tag{10}$$

$$v(z)[y(z) - b(z)] = 0 \quad \text{a.e. on } E_n, \tag{11}$$

$$w(z)[c(z) - y(z)] = 0 \quad \text{a.e. on } E_n. \tag{12}$$

One use of the complementary slackness conditions is given by the following property, Property 5. We note that parts a) and b) of the proof of Property 5 exactly parallel similar proofs in the duality theory of linear programming; it is part c) of the proof, which relies on Property 3 or Property 4, which is novel.

Property 5: Suppose y is a feasible solution to P-1 and (u,v,w) is a feasible solution to D-1. If P-1 satisfies the constraint qualification or if f is differentiable at u, then y is a solution to P-1 and (u,v,w) is a solution to D-1 if and only if y and (u,v,w) satisfy the complementary slackness conditions.

Proof: a) If y is any feasible solution to P-1 and (u,v,w) is any feasible solution to D-1, it will first be established directly that $g(y) \leq h(u,v,w)$. Given the hypotheses,

$$0 \leq u_i \, [a_i - \int_{E_n} q_i(z) \, y(z) \, dz], \quad i = 1, \ldots, m \tag{13}$$

so summing the inequalities (13) gives

$$0 \leq a\,u - \int_{E_n} uQ(z) \, y(z) \, dz.$$

Thus

$$g(y) = \int_{E_n} p(z)y(z) \, dz \leq \int_{E_n} [p(z) - uQ(z)] \, y(z) \, dz + a\,u.$$

Since

$$\int_{E_n} [p(z) - uQ(z)]y(z) \, dz = - \int_{E_n} v(z)y(z) \, dz + \int_{E_n} w(z)y(z)dz,$$

it follows that

$$g(y) \leq a\,u - \int_{E_n} v(z)y(z) \, dz + \int_{E_n} w(z) \, y(z) \, dz. \tag{14}$$

But

$$-v(z) \, y(z) \leq -v(z)b(z) \quad \text{for almost all } z \in E_n \tag{15}$$

and

$$w(z) \, y(z) \leq w(z) \, c(z) \quad \text{for almost all } z \in E_n, \tag{16}$$

so, on integrating the inequalities (15) and (16) over $E_n$ and using the inequality (14) it follows that $g(y) \leq h(u,v,w)$.

b) Given feasible solutions y and (u,v,w) to P-1 and D-1 respectively, it is an immediate consequence of part a) of the proof that y and (u,v,w) will be solutions to P-1 and D-1 respectively if $g(y) = h(u,v,w)$. When the complementary slackness conditions hold, all inequalities of part a) of the proof hold as equalities, so it follows that $g(y) = h(u,v,w)$.

c) If y is a solution to P-1 and (u,v,w) is a solution to D-1, then Property 3 or Property 4 implies $g(y) = h(u,v,w)$.

If the complementary slackness conditions do not hold, then at least one of the inequalities in (13) does not hold as an equality, or the inequalities (15) or (16) do not hold as equalities on a set of positive measure, so that essentially the same approach as in part a) of the proof implies that $g(y) < h(u,v,w)$, which gives a contradiction. Thus the complementary slackness conditions hold.

An approach quite similar to that used to establish Property 5 can be used to establish the following two properties.

Property 6: Suppose P-1 satisfies the constraint qualification. Then any feasible solution y to P-1 is a solution to P-1 if and only if there exists a feasible solution (u,v,w) to D-1 such that y and (u,v,w) satisfy the complementary slackness conditions.

Property 7: If (u,v,w) is a feasible solution to D-1 and f is differentiable at u, then (u,v,w) is a solution to D-1 if and only if there exists a feasible solution y to P-1 such that y and (u,v,w) satisfy the complementary slackness conditions.

Notice that Property 6 gives necessary and sufficient conditions for a solution to P-1 which may not at first appear to be equivalent to the customary necessary and sufficient conditions for a solution to P-1. However, the two sets of conditions are, of course, equivalent, and the equivalence may be established in a self-contained fashion. Prior to establishing the equivalence, it is useful to recall the following definitions, which hold for every $u \in E_m^+$:

$$S_1'(u) = \{z \in E_n: p(z) - uQ(z) < 0\},$$

$$S_1''(u) = \{z \in E_n: p(z) - uQ(z) = 0\} = S_1(u) - S_1'(u).$$

Property 8: Let y be a feasible solution to P-1. Then the following two statements are equivalent:

a) There exists a feasible solution (u,v,w) to D-1 such that y and (u,v,w) satisfy the complementary slackness conditions.

b) There exists a $u \in \mathbf{E}_m^+$ such that

$$y(z) = b(z) \quad \text{a.e. on } S_1'(u),$$

$$y(z) = c(z) \quad \text{a.e. on } S_2(u),$$

$$u_i \, [a_i - \int_{E_n} q_i(z) \, y(z) \, dz] = 0, \; i = 1, \dots, m.$$

Proof: Suppose the first statement is true. Then $g(y) = h(u,v,w)$ which implies $(u,v,w)$ is a solution to D-1, so, by Proposition 4, we may assume that $v(z) \, w(z) = 0$ a.e. on $E_n$. It then follows from the constraint (3) of D-1 that $v(z) \geqq uQ(z) - p(z) > 0$ a.e. on $S_1'(u)$, and that $w(z) = 0$ a.e. on $S_1'(u)$. The complementary slackness condition (11) then implies

$$y(z) - b(z) = 0 \quad \text{a.e. on } S_1'(u).$$

Further, it follows from the constraint (3) of D-1 that $w(z) = p(z) - uQ(z) > 0$ a.e. on $S_2(u)$ and that $v(z) = 0$ a.e. on $S_2(u)$, so that the complementary slackness condition (12) implies

$$c(z) - y(z) = 0 \quad \text{a.e. on } S_2(u).$$

Since $u \in E_m^+$, it follows that the second statement is true.

Next, suppose the second statement is true. Define $v(z) = uQ(z) - p(z) > 0$ on $S_1'(u)$ and $w(z) = 0$ on $S_1'(u)$. Clearly

$$v(z)[y(z) - b(z)] = 0 \quad \text{a.e. on } S_1'(u),$$

and

$$w(z)[c(z) - y(z)] = 0 \text{ a.e. on } S_1'(u).$$

Define $v(z) = w(z) = 0$ for all $z \in S_1''(u)$; then certainly

$$v(z)[y(z) - b(z)] = 0 \text{ a.e. on } S_1''(u),$$

and

$$w(z)[c(z) - y(z)] = 0 \text{ a.e. on } S_1''(u).$$

Finally, define $v(z) = 0$ and $w(z) = p(z) - uQ(z) > 0$ for all $z \in S_2(u)$; then certainly

$$v(z)\,[y(z) - b(z)] = 0 \quad \text{a.e. on } S_2(u),$$

and

$$w(z)\,[c(z) - y(z)] = 0 \quad \text{a.e. on } S_2(u).$$

Note that the definition of v and w on $E_n$ implies that (u,v,w) is a feasible solution to D-1. Evidently y and (u,v,w) now satisfy the complementary slackness conditions, and so the first statement is true.

## Duality When the Problem P-1 has Equality Integral Constraints

When the inequality constraints (1) in the statement of problem P-1 are replaced by equality constraints, duality results are readily obtainable by simple modifications to the foregoing results. The symbol $E_m$, Slater's constraint qualification is replaced by a relative interior point constraint qualification due to Dantzig and Wald [10], and used by Lehmann [19] in obtaining necessary conditions for a solution to the problem P-1 having equality constraints. Lehmann shows that if there exists at least one feasible solution to the problem P-1 then there exists a solution to the problem, so that when P-1 sattisfies the relative interior point constraint qualification the existence of solution to both the problems P-1 and D-0 is again assured by Lemma 1. The complementary slackness conditions may be simplified by eliminating the equations (10) which are trivially satisfied. With the above mentioned changes, it is readily established that all of the previous lemmas, propositions, and properties still hold.

## Concluding Remarks

In this paper it has been demonstrated that the duality theory of linear programming as it applies to the bounded variable linear programming problem is applicable, with appropriate modifications, to the fundamental generalized Neyman-Pearson problem of Statistics.

The adoption of a linear programming point of view permits insights which would perhaps not be obtained otherwise into the generalized Neyman-Pearson problem. In addition to any insights which may follow from the previously established results, some insight may also be gained as to why closed form solutions to versions of the problem P-1 appear to be so difficult to obtain for the case $m > 2$. While closed form solutions to variations of the problem P-1 have been obtained for the case $m = 2$, typical of which are the results of Rustagi [25] and of Karlin and Studden [16], the fact that closed form solutions are so far unavailable to the conceptually simpler bounded variable linear programming problem, and that virtually all of the interest in the bounded variable problem has been concerned with developing either theory for the problem or algorithms for finding solutions to the problem suggests that closed form solutions to P-1 for the case $m > 2$ will not be easily obtained. Other research questions which may be more tractable include relaxing the differentiability assumption for f needed for Lemma 2, and developing direct statistical interpretations of duality theory involving the problem D-0.

## Acknowledgments

The author would like to acknowledge contributions which two former students of his have made to this paper. The work of A. Yoshimura [30], which established the equivalence of a Lagrangian dual to the dual of the bounded variable linear programming problem suggested that it should be possible to establish the equivalence of the problems D-0 and D-1, although much of the analysis for the bounded variable problem does not carry over directly. Proposition 5 is due to H. Meeks.

## References

1. A. Ben-Israel, "Linear Equations and Inequalities on Finite Dimensional Real or Complex Vector Spaces: A Unified Theory," Journal of Mathematical Analysis and Applications, 27, 2, 367-389, (1969).
2. A. Charnes and C. Lemke, "Computational Theory of Linear Programming, The Bounded Variables Problem," ONR Research Memorandum No. 10, Graduate School of Industrial Administration, Carnegie Institute of Technology, Pittsburgh, Pennsylvania, (1954).
3. A. Charnes, W. Cooper, and B. Mellon, "Dyadic Models and Sub-Dual Methods," Naval Research Logistics Quarterly, 8, 1, 1-24, (1961).

4. A. Charnes and W. Cooper, Management Models and Industrial Applications of Linear Programming, John Wiley & Sons, Inc. New York, (1961).
5. H. Chernoff and H. Scheffé,"A Generalization of the Neyman-Pearson Fundamental Lemma," Annals of Mathematical Statistics, 23, 2, 213-225, (1952).
6. G. Dantzig, "Upper Bounds, Secondary Constraints, and Block Trangularity in Linear Programming," Econometrica, 23, 2, 174-183, (1955).
7. G. Dantzig, Linear Programming and Extensions, Princeton University Press, New Jersey, (1963).
8. G. Dantzig, and R. Van Slyke, "A Generalized Upper-Bounded Technique for Linear Programming," in Proceedings of the IBM Scientific Computing Symposium on Combinatorial Problems, IBM, White Planes, New York, (1966).
9. G. Dantzig and R. Van Slyke, "Generalized Upper Bounding Techniques", Journal of Computer and System Science, 1, 3, 213-226, (1967).
10. G. Dantzig, and A. Wald, "On the Fundamental Lemma of Neyman and Pearson," Annals of Mathatical Statistics, 22, 1, 87-93, (1951).
11. K. Eisemann, "The Primal-Dual Method for Bounded-Variables," Operations Research,12, 1, 110-121, (1964).
12. R. Francis and G. Wright, "Some Duality Relationships for the Generalized Neyman-Pearson Problem," Journal of Optimization Theory and Applications, 4, 6, 394-412, (1969).
13. D. Gale, H. Kuhn, and A. Tucker, "Linear Programming and the Theory of Games," in Activity Analysis of Production and Allocation, T. C. Koopmans (ed.), John Wiley & Sons, Inc., New York, (1951).
14. D. Gale, The Theory of Linear Economic Models, McGraw-Hill Book Company, New York, (1960).
15. S. Karlin, Mathematical Methods and Theory in Games, Programming, and Economics, Addison-Wesley Publishing Company, Reading, Massachusetts, (1959).
16. S. Karlin and W. Studden, Tchebycheff Systems, with Aplications in Analysis and Statistics, Interscience Publishers, New York, (1966).
17. O. Krafft,"Programming Methods in Statistics and Probability Theory," Nonlinear Programming, J. B. Rosen, O. L. Mangasarian, and K. Ritter, (eds.), Academic Press, New York, (1970).

18. H. Kuhn, "On a Pair of Dual Nonlinear Programs," Nonlinear Programming, J. Abadie (ed.), John Wiley & Sons, Inc., New York, (1967).
19. E. Lehmann, Testing Statistical Hypotheses, John Wiley & Sons, Inc., New York, (1959).
20. D. Luenberger, Optimization by Vector Space Methods, John Wiley & Sons, Inc., New York, (1969).
21. H. Meeks and R. Francis, "Duality Relationships for a Nonlinear Version of the Generalized Neyman-Pearson Problem,"a working paper, The Dept. of Industrial Engineering, Iowa State University, Ames, Iowa, (1971).
22. J. Neyman and E. Pearson, "On the Problem of the Most Efficient Tests of Statistical Hypotheses," Philosophical Transactions of the Royal Society of London, Series A, 231, 289-337, (1933).
23. J. Neyman and E. Pearson,"Contributions to the Theory of Testing Statistical Hypotheses," Statistical Research Memoirs, 1, 1-37, (1936), and 2, 25-57, (1938).
24. J. Neyman and E. Pearson, Joint Statistical Papers, University of California Press, Berkeley, California, (1967).
25. J. Rustagi, "On Minimizing and Maximizing a Certain Integral with Statistical Applications," Annals of Mathematical Statistics, 28, 2, 309-328, (1957).
26. M. Slater, "Lagrange Multipliers Revisited: A Contribu- to Nonlinear Programming," RM-676, The Rand Corporation, Santa Monica, California, (1951).
27. R. Van Slyke and R. Wets, "A Duality Theory for Abstract Programs with Applications to Optimal Control Theory," Journal of Mathematical Analysis and Applications, 22, 3, 679-706, (1968).
28. D. Wagner, "Nonlinear Functional Versions of the Neyman Pearson Lemma," SIAM Review, 73, 2, 52-65, (1969).
29. H. Wagner, "The Dual-Simplex Method for Bounded Variables," Naval Research Logistics Quarterly, 5, 3, 257-261, (1958).
30. A. Yoshimura, "An Unconstrained Dual to the Bounded Variable Linear Programming Problem," unpublished Master's Thesis, The Ohio State University, Columbus,Ohio, (1969).

# STATISTICAL CONTROL OF OPTIMIZATION

H. O. Hartley and R. Pfaffenberger*

Texas A & M University

## 1. Introduction

This paper provides a statistical solution to a very general problem in non-linear programming. Specifically, it does not make any assumptions about the convexity (or, for that matter, quasi-convexity or pseudo-convexity) of any of the functions involved in the problem. On the other hand it does not provide a guaranteed global optimum of the objective function, but is confined to:

(a) The computation of a feasible and 'near optimum' solution.

(b) The computation of confidence limits between which the global optimum is guaranteed to lie with a specified statistical confidence coefficient.

There are many problems in non-linear optimization in which an exhaustive search for the global optimum would lead to an astronomical amount of computing effort. On the other hand the insistence on finding the guaranteed global optimum is often not an absolutely essential part of the problem and a comparison of (a) with (b) will often lead to a 'good enough' solution quite acceptable for many applications of optimization. The present approach differs from the methods published in the literature and its relation to the limited number of papers on the subject is outlined in Section 8.

*This paper is a sequel to a dissertation by Roger Pfaffenberger now at Pennsylvania State University.

## 2. Mathematical Formulation of the Problem

The general non-linear programming problem here considered may be described as the numerical determination of

$$g^* = \sup_{x \in S} g(x) \tag{1}$$

where x is an n vector, g(x) is a function of x, usually assumed to be at least continuous and S is a subspace of the Euclidean n space $E_n$, usually assumed to be finite and closed. In addition to $g^*$, it is commonly required that an n vector $x^* \in S$ be determined such that

$$g^* = g(x^*) \tag{2}$$

In most cases, the subspace S of $E_n$ is defined by a finite number of inequalities. That is,

$$x \in S \text{ iff } f_i(x) \leq L_i, \quad i = 1,2, \ldots ,m \tag{3}$$

where the $f_i(x)$ are assumed to be at least continuous. Further, the subspace S of $E_n$ is assumed to have the following properties:

$$S \text{ is bounded and closed} \tag{4}$$

and

$$S \text{ is "n-dimensional"} \tag{5}$$

By (5), it is meant that for any $x \in S$ and any neighborhood of x defined by $\overline{yx} < \epsilon$, there exists an n-dimensional cube of points y entirely within S. Usually spaces defined by a set of inequalities as in (3) have the properties (4) and (5). There are trivial exceptions to (4) which can, however, be remedied by the addition of upper bounds not affecting the practical significance of the problem. A notable exception to (5) arises from a non-linear equality constraint which can be put in the form

$$f_1(x) \leq 0 \text{ and } f_2(x) = -f_1(x) \leq 0 \tag{6}$$

which implies that $f_1(x) = 0$ and usually violates (5). However, the procedures that shall be developed can be generalized to situations in which S comprises a finite number of recognizable lower dimensional subsets.

The confidence interval and point estimator for g* are designed specifically for the mathematical programming problem delineated above. For the methods developed in this research, two additional assumptions are required:

(i) Given any $x \in E_n$, it is possible to test whether or not $x \in S$ by a monitoring algorithm. (7)

(ii) The algorithm used to solve the mathematical programming problem is an ascent algorithm.. (8)

The first assumption is necessary for the purposes of implementing the methods on a computer. In the case where S is defined as in (3), the monitoring algorithm would require the evaluation of the $f_i(x)$ and comparison with the $L_i$.

An ascent algorithm is defined for the purposes of this research as a numerical procedure involving the following steps:

(i) The selection of a starting point $x_o \in S$. (9)

(ii) A recurrence algorithm or a Markov process which yields $x_{i+1} \in S$ as a unique function of $x_i$ symbolized by

$$x_{i+1} = h(x_i) \tag{10}$$

which has the property that

$$g(x_{i+1}) \geq g(x_i) \tag{11}$$

(iii) A stopping rule terminating the process if for r consecutive values of $x_i$,

$$g(x_{i+1}) \leq g(x_i) + \varepsilon \tag{12}$$

where $\varepsilon$ and r are parameters at the disposal of the analyst.

The value of $\varepsilon$ in (12) would normally be chosen sufficiently small to be of negligible practical significance for the applications of the objective function. It follows from (4) that $g(x)$ has a maximum $g^*$ and a minimum $g^-$ in S, and from (11) and (12) that the ascent algorithm terminates after a finite number of steps for which an upper bound is given by $(g^* - g^-)/\epsilon$.

Most of the well known ascent algorithms such as the method of steepest ascent, the gradient projection method (Rosen (1960) and (1961)), the methods of Fiacco and McCormick(1964) and Hartley and Hocking (1963) can be put into the form of an ascent algorithm as defined above provided that $g(x)$ and $f_i(x)$, $i = i,2, \ldots ,n$ have first derivatives.

## 3. Stochastic Process Generating the Starting Points

A stochastic process for selecting the starting points $x \in S$ is now developed. For convenience, only processes are considered which result in ordinate density functions $d(x)$ with the following property:

$$d(x) \text{ is piecewise continuous in } S. \tag{13}$$

By this it is meant that S can be subdivided into a finite number of mutually exclusive and exhaustive subsets $S_\lambda$ such that $d(x)$ is continuous in every $S_\lambda$. It is further assumed that for every n-dimensional cube $c$ in S

$$\int_c d(x)\, dx > 0. \tag{14}$$

Unless there is prior information available on the distribution $d(x)$ of starting points, a convenient and natural choice is the uniform distribution. That is,

$$d(x) = 1/\text{vol}, \text{ where vol} = \int_S dx. \tag{15}$$

A convenient algorithm for generating random samples from (15) is to first generate random coordinates $x_o$ for a vector $x_o$ within an n-dimensional cube $\alpha \leq x_o \leq \omega$ containing S (possible because of (4)) and then reject $x_o$ if it is not in S (accomplished with the help of (7)). If S is a hyperrectangle, no rejections are required, of course.

Experience with numerous examples has indicated that completely random selection of starting points as described above will be extremely wasteful of computer effort. Various devices of unequal probability selection may be employed in situations in which S can a priori be subdivided into strata such that certain strata have 'better chances' of leading to large terminal values than others. Efficient methods of constructing such strata will vary from problem to problem; in most instances, the following method of stratified sampling by rejection should prove to serviceable:

Select a set of s strata boundaries

$$-\infty = L_o < L_1 < L_2 < \dots < L_h < \dots < L_s = \infty \tag{16}$$

generating a strata $S_h$ of the form

$$x_o \in S_h \text{ if } L_{h-1} \leq g(x_o) < L_h;\ h = 1,2, \dots ,s. \tag{17}$$

Different sampling rates are now used within the s strata by the following two step acceptance procedure:

(i) Draw a random $x_o$ from S and compute $g(x_o)$.

(ii) If $g(x_o)$ falls into the $h^{th}$ stratum of (17), accept with probability $r_h$ the point $x_o$ as a starting point for the application of the ascent algorithm.

The $L_h$ and $r_h$ parameters are at the disposal of the analyst. Keeping in mind the general comments above with reference to the $r_h$, the precise values of $L_h$ and $r_h$ would normally be ascertained from 'trial ascent runs' or from previous experience with the specific problem. The theoretical properties of the procedures which give the estimates of the global optimum apply to any set of values chosen for $L_h$ and $r_h$, but the computational efficiency depends on a judicious choice of these parameters.

In the examples of Section 6, the values of $L_h$ and $r_h$ were chosen from trial runs. That is, a number of random starting points were chosen (usually 5) and each was inputed into the ascent algorithm with the terminal points and functionals being recorded. The distribution of terminal functional values was used to specify the values of $L_h$ and $r_h$. Unless prior information is available, it is suggested that the strata values and probabilities be chosen in this manner.

The ordinate distribution ($F(x_o)$,say) resulting from the stratified sampling process of selecting $x_o$ points will be piecewise continuous in the sense of (13). This distribution of $x_o$ starting values will induce an associated single variate distribution of $g_o = g(x_o)$ values which, because of (4), (5) and (11) will be a finite range distribution with upper end point $g^*$, the global optimum. It is now theoretically possible to estimate $g^*$ from a stratified sample of n starting points $x_{ot}$, $t = 1,2, \ldots ,n$ and associated functional values $g_{ot} = g(x_{ot})$. However, such an algorithm would constitute a mere random search procedure and clearly not utilize the ascent algorithm. Moreover, experience has indicated that very large samples n would normally be required to estimate a confidence interval for $g^*$ of adequate precision by this method. It is necessary, therefore, to use the information acquired in the ascent iterates in the development of a confidence interval. The next section deals with describing statistics that are functions of the iterate information and will later be used in the estimation procedure.

## 4. The Statistics Computed for the Ascent Paths

Denote by $I(x_{ot})$ the terminal iteration resulting from employing the ascent algorithm on the t-th starting values, $x_{ot}$. Further, denote by $x_{it}$ the i-th iterate of the t-th starting value $x_{ot}$ where $t = 1,2, \ldots ,n$, and $i = 1,2, \ldots ,I$, and I is a conveniently chosen index of truncation. For any starting value $x_{ot}$ for which the stopping rule terminates the sequence $x_{it}$ at $I(x_{ot}) < I$, define

$$x_{it} = x_{I(x_o),t} \quad \text{for } i = I(x_o), \ldots ,I. \tag{18}$$

Further, let

$$g_{it} = g(x_{it}) \text{ for } t = 1,2, \ldots ,n \tag{19}$$

$i = 1,2,\ldots ,I$ and define the statistic

$$g_i = g(x_i). \tag{20}$$

Finally, consider the statistic

$$g = \sum_{i=1}^{I} c_i g_i + u \tag{21}$$

where the $c_i$ are weight coefficients with the property

$$\sum_{i=1}^{I} c_i = 1 \tag{22}$$

and u is a symmetrical beta-variable drawn from

$$f(u) = \Gamma(2m+4)\,\Gamma(m+2)^{-2}\,(u/\delta)^{m+1}\,(1-u/\delta)^{m+1}\,\delta^{-1}. \tag{23}$$

The $c_i$ and the beta parameters m and $\delta$ are parameters of the algorithm. The theory developed in Section 5 applies to <u>any</u> choice of these parameters, but the $c_i$ should normally increase with the i and $\delta$ would normally be chosen sufficiently small so as to be practically negligible compared with the objective function g(x) and m is defined in (29).

Clearly, the statistic g in (21) has a continuous c.d.f. because of the addition of the u component and the boundedness of (21). Moreover, because of (4), (5), (10), (11), (22), and (23), the statistic g has a finite range distribution with upper limit $g^* + \delta$. It is the purpose of Section 5 to show how the n sample of g values

$$g_t = \sum_{i=1}^{I} c_i g_{it} + u_t, \quad t = i,2, \ldots ,n \tag{24}$$

can be used to obtain a point estimate and a confidence

interval for g*.

Before proceeding with this we define the 'approximate solution' to the non-linear optimization problem to be the terminal input vector $x^+$ which yields the largest objective function value and is symbolized by

$$x^+ = x_{I(x_{ot}')} \text{ iff } g(x_{I(x_{ot}')}) = \max_t g(x_{I(x_{ot})}). \qquad (25)$$

In the first example problem of section 6, the global maximum is attained; that is, g(x) = g* and x* is not unique.

It should be apparent that the stratified sampling procedure does not alter properties of the steepest ascent algorithms to converge to g* as $n \to \infty$ since every starting point $x_{ot} \in S$ has a non-zero probability of being incorporated into the statistic $g_t$ (every strata probability $r_h$ in the stratified sampling scheme must be greater than zero). That is, it is easy to show that for our stratified sampling procedure and with the definition (25)

$$\operatorname{pr}\lim_{n \to \infty} g(x^+) = g^* .$$

## 5. The Confidence Interval for g*

The $g_t$ computed from (4) for each path in accordance with the procedure in Section 4 represent a random n-sample drawn from a finite range distribution with upper limit g*. Our problem now is to compute from the $g_t$ an upper confidence limit for g*. A lower confidence point is not required since it is known with certainty that

$$g(x^+) \leq g^* \qquad (27)$$

Denoting the (unknown) c.d.f. of g by

$$p(G) = \Pr\{g \leq G\} \qquad (28)$$

this problem is essentially unsolvable if no assumptions are made about p(G). The analytic properties of p (G) will depend on

(i) The smoothness assumptions made on the functions

$g(x)$ and $f_i(x)$ in (1) and (3).

(ii) The stochastic process of selecting the starting points $x_{ot}$.

(iii) The functional form of the ascent algorithm (10).

If we denote by $\pi(v)$ the c.d.f. of $\Sigma c_i g_i$, that is if we introduce $\pi(v) = \Pr\{\Sigma c_i g_i \leq v\}$ then it follows from (21) and (23) that

$$p(G) = \int_{G-\delta}^{G} f(G - v)\,\pi(v)dv. \tag{29}$$

MOreover, because of the above assumptions (i) to (iii) it follows that p(G) has continuous derivatives up to order m+1 and hence permits a Taylor expansion at $G = G^*$ of the form

$$P(G) = \sum_{j=1}^{m} \alpha_j (G - G^*)^j + 1 \tag{30}$$

where $G^*$ is the upper limit of the distribution of g and hence is given by

$$G^* = g^* + \delta. \tag{31}$$

Consider now a sample of n values of the statistic g given by (21) and denote by $g_t$ the order statistics in the sample of n values. Clearly the probability transforms $p(g_t)$ are the order statistics from a uniform distribution and the k largest $p(g_t)$ (for $t = n - k + 1, \ldots, n$) follow the distribution of the k largest order statistics from a uniform distribution.

We now make the following "smoothness assumptions" about p(G). We assume that for a suitably chosen k with $m < k \leq n$ the transforms $P(g_t)$ (using the Taylor expansion (30) in place of $p(g_t)$) for $t = n - k + 1$ follow the joint distribution of the k largest order statistics from a uniform distribution and that the $g_t$ lie with probability $1 - \epsilon$ within a range $\underline{G} \leq G \leq G^*$ for which $\frac{dP}{dG} \geq 0$.

These assumptions would of course be violated if the Taylcr Taylor expansion (30) differs appreciably from p(G) but the assumption is monitored by the test described below. In what follows we proceed under these assumptions.

The upper confidence points for G* will now be obtained with the help of a new non-parametric goodness of fit criterion specially developed to have high power for the detection of departures of a c.d.f., p(G), in the upper tails: Assuming that the sampled $g_t$ in (24) are in ascending order of magnitude and that the k highest order statistics $g_{n-k+1}, \ldots, g_n$ fall within the range $G \leq g \leq G^*$ in (30) we transform the $g_t$ to the 'uniform' order statistics

$$z_t = P(g_t) = \sum_{j=1}^{m} \alpha_j (g_t - G^*)^j + 1. \tag{32}$$

Now from well known properties of the uniform order statistics we have

$$Ez_t = \frac{t}{n+1} = \mu_t \text{ (say)} \tag{33}$$

and

$$\text{Cov } z_t z_\tau = \frac{t(n - \tau + 1)}{(n+1)^2(n+2)} = v_{t\tau} \quad \text{(say)} \quad t \leq \tau. \tag{34}$$

Using these we form as a goodness of fit criterion the quadratic form in the differences

$$(z_t - \frac{t}{n+1})$$

with the inverse of the variance-covariance matrix in the form

$$S^2(k,n) = (z - \mu)' V^{-1}(z - \mu) \tag{35}$$

where z and $\mu$ are the k-vectors with elements $z_t$ and $\mu_t$ given by (32) and (33) and V has elements $v_{t\tau}$ given by

(34). Using these values (35) can be simplified to

$$S^2(k,n) = (n+1)(n+2)\left\{\frac{z^2_{n-k+1}}{n-k+1} + \sum_{t=n-k+1}^{n+1}(z_t - z_{t-1})^2\right\} - (n+2) \qquad (36)$$

where $z_{n+1} = 1$.

An extensive study of the null-distribution of $S^2(k,n)$ was conducted by D. Lurie* who derived its exact moments and showed that the moment based Pearson Type approximations compared extremely well with massive Monte Carlo evaluations. He also provided Monte Carlo evidence that $S^2(k,n)$ is considerably more powerful than the customary non-parametric goodness of fit criteria for detecting departures in the upper tail. A short table of his percentage points is given in Appendix A.

Denoting these upper $100\alpha\%$ points of $S^2(k,n)$ by $S_\alpha^2$ a confidence region for the unknown parameters $\alpha_j$ and G* can be set up in the form

$$(n+1)(n+2)\left\{\frac{z^2_{n-k+1}}{n-k+1} + \sum_{t=n-k+1}^{n+1}(z_t - z_{t-1})^2\right\} - (n+2) \le S_\alpha^2 \qquad (37)$$

where the $z_t$ depend on the parameters $\alpha_j$, G* through equation (32). Writing therefore the confidence region (37) in the form

$$S^2(\alpha_j, G^*) \le S_\alpha^2$$

we note that $S^2(\alpha_j, G^*)$ is a positive definite quadratic function of the $\alpha_j$. Moreover since

$$\frac{dP}{dG} \ge 0$$

we known with certainty that the parameters $\alpha_j$ and G* satisfy the inequalities

*Doctoral Dissertation, Texas A & M University.

$$- \sum_{j=1}^{m} j \alpha_j (G - G^*)^{j-1} \leq 0, \qquad (39)$$

for a suitably chosen dense grid of G-values $\leq$ G*.

This makes it possible with the help of quadratic programming to:

(a) Determine whether the confidence region defined by (38) and (39) is empty in which case the composite polynomial model is rejected by this monitoring test.

(b) In case the confidence region is not empty to determine the 'largest value' of G* in the confidence region.

Specifically for any given G* the function $S^2(\alpha_j, G^*)$ is a positive definite quadratic form in the $\alpha_j$ and the restrictions (39) are linear in the $\alpha_j$. Using then the quadratic programming algorithm by Wolfe (1959) it is possible to compute

$$S^2(G^*) = \inf_{\alpha_j} . \; S^2(\alpha_j, G^*) \qquad (40)$$

subject to (39), for suitably chosen dense sequence of G*-values. Parametric programming techniques are ideally suited to the purpose. If the tabled

$$S^2(G^*) \geq S_\alpha^{\;2}$$

for all G* we have case (a) i.e. the rejection of the model by our monitoring test. It there is at least one G* with

$$S^2(G^*) < S_\alpha^{\;2}$$

then there must exist a largest G* say $\overline{G}(\alpha)$ with

$$S^2(G^*) \leq S_\alpha^{\;2}.$$

Clearly, no value of G* $\geq \overline{G}(\alpha)$ can possible lie in the confidence region (38). This value $\overline{G}(\alpha)$ therefore represents a conservative upper confidence point for $G^*$ with a confidence coefficient of at least $1 - \alpha$. Accordingly $\overline{G}(\alpha) - \delta$ represents the upper confidence point for the global maximum $g^*$.

We have proved the consistency of $\overline{G}(\alpha)$ in the sense that

$$\operatorname*{pr\,lim}_{n \to \infty} \overline{G}(\alpha) = G^* \tag{41}$$

However, for finite n it is conceivable that for samples occurring with small probabilities $\overline{G}(\alpha) = +\infty$. Nevertheless, a judicious choice of all algorithm parameters in Section 4 has, in the numerous examples considered, not only resulted in finite $\overline{G}(2)$ but confidence limits close to the true values of $G^*$. In view of the consistency result, dissappointingly large values of $\overline{G}(\alpha)$ would induce the analyst to increase his sample size n by adding ascent paths.

## 6. Examples

We confine ourselves to two examples in the optimization of second order response surface designs. With these designs, it is assumed that the output y from inputs $V_1, V_2, \ldots, V_r$ is given by

$$y = \beta_o + \sum_{t=1}^{r} \beta_t V_t + \sum_{s,t} \beta_{ts} V_t V_s + e \tag{42}$$

where the $\beta$'s are unknown coefficients and e is an error term.

The problem is to estimate the $\ell$-vector

$$\gamma' = (\beta_o; \beta_1, \ldots, \beta_r; \beta_{11}, \beta_{12}, \ldots, \beta_{rr})' \tag{43}$$

from experimental data consisting of N individual "trials". For each of the N trials, an r-vector $V_\theta$ with elements $V_{\theta 1}, V_{\theta 2}, \ldots, V_{\theta r}$ is selected and the corresponding scalar output $y_\theta$ is measured. Let $b_\theta$ be the $\ell$-vector of zero, first and second order terms for the θ-th trial; that is,

$$b_\theta = (1;\ V_{\theta 1},\ \ldots\ ,V_{\theta r};\ V_{\theta 1}^2, V_{\theta 1}, V_{\theta 2},\ \ldots\ ,V_{\theta r}^2). \quad (44)$$

The N x ℓ design matrix B may be expressed as

$$B' = (b_1, b_2,\ \ldots\ ,b_N)' \quad (45)$$

and the linear model describing the N trials may now be written as

$$Y = B\gamma + e \quad (46)$$

where Y is the N x 1 vector of responses, B is the N x ℓ design matrix, $\gamma$ is the ℓ x 1 vector of unknown coefficients and e is an N x 1 vector of error terms.

Assuming the linear model give by (46), it follows from least squares that

$$\hat{\gamma} = (B'B)^{-1}B'Y. \quad (47)$$

It is also known that $\hat{\gamma}$ has the properties that

$$\text{(i)} \qquad E(\hat{\gamma}) = \gamma \quad (48)$$

and

$$\text{(ii)} \qquad E(\hat{\gamma} - \gamma)(\hat{\gamma} - \gamma)' = (B'B)\sigma^2 \quad (49)$$

if it is assumed that $E(e) = 0$ and $Var(e) = \sigma^2 I$.

Harltey and Ruud (1969) have chosen to select the design of the experiment (the selection of the N input vectors $V_\theta$) by minimizing the generalized variance of $\hat{\gamma}$,

$$|(B'B)^{-1}|.$$

It is assumed in their paper that the space of experimentation is the convex intersection S of R linear inequalities.

Formally, the mathematical programming problem is:

$$\text{Minimize} \quad |(B'B)^{-1}| \quad (50)$$

Subject to:

$$AV_\theta \leq L; \quad (\theta = 1,2, \ldots ,N) \tag{51}$$

where A is an R x θ matrix of given coefficients, L is an R-vector of upper bounds and $V_\theta$ is in S.

Although the constraints (51) are linear(convex), the objective function is not convex and we have, therefore, a non-linear optimization problem. Notice that the dimensionality of the constraint space is N x r, where r is the number of elements in V. For most problems, the number of variables is sufficiently large to preclude the utilization of scanning procedures.

Two examples of the optimization of second order response surface designs will be discussed, both of which involve two input variables, $V_1$ and $V_2$. In the first, six points in the $V_1$ $V_2$ space are chosen optimally and in the second, eight points are selected. The space of experimentation is a rectangle.

Details of the algorithm parameters are given in Pfaffenberger (1971) and are not repeated here. However, it should be stated that the sample size for random starts was n = 20 and the degree of polynomial postulated m = 4 in both cases. The objective function g(x) to be maximized was log $|B'B|$.

The results of the computation of upper confidence points for g* are give in the table below where they are compared with the true values of g* known from other considerations.

| | Global Maximum g* | Confidence Points G(.05) | $\overline{G}$(.01) |
|---|---|---|---|
| Six-Point Example | $1.75 \times 10^7$ | $1.86 \times 10^7$ | $1.99 \times 10^7$ |
| Eight point Example | $15.46 \times 10^7$ | $17.20 \times 10^7$ | $18.06 \times 10^7$ |

It must, of course, be remembered that these confidence points have been computed from a very modest sample of $n = 20$. The G values for the six-point example are most reasonable in the light of the sample size. The eight-point G values, however, suffer from samples of this size. If n is increased to 30, $\bar{G}(.05) = 16.27 \times 10^7$, for example.

## 7. Limitations and Extensions

The task of making a statement about the maximum of a completely general mathematical function g(x) of an n-vector x is obviously an unsolvable problem. For even if 'continuity' or 'smoothness conditions' are made there is always the possibility of a very 'narrow spoke' which may escape any scanning operation even if this is supposedly exhaustive. The problem is related to the well known futility of claiming that formulas in finite difference calculus (such as interpolation formulas) are applicable to 'any function'. Definite statements about the errors involved can only be made if the smoothness assumptions are made quantitative, for example by postulating an upper bound for the modulus of a higher derivative of the function.

In the present statistical approach 'quantitative' smoothness assumptions are made by postulating that the c.d.f. can be represented by a polynomial of specified degree m, in a 'tail range' $G \leq g \leq G^*$ as well as in the assumption that n is sufficiently large that a specified number of k upper order statistics will lie in this range with probability $1 - \epsilon$. Our assumptions will however be satisfied for some value of m given n and k. It appeared therefore of interest to us to modify our assumptions to cover a wider situation when for given n,k there is available from experiences encountered in 'related optimization problems' a data based prior distribution $\phi(m)$ (say) on the degree of the polynomial m. The confidence statement can then be modified to apply to the posterior distribution computed from $\phi(m)$ and the data likelihood. Some results of such computations will be found in Pfaffenberger [1971] and are not reproduced here.

## 8. Survey of Literature

It is well known that the special case of 'convex programming' in which the functions $-g(x)$, $-f_i(x)$ in (1) and (3) are convex is well established as a workable problem and well documented by available computer codes. However, for non-convex problems the literature is more restricted.

The current procedures for solving the non-convex problems are essentially of two kinds:

(i) Exhaustive scanning methods of the feasible x-space which are computationally manageable only for moderately small dimensionality of the constraint space. Hooke and Jeeves (1961) summarize the literature on scanning methods and discuss in some detail the consequences of the vastness of hyperspaces (see also Spang (1962) for relevant comments).

(ii) Application of convex programming algorithms to the non-convex problem which usually guarantees only the attainment of local optima (see, for example, Fiacco and McCormick (1964) and Rosen (1960)). More recently, attempts have been made to take advantage of structural conditions of the objective function and/or the constraint space. Hartley and Hocking (1968) have identified a number of conditions which enable them to use convex procedures in computing the optimal solution.

Inasmuch as an increasing number of important problems are falling into the non-convex class, techniques are continually being devised to at least conduct the search or scanning procedures in a more efficient way. Seceral such techniques will now be briefly discussed:

Since in the general non-convex case, the local optimum reached by a convex algorithm will depend upon the starting point, Reiter and Sherman (1965) suggest taking a number of starting points, say k, each drawn from an assumed initial distribution of the variables over the constraint space, and allowing each to move to a terminal point (a local optimum) by means of the optimization algorithm being employed. The resulting local optima are bounded by a hyperrectangle which, hopefully, will be a small subset of the constraint space. For the next N - k starting points, this hyperrectangle is given a higher probability of being sampled. In a later paper, Reiter and Rice (1966) apply this procedure to 39 test problems in an effort to demonstrate its merit in resolving non-convex mathematical programming problems.

Very recently, Graves and Whinston (1970) have reported a technique yielding the global solution to the general non-convex problem. However, the technique involves the repeated evaluation of n-variable integrals over subspaces of $E_n$ and since such integrals cannot in general be obtained analytically, their evaluation by numerical integration would computationally be equivalent to n-dimensional scanning of the integrand functions with an added uncertainty about the remainder term. Moreover, the method described only applies to unrestrained optimization.

Clough (1969) has developed a procedure which gives a confidence interval for the global optimum based on asymptotic extreme-value sampling theory as developed by Fisher and Tippett (1927). The extreme-values that Clough uses are, of course, the local optima resulting from repeated trials on a specific mathematical programming problem. The apparent difficulty with this technique is the unusually large sample sizes required before precise statements can be made using the extreme-value theory. Reiter and Rice (1966) note that, if their procedure utilizing the hyperrectangles as described above is used the distribution of local optima appears to conform to a beta distribution in the example problems they have considered. By fitting a beta distribution to these points, they suggest it may be possible to use the fitted distribution to predict the global optimum. At this time, they have apparently not considered this extension.

## References


1. Clough, D. J., (1969) "An Asymptotic Extreme Value Sampling Theory for the Estimation of a Global Maximum", Canadian Operations Research Society Journal, 7, 105-115.
2. Fiacco, A. V. and G. P. McCormick, (1964) "Computational Algorithm for the Sequential Unconstrained Minimi-

zation Technique for Nonlinear Programming", Management Science, 10, 601-617.
3. Graves, G. and A. Whinston, (1970) "An Algorithm for Non-convex Programming", Seminar Manuscript by courtesy of authors.
4. Hartley, H. O. and R. R. Hocking, (1963) "Convex Programming by Tangential Approximation", Management Science, 9, 600-612.
5. Hartley, H. O. and R. R. Hocking, (1968) "Non-convex Programming", Technical Report No. 6, Project Themis Contract. N00014-68-A-0140, Project NRO 47-700.
6. Hartley, H. O. and P. G. Ruud, (1969) "Computer Optimization of Second Order Response Surface Designs", Statistical Computations, Proceedings of a Conference held at the University of Wisconsin, Madison, Wisconsin, April 29-30, 1969, 441-462.
7. Hooke, R. and T. A. Jeeves, (1961) "Direct Search Solution of Numerical and Statistical Problems", Association for Computing Machinery, 8, 212-229.
8. Reiter, S. and Rice, (1966) "Discrete Optimizing Solution Procedures for Linear and Non-linear Integer Programming Problems", Management Science, 12, 829-850.
9. Reiter, S. and G. Sherman, (1965) "Discrete Optimizing", J. of the Soc. for Industrial and Applied Mathematics, 13, 864-889.
10. Rosen, J. G. (1960) "The Gradient Projection Method for Non-linear Programming, Part I. Linear Constraints", Journal of the Society for Industrial and Applied Mathematics, 8, 181-217.
11. Rosen, J. G., (1961)"The Gradient Method for Non-linear Programming, Part II. Nonlinear Constraints", Journal of the Society for Industrial and Applied Mathematics 9, 514-532.
12. Spang, H. A., (1962) "A Review of Minimization Techniques for Nonlinear Functions", Journal of the Society for Industrial and Applied Mathematics, 4, 343-365.
13. Wolfe, P., (1959) "The Simplex Method for Quadratic Programming", Econometrica, 27, 382-398.

## Appendix A

### Critical Values for $S^2(k,n)$
$\alpha = .05$

| n \ k | 5 | 10 | 15 | 20 | 25 | 30 | 35 | 40 | 45 | 50 |
|---|---|---|---|---|---|---|---|---|---|---|
| 10 | 13.46 | | | | | | | | | |
| 15 | 14.42 | 21.85 | | | | | | | | |
| 20 | 14.87 | 23.12 | 29.38 | | | | | | | |
| 25 | 15.14 | 23.88 | 30.72 | 36.52 | | | | | | |
| 30 | 15.31 | 24.37 | 31.60 | 37.84 | 43.41 | | | | | |
| 35 | 15.43 | 24.74 | 32.22 | 38.78 | 44.68 | * | | | | |
| 40 | 15.52 | 25.00 | 32.70 | 39.47 | 45.64 | 51.34 | * | | | |
| 45 | 15.60 | 25.20 | 33.05 | 40.01 | 46.39 | 52.31 | 57.85 | * | | |
| 50 | 15.65 | 25.35 | 33.34 | 40.43 | 46.97 | 53.06 | 58.82 | * | * | |
| 60 | 15.72 | 25.61 | 33.76 | 41.07 | 47.82 | 54.18 | 60.20 | 65.99 | * | * |
| 70 | 15.78 | 25.77 | 34.07 | 41.51 | 48.43 | 54.97 | 61.19 | 67.17 | 72.92 | * |
| 80 | 15.83 | 25.89 | 34.29 | 41.85 | 48.88 | 55.55 | 61.91 | 68.05 | 73.96 | 79.76 |
| 90 | 15.86 | 26.00 | 34.46 | 42.10 | 49.22 | 56.00 | 62.48 | 68.73 | 4.79 | 80.65 |
| 100 | 15.88 | 26.07 | 34.59 | 42.31 | 49.51 | 56.36 | 62.91 | 69.26 | 75.40 | 81.42 |

### Critical Values of $S^2(k,n)$
$\alpha = .01$

| n \ k | 5 | 10 | 15 | 20 | 25 | 30 | 35 | 40 | 45 | 50 |
|---|---|---|---|---|---|---|---|---|---|---|
| 10 | 21.47 | | | | | | | | | |
| 15 | 24.21 | 31.99 | | | | | | | | |
| 20 | 25.69 | 34.89 | 41.20 | | | | | | | |
| 25 | 26.60 | 36.72 | 43.94 | 49.70 | | | | | | |
| 30 | 27.23 | 37.99 | 45.83 | 52.26 | 57.75 | | | | | |
| 35 | 27.66 | 38.90 | 47.22 | 54.12 | 60.12 | * | | | | |
| 40 | 28.00 | 39.59 | 48.27 | 55.54 | 61.95 | 67.88 | * | | | |
| 45 | 28.27 | 40.14 | 49.10 | 56.67 | 63.38 | 69.44 | 75.01 | * | | |
| 50 | 28.48 | 40.59 | 49.77 | 57.57 | 64.53 | 70.86 | 76.70 | * | * | |
| 60 | 28.81 | 41.25 | 50.77 | 58.95 | 66.28 | 73.02 | 79.26 | 85.17 | * | |
| 70 | 29.04 | 41.73 | 51.52 | 59.92 | 67.54 | 74.56 | 81.12 | 87.32 | 93.21 | * |
| 80 | 29.21 | 42.09 | 52.06 | 60.69 | 68.49 | 75.71 | 82.50 | 88.94 | 95.11 | 101.00 |
| 90 | 29.35 | 42.36 | 52.49 | 61.27 | 69.23 | 76.61 | 83.60 | 90.20 | 96.58 | 102.69 |
| 100 | 29.46 | 42.60 | 52.84 | 61.74 | 69.83 | 77.36 | 84.45 | 91.24 | 97.71 | 103.98 |

* A Pearson's Type IV distribution.

# CURRENT CAPABILITIES IN MATHEMATICAL PROGRAMMING

Philip Wolfe

IBM Research

Abstract

An effort will be made to present information useful to one who has a specific optimization problem to solve or is concerned with the development of efficient procedures for the solution of classes of problems. We can now make estimates of the difficulty of solving some types of problems--for example, large-scale linear programming problems--and indicate the bounds of practicality, in terms of problem size and complexity and presence or absence of constraints, of other classes of optimization problems. Results from comparisons of nonlinear programming techniques will be presented, and the directions in which fruitful research is currently being done indicated. Distinction will be made among those procedures which are interesting or promising, those which are well understood, and those for which reliable and usable computer routines exist.

# PATTERNS AND SEARCH STATISTICS*

Allen Klinger

Computer Science Department
University of California, Los Angeles

## 1.0 Introduction

The nature of this paper is expository and the goal is to outline many problems which exist in the planning and execution of a search of a small two-dimensional visual field or pattern. Most of the paper discusses one particular application and the work presented is descriptive of approaches to that application. These, of necessity, constitute ideas rather than general results. No theorems are stated and whatever tables and figures are presented constitute trial cases needed to illustrate a conceptual approach. This material thus constitutes a working record of an exploration into a rich problem area. It is hoped that the record will stimulate intensive efforts in this fundamental area of pattern perception by automatic means. The paper itself reflects the interest and enthusiasm of my students in discussing this work.

In situations which arise in artificial intelligence initial null and alternative hypotheses must be modified as information is obtained. One modification concerns the establishment of other hypotheses to which the observed last datum also contributes and a decision regarding which pair of null and alternative hypotheses to emphasize in further testing. For example, specific instances could oc-

*Research sponsored by the Air Force Office of Scientific Research, Office of Aerospace Research, USAF, under Grant No. AFOSR 70-1915 and the National Science Foundation under Grant No. NSF-GK-4827. The United States Government is authorized to reproduce and distribute reprints for Governmental purposes notwithstanding any copywright notation herein.

cur when a computer is controlling an object and seeking to cause it to reach a goal such as moving a robot through a doorway. Let us suppose that at each point on its progress towards the doorway it can view only one of five points: three points colinear with its trajectory, N,M, F (near, middle and far), and two points ten degrees on either side of M, L, and R (left and right). It can also save its battery by "not looking" and accept or reject a previous hypothesis such as "at wall", "at corner", or "at doorway". Now replace "at" with "near" or "on direct line with", and let the robot be seeking to guide its future course by the outcome of the test of these hypotheses

$$\{H_o^1, H_1^1\}, \quad \{H_o^2, H_1^2\}, \text{ and } \{H_o^3, H_1^3\}.$$

The past data is available as a light intensity value at a time (robot coordinate position) i = 0,1, ... ,n and at a point relative to the robot, i.e., $N_i$, $M_i$, $F_i$, $L_i$, or $R_i$. Let us call the history X,Y the pair of two (n + 1)-dimensional vectors and the <u>principal hypothesis</u> $H_1^j(n)$ the j-th alternate hypothesis at stage n; $X_n$ is a random variable of light intensity observed at a location specified by $Y_n$. The goal is a rule for j(n + 1)-selection of $H_o^j(n + 1)$. and $Y_{n+1}$ a next stage search location based on the history.

Clearly $Y_{n+1}$ will depend on the choice of next principal hypothesis. We will employ the ideas of this formulation later.

The robot example is an instance of "artificial intelligence", loosely, the use of computers to accomplish complex tasks which can be done by natural systems, e.g., human beings. Another instance is the processing of pictorial data. Computers have been programmed to classify chromosomes, alphanumeric characters, electroencephlograms, and to enhance the quality of pictures taken by spacecraft. The theory associated with these cases and related problems is known as "pattern recognition"; as the robot example shows, pattern recognition is a central problem area in artificial intelligence. In processing pictorial data the key computer problem is reducing the volume of light intensity values or grey levels at different points in the scene to a limited number of descriptive terms associated with objects. Examples are 1) "curved line", "circle", "straight line at 45°", etc., a set of "<u>features</u>" useful for distinguishing the numerals 4 and 9, 2) the actual slope intercept equations of a straight line approximation to the outlines of the elementary bodies in the scene, or the "line drawing", and 3) the linear representation of each enclosed region in the scene, called the "skeleton".

Sequential analysis, Wald [1], is a useful tool in processing a pattern history. Fu [2], [3], has applied sequential decision procedures and dynamic programming to pattern recognition. He considers "successive observations or feature measurements" ([3], p. 46) and uses the outcome to decide whether to continue or to classify the pattern. The distinction between this paper and the work presented by Fu is that we distinguish between observations and features. Here an observation will be a point evaluation. It will contribute to many hypotheses or features.

The point evaluations we will make will be of signal intensity or grey-level at a location in a two-dimensional grid. The grid corresponds to a digitized picture or scene. We could be concerned with minimizing the number of searched locations needed to find "objects" in the scene or minimizing a cost of taking observations. Total time until the scene is understood might be increased if a search frequently alternated between ends of a disc-stored scan of a picture. Likewise robot battery power could be quickly expended by large physical movements of any sensor. However, analysis of "minimum number of locations searched" will suffice for our purposes.

A specific instance of such a search can be based on the mathematical recreation game "Patterns" (Sackson [4], [5]). The game involves a 6 x 6 grid with each location (cell) containing one of four symbols. The goal is to guess the pattern accurately after as few queries as possible regarding the contents of specific cells: the score obtained is the number of correct guesses minus the number of incorrect guesses; the extreme scores attainable, maximum and (negative) minimum, become smaller in magnitude as more information is sought. At any stage the player can make an additional query and obtain the exact cell contents or a decision to terminate the search and guess the pattern. The "player" is rewarded when he establishes a hypothesis and successfully guesses the unsearched part of the pattern;

the "Patterns" are constructed by a "designer" whose goal is a figure scene which can be found by some skillful choice of hypotheses and inductive conclusions but which will be obscure otherwise. The "player" seeks to recognize the pattern after examining only a small part of it; this coincides with the general goals of the artificial intelligence research problems described above.

The following terminology will be used. A pattern will be denoted by P; when dealing with Sackson's game we will consider P(6,6), a 36 cell matrix with cell symbols p(i,j) ranging over 0,+,.˙., or *. A pattern hypothesis will be a matrix M(m,n) of m rows and n columns which contains p(i,j) obtained by search and conjectured values completing the array. A feature will be represented by a set of cells to be searched; the set will be denoted by L and will be referred to as a likelihood region. The latter term reflects the possible modification of an hypothesis: e.g., the final "feature" may be a subset of the postulated feature; the currently postulated feature is the likelihood region L under test. L may be a row, column, or diagonal such as M(1,6), M(6,1) or $\{p(i,i) | i = 1,2, \ldots ,6\}$ or a set of cell-pairs such as $\{p(i,j),p(j,i) | i \neq j, i = 1,2, \ldots ,6, j = 1,2, \ldots ,6\}$ to test an hypothesis of symmetry with respect to a diagonal. When considering sequential procedures within a likelihood region L we will use $x_i$ and $y_i$ to denote the i-th cell or cell-pair sample. Subscripted x will always denote query of a single cell at iteration i; $y_i$ represents a binary predicate or truth value of the conjecture that the cell-pair p(i,j), p(j,i) chosen at iteration i contains only one symbol. The symmetry conjecture tested via a sequence of y values (e.g. 10111 from query of 10 cells) is a partial pattern hypothesis. Predicates t, v, and w associated with other partial pattern hypotheses involve groups of cells-nine cells for v. "False values" acquired during search of fewer cells will be incorporated in the decision rule for j(n + 1) to choose the (n + 1)st principal hypothesis (for example, we may find two different symbols in a nine cell group where 1,2, ... , or 7 cells remain unqueried so that $v_i = 0$ is noted). If the predicate is found "false" based on fewer cell samples than needed for "true" its associated partial pattern hypothesis will be rejected or modified, and the list of test hypotheses will change. "True values" can occur as well. Prior cell samples can combine with the current queries which are located to test the principal hypothesis, to yield a true value of a $y_i$ or another predicate. In some cases this will cause a change of principal hypothesis or acceptance of a non-principal hypothesis. Where no confusion would result we abbreviate "partial pattern hypothesis" to "pattern hypothesis".

Two types of likelihood regions of particular importance are the arrays of sizes 2 x 2 and 3 x 3. Figure 1 illustrates the reduction of P(6,6) into subblocks based on each of these. That is, P(6,6) consists of four 3 x 3 subblocks or nine 2 x 2 subblocks, and each formulation involves the uses of the other subblock type on a coarser grid: we have P(2,2) with P(3,3) cells or P(3,3) with P(2,2) cells; that is, the cells or constituent elements are themselves arrays of cells at a higher level of resolution. The importance of P(2,2) is that it is a symmetry-locating feature: horizontal, vertical, upward diagonal or downward diagonal symmetry axes could be conjectured based on a coarse-grid two-dimensional binary search. P(3,3) contains information regarding the eight neighboring cells of its midpoint, hence it is useful for choosing subsequent search directions.

Hierarchical methods for search of practical digitized representations of scenes can be based on recursion via subblock storage. Here coarse-grid cell values would indicate average grey level intensity over the cells in a given subblock. By "practical" we seek to differentiate the "Patterns" example from stored picture data where the order of magnitude of the number of cells ranges from 1,000 to 1,000,000. Realistic assembly language data structures for representing pictorial data could well be based on list structures with average grey level over an array stored in a list head and actual intensities stored in nodes following the head. For many processing algorithms fine cell search could be directed by coarse cell grey value in list and sublist heads. Scanners which yield the digitized picture data (grey levels by cell) can be envisioned to produce average or cumulative intensity (grey level) over subblocks: the subblock then becomes a Coarse cell. A clocked scan with a cumulative intensity count could be the basis of such a system.

The main body of the paper consists of three sections. Section 2. discusses "Statistical Procedures for "Patterns";" Section 3., "Pattern Complexity", and Section 4., "Recursion, Storage, and Scene Search".

## 2.0 Statistical Procedures for "Patterns"

Let x, the contents of a particular cell in the patterns game, be a discrete random variable taking on the values +, *, 0, .˙. with probability distribution $f(x|a,b,c)$, $p(x = +) = a$, $p(a = *) = b$, $p(x = 0) = c$ and $p(x = .˙.) = 1 - (a + b + c)$. Then a null hypothesis of $a_o = b_o = c_o = \frac{1}{4}$ can be compared to an alternate simple hypothesis over either the entire array or a subarray. Specifically, we could test the hypothesis that only a single symbol appears (e.g., $a_1 = 1$, $b_1 = c_1 = 0$) or that one symbol does not appear (e.g., $a_1 = 0$, $b_1 = c_1 = 1/3$) in the total pattern or some geometric feature. If we extend the domain from "cells x" to cell-tuples# related geometrically", we obtain a sample space or likelihood region L for symmetry and subblock structure test hypotheses. Let y be the truth value of the predicate "same symbol in all cells in the cell-tuple". Then y = 1 corresponds to x = s the single symbol we believe present, so that similar statistical tests can be used to make decisions regarding uniformity of symbols in regions and symmetry over the entire array.

If the query of any cell in P(6,6) corresponded to an independent random trial and symbols were identically distributed according to f with parameters (a,b,c) having either subscript 0 or 1 we could employ sequential methods for making decisions to accept $H_o$, accept $H_1$, or continue sampling cells. The sequential probability ratio test (SPRT) could then be used to design decision procedures in accord with desired error probabilities $\alpha = p(H_1|H_o)$ and $\beta = p(H_o|H_1)$. Typical tests would be between a pattern or nonpattern in a given likelihood region:

$$H_o = H^L_{np} \text{ and } H_1 = H^L_p$$

or more specifically, $H_1 = H^L_s$, the hypothesis that region L contains or omits only symbol s. However, both in "Patterns" and in

#Cell-pair, cell-triple, etc.

actual picture processing applications cell samples are not independent because there is a regular pattern. Design of an SPRT for each possible history (sampling and observation sequence) to account for dependence would be highly involved for the 36-cell case. Likewise the small samples in "Patterns" and our goal of introducing recursion for analogous real world contexts leads us to present only heuristic rules for sequential decisions.

## 2.1 Problem Reduction

One basic technique in solving a problem is to solve a series of simple subproblems [6]. In "Patterns" the first type of problem reduction is geometric: instead of the entire field of 36 cells we can search subblocks-smaller arrays which are similar to the original array. A second type of problem reduction is symbolic: we decide that the area being searched has fewer symbols than the entire alphabet. Symbolic reduction can be accomplished by accepting a symbol hypothesis for a likelihood region. The assumed probabilities of different symbols, $p(x = s)$, $s = +$, 0, *, $\dot{\cdot\cdot}$, when investigating a likelihood region and the area remaining to be searched are variable quantities which may differ at each stage in a sequential search. Thus statistical methods for symbolic problem reduction involve calculation of expected error probabilities under the stopping condition accept $H_p^L$ given f and "size unsearched part of L".

In any automatic mode for pattern search the two aspects of problem reduction will interact. Likelihood areas will be chosen based on geometric considerations and on actual symbol observations. One way to do this is to establish a prior set of geometric features and maintain a list with total number of queries and total symbols of each type, by feature. A ranking of the list would be based on this information and a likelihood region chosen as the highest ranked feature. The list could grow as pattern information is obtained. New features would be derived by algorithms which associate like-symbol cells with one another. For example when $p(1,1) = p(2,2) = p(1,3)$ is known by a completed search three symbol-chain features could be derived:

$$E=\{p(i,j)\,|\,i + j \text{ is even}\},$$

$$E(2,6) = \{p(i,j) \mid i+j \text{ is even, } i = 1,2\}, \text{ and}$$

$$E^1 = \{p(i,j) \mid i + j \text{ is even and } i \neq j \text{ or } i=j=1,2,5,6\}$$

(even, even in top two rows, even in outer rings-central 2 x 2 subblock deleted). The list could be modified when a found symbol causes a hypothesis to be reduced in scope: a likelihood region of size six with two s symbols could be changed to one of size four if an "ns" extreme cell were found.

## 2.2 Symbol Sequential Tests

Suppose we have L an established likelihood region in the list, and a sequence $\underline{x} = \{x_1, x_2, \ldots, x_q\}$ of length q of sampled cells $x_i$ in L. The relative frequencies of the symbols in the sequence, the sequence length, and the size of L are the significant variables for a symbol hypothesis test. We decide that L is marked only with symbol s when q is Q, a substantial fraction of the size of L and $\underline{x}$ contains only s. Another type of symbol test would compare the relative frequencies over L with those over the entire pattern P. Control chart methods could detect nonuniformities in L as compared to the overall pattern.

An SPRT-like procedure could be based on setting $(a_1, b_1, c_1), \alpha$, and $\beta$ as well as Q empirically to meet reasonable search goals. When $q < Q$ we compute $Z_i = f(x_i, a_1, b_1, c_1)/(f(x_i, a_o, b_o, c_o)$ given by:

$$Z_i = \log \frac{a_1}{a_o} \quad \text{if } x_i = +$$

$$\log \frac{b_1}{b_o} \quad \text{if } x_i = * \text{ , } \log \frac{c_1}{c_o} \quad \text{if } x_i = 0,$$

$$\log \frac{1 - (a_1+b_1+c_1)}{1 - (a_o+b_o+c_o)} \quad \text{if } x_i = \therefore \text{ ;}$$

so that $\sum_{i=1}^{q} Z_i = m^+\log \frac{a_1}{a_o} + m^*\log \frac{b_1}{b_o} + m^0\log \frac{c_1}{c_o} \quad +$

$$m^{\cdot\cdot}\log \frac{1-(a_1+b_1+c_1)}{1-(a_o+b_o+c_o)} .$$

where $m^s$ = no. of occurrences of "s" in sample sequences of length q. Then with $A = \beta/(1-\alpha)$ and $B = (1-\beta)/\alpha$ we

$$\text{accept } H^L_{np}(H_o) \text{ if } \sum_{i=1}^{m} Z_i \leq \log A$$

$$\text{accept } H^L_s(H_1) \text{ if } \sum_{i=1}^{m} Z_i \geq \log B$$

and take another sample $x_{m+1}$ otherwise. Acceptance of the presence of symbol s in region L will cause that symbol to be inserted in all nonqueried cells of that region; the induction result will be the pattern hypothesis for region L: all cells contain s. For the case where $q = Q$ we decide arbitrarily in favor of $H^L_{np}$ and reject the symbol hypothesis, unless $m_s = q = Q$.

For this procedure $(a_o, b_o, c_o)$ could be $a_o = b_o = c_o = 1/4$ or the relative frequencies over P at the current iteration. The former would be used when $L = P$ and overall symbol reduction were the goal of the test procedure.

## 2.3 Nonsymbol Sequential Tests

Two types of information which aid in geometric problem reduction are symmetry and structure. We will employ sequential tests to determine the presence of these types of information using $H_p$ in place of $H_s$. Table 1 lists several symmetries[#] ("pattern" p = V,H,U,D,C,T) while likelihood regions and structure hypotheses are listed in Table 2. The regions are given descriptive names which make them self-explantory. Reference to Figure 1 or Sections 3 and 4 can relate them to the complementary concept "pattern complexity" and the role of recursion.

In accord with the interaction between geometric and symbolic problem reduction described in 2.1, the basic hypotheses in Tables 1 and 2 will be augmented by complemen-

[#]See Polya [7], Vol. I, pp. 89 and 236, prob. 18.

tary hypotheses. For example, a regular antisymmetry hypothesis will be established when two pairs of cells related by symmetry have the same symbol pairs(e.g.(+,0),(+,0)): these will be denoted by ap(aU,aV,...). Thus, the nonsymbol tests are the basis of a set of features which are symbol dependent. (Subsequently, p. 20, we present several examples from [4] and [5] which illustrate this: in Figure 4, examples 7 and 11 involve antisymmetry based on .˙. opposite to 0.) This basis and the symbols in the pattern history constitute an input to an algorithm to describe new features or test hypotheses to be added to the current list.

A test of $H_p$ against $H_{np}$ will require sampling pairs, triples, or four cells related by the particular structure or symmetry. Each sample will confirm or disprove the hypothesis or conjecture; this will be indicated by the truth value at each sample point; i.e., like symbols in the cell-tuple imply truth value "one". Procedures for constructing sequential tests are similar to the symbol hypotheses in likelihood regions. A proportion of the available sample points is to be tested sequentially with acceptance occurring if all points have "true", rejection if one or two have "false". Data on the established tests are contained in Table 3, which is due to E. Purcell.

## 2.4 Heuristic Sequential Decisions

A decision rule which truncates an SPRT-like procedure at Q samples approximately equal to one-third the size of the likelihood region is summarized in Table 3. This choice of Q results in a highly cautious approach to search. Hypotheses are accepted only when a large proportion of the search is accomplished. The sequential nature of the search is used to allow for rejection of hypotheses both early in the process via the SPRT limit or at Q, when an inconclusive sample sequence x(y,t, ...) has been observed. This is illustrated in the hand-generated Table 4 where a search (of Example 5 of Figure 5 ) is described. The initial principal hypothesis is of symmetry with respect to the upward diagonal (U in Table 1); as a consequence all queries are of cell-pairs outside this diagonal $\mathcal{R}$ . As the search procedes comparison of cells queried at different iterations yields samples for other hypotheses. Table 4 shows only the other symmetry hypotheses V,H, ... ,

T; a table from a programmed example would include other test hypotheses. Acceptance of U and evalution of the number of cells remaining to be sampled until Q for these subsidiary hypotheses causes C to be the next principal hypothesis. The process is illustrated by Figure 2 which shows the queries at search iterations 3,6, and 9 in Table 4.

A conservative procedure should insure an accurate guess of the final pattern. This is the case in the search example considered here. Conjectures made based on the accepted hypotheses of Table 4 yield the patterns of Figure 3. When the three hypotheses U, C, and $\beta$ are accepted at iteration 9 in Table 4, induction yields five symbols guessed for the fifteen queried and eight guessable from eight more queries. This is on the basis of "no guesses based on guesses" -we will not use two accepted hypotheses in cascade for induction. Further iterations result in acceptance of hypotheses D, $h_1$, and $h_6$ so that ultimately a pattern hypothesis M(6,6) is achieved after only six additional queries - a total of twenty-one queries resulting in a perfect replica of P(6,6). The example thus illustrates the concept of hypothesis lists and their effectiveness in correlating search statistics.

## 3.0 Pattern Complexity

To gauge the effectiveness of an heuristic search procedure we need a normalized set of test examples. Accomplishing the normalizations leads to the question "How much information is in a two-dimensional array?" A measure of this must be based on the two ideas introduced above, symbol frequency and pattern geometry. In this section we will describe several examples of "patterns" P(6,6) and some possible measures of <u>pattern complexity</u>. In developing these measures we will refer to the examples. These will illustrate some of the subarrays or likelihood regions which were noted in the previous section.

### 3.1 Symbol Uniformity

If the least complex pattern is a uniform placement of the same symbol over the array, one measure of pattern complexity could be based on the number of symbols in a region.

Thus in a subblock of number type (2 x 2 array) the most complex pattern has (all four symbols) complexity 4, and the least complex has complexity 1. We extend this to subblocks of letter type. Then a measure of the total pattern complexity can be based on this subblock measure; we will denote the former by K and the latter by k(i), $i \in \{1, \ldots, 9, a, \ldots, d\} = S$. Then we will use:

$$K_1 = \frac{1}{13} \sum_{i \in S} k(i)$$

This measure is in accord with our perception of pattern complexity as shown by the set of examples in Figure 4. The three examples are increasing in complexity in the same order as $K_1$. Two additional examples based on arrays other than the thirteen subblocks S, one whose complexity is intuitively high and one whose complexity is intuitively low, and their $K_1$ values are given in Figure 5.

### 3.2 Geometric Structure

Example 4 is of high complexity, constructed by an algorithm based on the rings: 12 elements, $\beta$: 20 elements, $\gamma$. Each ring is a cycle of all four symbols, followed by a first-to-last shift to begin a new cycle; the middle ring $\beta$ starts with a different cycle, than the outer ring $\gamma$. Example 5 is low in complexity: two types of structure present are the rings of + and 0 and the low number of different symbols per line in many of the horizontal and vertical lines.[#] However, because $K_1$ emphasizes order on a local scale, although this pattern is less complex than example 2, it has a higher $K_1$ value. Consequently, we will base a new complexity measure $K_2$ on an expanded group of subsets T: the previous 13 subblocks, the four corners ($\alpha$), the two rings, ($\beta$, $\gamma$), the "mid-line" set $\delta = \{a7, a3, d7, d3, b1, b9, c1, c9\}$, and fourteen 6-element lines (6 each horizontal, $h_1, \ldots, h_6$, and vertical, $v_1, \ldots, v_6$; 2 diagonal, $\ell, \pi$). [i.e., $T = S \cup \{\alpha, \delta, \beta, \gamma, h_1, \ldots, h_6, v_1, \ldots, v_6, \ell, \pi\}$]. We will weigh these in proportion to the number of elements in each subset so that a large subset with few symbols lowers complexity, thus

[#] This comment was made by J. Pearl.

$$K_2 = \sum_{i \in T} w(i)\, k(i)$$

The number of subsets with various sizes is also given in Table 5, which gives the weight per subset of each type.

### 3.3 Relative Order

The complexity of the five example patterns under $K_2$ is given by Table 6: Example 5 is seen less complex than example 2 under $K_2$.

A comparison of the local/global character of $K_1$ and $K_2$ can be given. Note that in $K_1$ all weight is given to the 2 x 2 and 3 x 3 subblocks, while in $K_2$ only 0.36 weight is given to these local measures. $K_2$ also weights linear measures with the relatively high value of 0.42, and reserves 0.8 weight for subsets of size greater than 4. This tends to raise the numerical value of complexity on the 1-4 scale. However, we are concerned with an ordering rather than an absolute value. We have these as:

$$1 < 5 < 2 < 3 < 4,$$

for the complexity ranking under $K_2$, and

$$1 < 2 < 5 < 3 < 4$$

for the complexity ranking under $K_1$.

Finally, a third measure of complexity could be obtained by removing the 2 x 2 subblocks which lie entirely in a 3 x 3 subblock. Then useful new features to replace these (i.e., 1, 3, 7 and 9) are four 2 x 2 subblocks which test symmetries: $w = \{1b, 1c, 3a, 3d\}$, $x = \{1b, 1c, 7a, 7d\}$, $y = \{1b, 1c, 9b, 9c\}$, and $z = \{3a, 3d, 7a, 7d\}$. Respectively, these test left/right, up/down, origin/upward-diagonal, and origin/downward-diagonal symmetries. We note that the mid-line set already tests all these symmetries occurring simultaneously. A complexity measure $K_3$ uses these sets and is also described by Table 1, where the needed weights are given. The set T is replaced by

$$T' = \{\alpha, \delta, \beta, \gamma, h_1, \ldots, h_6, v_1, \ldots, v_6, \ell, r, w, x, y, z,$$

2,4,5,6,8,a,b,c,d}

in evaluating $K_3$.

A direct conparison of $K_2$ and $K_3$ is given in Figure 6. Six patterns (Sackson [4], p. 157) are presented with their complexities under $K_2$ and $K_3$. We see that the two values are almost the same in each case. The order under $K_2$ is

$$7 < 10 < 6 < 11 < 8 < 9;$$

that under $K_3$ is

$$7 < 10 < 6 < 8 < 11 < 9.$$

The only ordering change, examples 8 and 11, reflects the presence of central and other symmetries in 8 and their absence in 11; this change seems in accord with our intuitive notion of complexity.

Sets of likelihood regions can be used for accumulation of pattern search statistics. Thirty-five subsets were described and listed in T or T'. All cells are members of at least five of these: a 2 x 2, a 3 x 3, a row, a column, and a ring or diagonal line. Twenty-four cells are members of six, seven, or eight subsets. (Those in six:

{1d,3c,7b,9a,9b,9c,a7,a3,d7,d3,b1,b9,c1,c9};

seven:

{1a,3b,7c,9d,3a,3d,7a,7d}; and

eight:

{1b, 1c}.)

Suppose the number of queries and the relative frequencies are recorded by subset in either a list like T or T' or one based on the current hypotheses. Then if an unknown cell were in several subblocks with similar symbol frequencies (e.g. zero+ , zero *) a conjecture could be established (not+ , not *).

Clearly, any specific set of likelihood regions such as T or T' is incapable of representing all types of patterns.

E. Purcell has described three algorithmic patterns of similar constructions but widely different visual complexity, which are presented in Figure 7. No general theory of two-dimensional pattern complexity as information has been intended in this section. Our only concern has been in relating visually discernable patterns to one another through the relative frequencies of symbols in different geometric regions. We now turn to an application of features in T' to search and induction to eliminate further search.

### 3.4 Limited Search

In section 2.4 a conservative approach to guessing the pattern was applied to example 5 of Figure 5 which had complexity $K_2 = 2.14$. After 21 queries the pattern was found with no errors for a game score 15. Here we will describe query of some features in T' used to make a limited search of pattern example 10 in Figure 6 which had $K_2 = K_3 = 2.12$. A single symmetry feature, such as the four cells of x (two cell-pairs or sample points from the likelihood region of symmetry H), when queried and found symmetrical establishes the corresponding partial pattern hypothesis. As a result eight and twelve query pattern hypotheses will be generated with comparable game score results to that for example 5. All the simulated results compare favorably to a "flub": a computer-program-recorded search made by a person who did not use the relative frequency information. We now describe the detailed rationale of this less cautious search.

As before it is logical to begin search by looking for symmetry. Hence six cells which comprise x and w are initially queried. At this point the pattern is a candidate for symbol reduction (six queries, no .˙., no *, equal + and 0). No symmetry was found so a further search of the remainder of y (eight queries, no .˙., no *, equal + and 0) adds strength to the "two-symbols only" conjecture and "locates" a new conjecture: symmetry with respect to the upward-diagonal $\nearrow$. An eight query hypothesis based on random generation of equiprobable + and 0 for thirteen cells can be based on these two conjectures and creation of like-symbol line segments of four cells whose end cells are known (query results). Figure 8 shows three such simulated hypotheses. Three other pattern guesses are shown in

Figure 9. The first is a full query of the eight cells in subblocks 1 and 9 followed by the guess that all numerical subblocks are the same as these two. Since three others in the actual pattern are the same and two more are rotations of these by $\pi/2$ (causing two errors each) this is a fortunate guess and a score of 16 is found. By comparison the random simulations did well - scores of 8, 12, and 10. Another subblock replication based on query of the twelve cells of subblocks 3, 5, and 7, and a guess that 5 repeated on the diagonal 1 and 9, while the antisymmetry of 3 and 7 with respect to that diagonal continued elsewhere and that off-diagonal 2 x 2's were like these two, also yielded a 16 score. (The illustrated choice of subblocks 2 and 6 was arbitrary; if these were instead taken to be like 3 and the antisymmetry kept four errors would result for a score of 10.) Both these subblock repetition cases of Figure 9 are favorable scores obtained after symbolic reduction to (0 and +). The observations of (eight queries, no .˙., no *, equal + and 0) and (twelve queries, no.˙., no *, equal + and 0) led to a pattern hypothesis with the same representation of + and 0 symbols in the unknown cells. When this reduction is not made the results are like the unfavorable score of -16 actually obtained by a player after twelve queries, the four symbol hypothesis shown in Figure 6.

Finally, note that example 10 contains five identical 2 x 2 subblocks. Tables 2 and 3 present test hypotheses for finding "all subblocks the same".

## 4.0 Recursion, Storage, and Scene Search

In this section we apply the subblock cell notation introduced in Figure 1. The first application is conceptual; we extend our notion that visual pattern perception depends on the regions described above. Consider the two subblocks as alternately used mechanisms of scanning a scene and searching a small portion of a scene. Natural systems employ both methods of searching a field of view. Examination of Figure 1a shows that at one level of organization P(6,6) consists of four major areas eaxh a 3 x 3 array, while at another perceptual level it is a 3 x 3 array of "coarse cells" each one a 2 x 2 - nine P(2,2) patterns. The concept we introduce is the use of this to enable recursive analysis of two- dimensional visual scenes, by alternately subdividing a scene into finer and finer 2 x 2 or 3 x 3 coarse cells.

### 4.1 Coarse Representations of a Pattern from a Man-made Object

In Figure 10 a 12 x 12 array is shown superimposed on a visual scene. An object is encoded according to a correspondence between the grey levels or signal intensities in each cell and the symbols used in "Patterns", and these values assigned according to the portion of a cell enclosed within the object by its boundary. The result is a pattern P(12,12) with 12.2 units of intensity over the entire array, divided into a = 4.9, b = 0.7, c = 1.0, and d = 5.6 units. The letters here signify the four coarse cells each of which is a P(6.6) array, and the numbers are based on the atomic cell symbol numerical values * = 1, + = 0.7, .˙. = 0.3 and 0 = 0.0.

If the pattern is to be stored for computer processing, the successive refinement of the overall array by subblocks of alternating type can be the basis of a useful data representation format. Figure 11 shows the data structure [8] which results when locations are assigned their aggregate intensity values over the entire coarse cell and when "a" signifies upper left quadrant, "a.1", the upper left ninth of "a", "a.1.b" the upper right quarter of "a.1", etc. This structure is not the most concise which could be created for the sparse matrix P(12,12), but it does use only 61 nodes to store the 144 cell pattern (72 would be needed for direct sequential storage with two cells per node). Furthermore the aggregate values of intensity and the format of Figure 11 will be useful for a scene search example which follows shortly. For alternate subdivision of a scene by 2 x 2 and 3 x 3 subblocks it is useful to have available the exact edge or side sizes and the total numbers of cells which can be evenly divided by this method over several iterations. This data is summarized in Table 7 up to 419,904-cell patterns.

### 4.2 Search by "Quadrant" and "Four/Nine/Four" Decomposition

Table 8 compares two searches of P(12,12) which had as

their common goal rapid detection of the cells which contained more than 80% of the signal intensity. The quadrant search is an alternative to the binary search "upper half? left half? etc." which yields the same result as "quadrant a" for the "yes,yes" answer. We assumed that all coarse cell subblocks' aggregate intensities were accessible, including P(12,12), and that no coarse cell would be refined unless 40% or more of the unacquired intensity were contained therein. Search within a subblock was sequential: a.a, a.b, a.c, a.d or a.d.1, a.d.2, a.d.3, ... , a.d.9 were typical in quadrant search. When a coarse cell value combined with full search of that cell in all but one place could yield the last value we did not search for it. When "backing up" - i.e., searching a previously skipped zone - we employ adjacency to known high intensity cells to direct the search.

Four/Nine/Four search compared very well to quadrant search for the example of Figure 10.

### 4.3 Problem Reduction Search Trees

Binary search is of some limited value in this connection. An application of the problem reduction approach based on and/or tree representations (Nilson [6]) to pattern hypothesis M for a pattern P(4,4) was made by Joan Ellen Katz Slottow. Grey levels were given by the numerical representation * = 4, + = 3, .˙. = 2 and 0 = 1, where as in Figure 10, * corresponds to maximum intensity (black cell) and 0 corresponds to a minimum intensity or white cell. Coarse squares are assigned grey level 4 if and only if all atomic cells therein are black. Coarse cell values are the average of the atomic cell values within the coarse cell rounded to the nearest integer. The analysis considers only three features:

1. The pattern is symmetric.

2. The pattern is inverted. Grey levels 1 and 4 and grey levels 2 and 3 will be considered opposites under inversion.

3. The pattern is neither symmetric or inverted.

The procedure as described by Mrs. Slottow follows.

When any coarse square is divided in half the grey levels of the resulting coarse squares are measured and a conclusion is reached.

1. If the resultant coarse squares have the same grey level, the pattern is judged symmetric and the problem of finding the atoms in any one of the two parts is substituted for the problem of finding the pattern in the original coarse square.

2. If the resultant coarse squares have opposite grey levels, the pattern is judged inverted and the problem of finding the atoms in any one of the two parts is substituted for the problem of finding the pattern in the original coarse square.

3. If neither 1 nor 2 holds, the pattern is judged neither symmetric nor inverted and both halves must be further investigated.

4. The search on any tree branch comes to an end when either a grey level of 1 or 4 or an atom is reached.

<u>Example</u>

M(4,4)

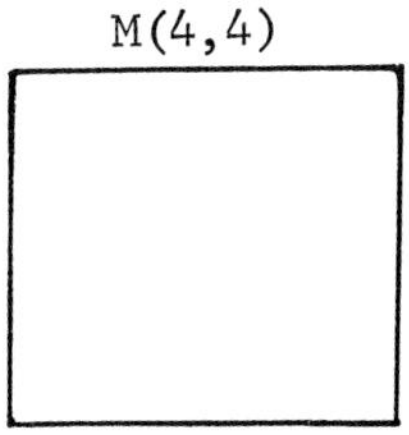

Divide the total pattern in half.

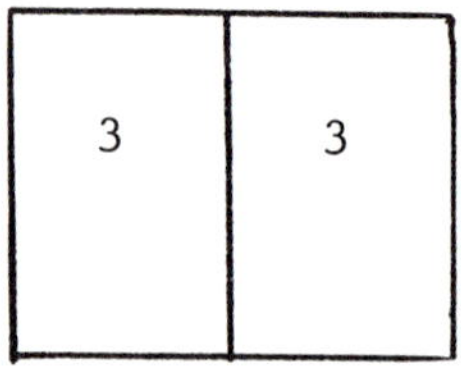


Since the grey values of the two halves are equal, the problem is reduced to the left half (Cost so far equals 2).

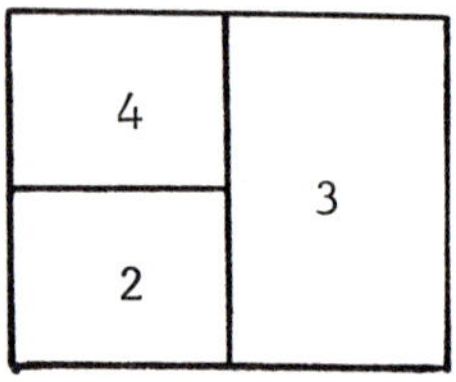


Since the left half is neither symmetric or inverted, both top and bottom must now be examined. However, since the top half has grey level 4 signifying that all atoms in it are black, only the bottom half remains unknown (Cost so far equals 4).

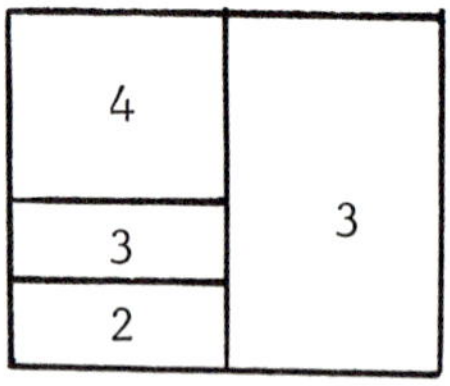


The coarse square on the left is assumed to contain two atoms of level 3. If all assumptions have been correct, the pattern has been discovered with a cost of 8. To make sure the assumptions really hold in this case, one atom at least of each coarse square of grey level 3 should now be examined. This will bring the total cost to 10 or greater depending on how many further complications are found when these tests are made. In the worst possible case, the left and right halves could turn out to be non-symmetric.

In the following reduction trees for the above problem, o nodes represent coarse squares and • nodes are terminal nodes representing atoms and coarse squares with grey levels 1 and 4. The initial problem is to assign appropriate values to the elements of M(4,4). M(a-b,c-d) as a subproblem statement means find all atoms in the matrix M whose first index varies from a to b and whose second index varies from c to d.

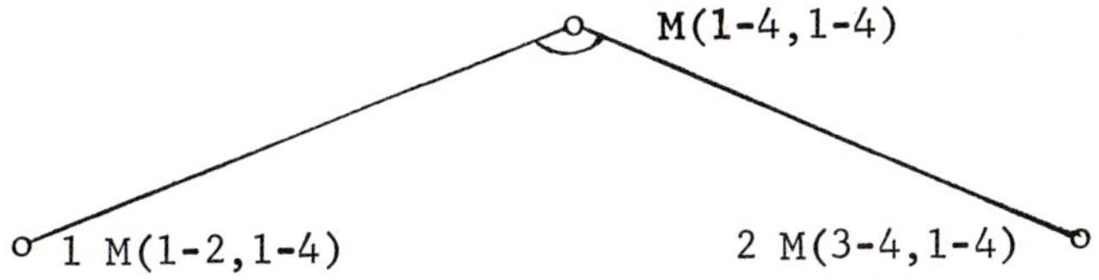


Because symmetry is assumed, branches 1 and 2 are replaced by one branch.

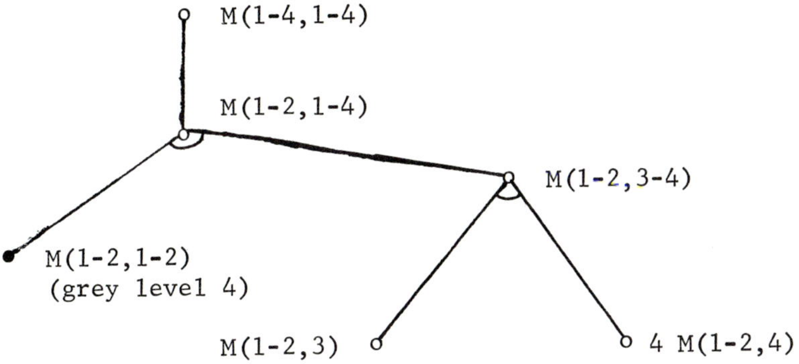


Because inversion is assumed, branches 3 and 4 are replaced by one branch.

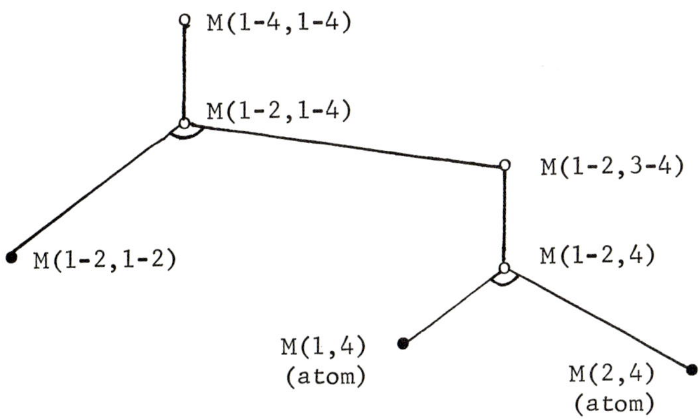

To summarize, in this particular method only a small number of features (symmetry and inversion) were tested. The method of division in half meant that only symmetry and inversion around specific lines would be found. Nevertheless, the general approach should be considered as a tool of some limited value in locating objects in a visual scene.

## 5.0 Conclusion

We have seen how statistical procedures and definitions aid in describing the information storage and retrieval problems which are at the root of practical pattern recognition for some scene analysis. Several concepts have been introduced. These include the notions: hypothesis list, principal hypothesis, and modification of hypotheses based on search results. Specific modifications included change of principal hypothesis, addition, deletion and revision of likelihood regions to create a new hypothesis list at each search iteration, as well as acceptance and rejection of principal or subsidiary hypotheses. Pattern complexity was introduced and quantified based on visual features. Several example patterns were presented. These were used in both statistically-based conservative search modes and less cautious search modes based on visual features and symbol relative frequencies. Comparable results were obtained for "patterns" of like complexity. Recursion based on sub-block storage was described and compared favorably to a modified binary search (quadrant search) in a specific example. Problem reduction and/or trees were used along with the related notions of coarse cell or average intensity representations of picture data. A data structure for a "realistic" example pattern over 144 elementary points were given.

Several ideas and approaches were presented concerning the problem of reducing and reconstructing pictorial patterned data. Much remains to be done to obtain a clear understanding of the merits and disadvantages of the techniques available.

Acknowledgement:

The author wishes to express his thanks to Mrs. Joan Ellen Katz Slottow and Mr. Edward Purcell for their permission to include descriptions of their work within this paper, to the Reports Group at UCLA for their aid in preparing this paper, to the Air Force Office of Scientific Research, and the National Science Foundation for sponsoring this endeavor, and to my many colleagues, students and friends who have discussed "Patterns" with me.

## References

[1] A. Wald, Sequential Analysis, Wiley, New York, 1949.

[2] K. S. Fu, Sequential Methods in Pattern Recognition and Machine Learning, Academic Press, New York, 1968.

[3] K. S. Fu, "Statistical Pattern Recognition", Adaptive Learning, and Pattern Recognition Systems, Academic Press, New York, 1970.

[4] S. Sackson, A Gamut of Games, Random House, 1969.

[5] M. Gardner, "Mathematical Games", Scientific American, Vol. 221, No. 5, November, 1969.

[6] N. Nilsson, Problem Solving Methods in Artificial Intelligence, McGraw-Hill, 1971.

[7] G. Polya, Mathematics and Plausible Reasoning: Vol. I Induction and Analogy in Mathematics, Vol. II, Patterns of Plausible Inference, Princeton University Press, Princeton, 1968 .

[8] D. Knuth, The Art of Computer Programming Vol. 1 Fundamental Algorithms, Addison-Wesley Publishing Company, 1968.

TABLE 1

List of Symmetry Hypotheses for P(6, 6)

| Name | Symbol | Analytic Description | Size of Sample Space | Cells in Sample Space | Cells per Trial |
|---|---|---|---|---|---|
| Vertical | V | $\forall_i\ p(i,j)=p(i,7-j), j=1,2,3$ | 18 | 36 | 2 |
| Horizontal | H | $\forall_j\ p(i,j)=p(7-i,j), i=1,2,3$ | 18 | 36 | 2 |
| Upward Diagonal | U | $\forall_{i,j}\ p(i,j)=p(7-j,7-i)$ | 15 | 30 | 2 |
| Downward Diagonal | D | $\forall_{i,j}\ p(i,j)=p(j,i)$ | 15 | 30 | 2 |
| Central | C | $\forall_{i,j}\ p(i,j)=p(7-i,7-j)$ (U & D) | 18 | 36 | 2 |
| Central, Vertical, & Horizontal | T | C & V & H (U & D & V & H) | 9 | 36 | 4 |

Table

| Pattern or Likelihood Region Tested | Size of Sample Space | Maximum Number of Points Sampled | Maximum Number of Cells Sampled | Number of Possible Sample Sequences of Maximum Length | Proportion of Non-Deciding Sample Sequences | Probability of Non-Deciding Sample Sequences | Random Variable (Composite or Simple) |
|---|---|---|---|---|---|---|---|
| Symmetry V, H, U, D, C; 6 Ring ε | 15, 18, 20 | 6 | 12 | 64 | .531 | .151 | y |
| | 15, 18, 20 | 6 | 12 | 64 | .3125 | .035 | y |
| | 15, 18, 20 | 6 | 12 | 64 | 0 | 0 | y |
| Rectangular Subarray 4 Ring β | 12 | 4 | 12 | 16 | .562 | .018 | t |
| | 12 | 4 | 12 | 16 | 0 | 0 | t |
| 2 x 2 Subarray | 4 | 1 | 9 | 2 | 0 | 0 | v |
| 3 x 3 Subarray | 9 | 3 | 12 | 8 | .625 | .031 | w |
| | 9 | 3 | 12 | 8 | 0 | 0 | w |
| 6 Cell Regions | 6 | 3 | 3 | 64 | 0 | 0 | x |
| | 6 | 3 | 3 | 64 | 0 | 0 | x |

TABLE 2

List of Symbol and Structure Hypotheses for P(6, 6)

| Name | Symbol | Size of Sample Space | Cells per Trial |
|---|---|---|---|
| (Uniform Symbol in Region) | (Likelihood Region L) | | |
| Downward Diagonal | ℓ | 6 | 1 |
| Upward Diagonal | ʎ | 6 | ↓ |
| Row i | hi | 6 | ↓ |
| Column j | vj | 6 | ↓ |
| Four Ring | $\beta$ | 12 | ↓ |
| Six Ring | $\gamma$ | 20 | ↓ |
| 2x2 Subblock | n(=1, 2, . . . , 9) | 4 | ↓ |
| 3x3 Subblock | m(=a, b, c, d) | 9 | ↓ |
| Rectangle | hij | 12 | ↓ |
| | vij | 12 | ↓ |
| (Structure) | | | |
| 2x2 Subarray | N | 4 | 9 |
| 3x3 Subarray | M | 9 | 4 |
| Rectangular Subarray | Hij, kl | 12 | 3 |
| | Vij, kl | 12 | 3 |

# 3

| Number of Such Tests Applicable to Game | Probability (1 or S) Given Null Hypothesis | Probability (1 or S) Given Pattern or Symbol Hypothesis | Range of A | Range of B | Range of $\alpha$ | Range of $\beta$ |
|---|---|---|---|---|---|---|
| 4 | $\frac{1}{4}$ | $\frac{3}{4}$ | $[\frac{1}{9}, \frac{1}{3})$ | (243, 729] | $[\frac{1}{820}, \frac{1}{364})$ | $[\frac{1}{9}, \frac{1}{3})$ |
| 4 | $\frac{1}{4}$ | $\frac{255}{256}$ | $[\frac{1}{192^2}, \frac{1}{192})$ | (2048, 4096] | $[\frac{1}{4096}, \frac{1}{2048})$ | $[\frac{1}{192^2}, \frac{1}{192})$ |
| 4 | $\frac{1}{4}$ | 1 | $0 + \varepsilon$ | (2048, 4096] | $[\frac{1}{4096}, \frac{1}{2048})$ | 0 |
| 2 | $\frac{1}{16}$ | $\frac{15}{16}$ | $[\frac{1}{225}, \frac{1}{15})$ | (225, 3375] | $[\frac{1}{3375}, \frac{1}{225})$ | $[\frac{1}{225}, \frac{1}{15})$ |
| 2 | $\frac{1}{16}$ | 1 | $0 + \varepsilon$ | (256, 4096] | $[\frac{1}{4096}, \frac{1}{256})$ | 0 |
| 1 | $\frac{1}{69536}$ | 1 | $0 + \varepsilon$ | $2^{16}$ | $2^{-16}$ | 0 |
| 1 | $\frac{1}{64}$ | $\frac{63}{64}$ | $\frac{1}{63^2}, \frac{1}{63}$ | (3969, 250047] | $[\frac{1}{63^3}, \frac{1}{63^2})$ | $[\frac{1}{63^2}, \frac{1}{63})$ |
| 1 | $\frac{1}{64}$ | 1 | $0 + \varepsilon$ | (4096, 262144] | $[\frac{1}{64^3}, \frac{1}{64^2})$ | 0 |
| 14 | $\frac{1}{4}$ | $\frac{3}{4}$ | $\frac{1}{3}$ | (9, 27) | $[\frac{1}{40}, \frac{1}{13})$ | $[\frac{4}{13}, \frac{13}{40})$ |
| 14 | $\frac{1}{4}$ | 1 | $0 + \varepsilon$ | (16, 64) | $[\frac{1}{64}, \frac{1}{16})$ | 0 |

TABLE 4

Sequential Search Based on Symmetry Hypotheses

| Pattern Example Iteration | Principal Hypothesis | Query | Sample Point Truth Value | Number of Sample Points V | H | U | D | C | T | Number of True Samples V | H | U | D | C | T | No. of Queries | Symbols Found 0 | ∴ | + | * | Symmetry, Symbol or Structure Hypothesis Accepted |
|---|---|---|---|---|---|---|---|---|---|---|---|---|---|---|---|---|---|---|---|---|---|
| 1 | U | 17 | | | | | | | | | | | | | | 1 | 1 | | | | |
| | | 10 | 1 | | | 1 | | | | | | 1 | | | | 2 | 2 | | | | |
| 2 | U | 27 | | | | | 1 | 1 | | | | | 1 | 1 | | 3 | 3 | | | | |
| | | 20 | 1 | | | 2 | 2 | 2 | | | | 2 | 2 | 2 | | 4 | 4 | | | | $\beta$ |
| 3 | U | 28 | | 1 | 1 | | | | | 1 | 1 | 2 | 2 | 2 | | 5 | 5 | | | | |
| | | 14 | 1 | 2 | 2 | 3 | 2 | 2 | | 2 | 2 | 3 | 2 | 2 | | 6 | 6 | | | | |
| 4 | U | 3 | | | | | | | | | | | | | | 7 | 6 | 0 | 1 | 0 | |
| | | 24 | 1 | 2 | 2 | 4 | 2 | 2 | | 2 | 2 | 4 | 2 | 2 | | 8 | 6 | 0 | 2 | 0 | |
| 5 | U | 7 | | | | | | | | | | | | | | 9 | 6 | 0 | 3 | 0 | |
| | | 35 | 1 | 2 | 2 | 5 | 2 | 2 | | 2 | 2 | 5 | 2 | 2 | | 10 | 6 | 0 | 4 | 0 | |
| 6 | U | 8 | | | | | | | | | | | | | | 11 | 7 | 0 | 4 | 0 | |
| | | 29 | 1 | 2 | 2 | 6 | 2 | 3 | | 2 | 2 | 6 | 2 | 3 | | 12 | 8 | 0 | 4 | 0 | U |
| 7 | C | 30 | 1 | 2 | 2 | 6 | 3 | 4 | | 2 | 2 | 6 | 3 | 4 | | 13 | 8 | 0 | 5 | 0 | |
| 8 | C | 23 | 1 | 3 | 3 | 6 | 4 | 5 | | 3 | 3 | 6 | 4 | 5 | | 14 | 9 | 0 | 5 | 0 | |
| 9 | C | 34 | 1 | 3 | 3 | 6 | 5 | 6 | | 3 | 3 | 6 | 5 | 6 | | 15 | 9 | 0 | 6 | 0 | C |
| 10 | D | 5 | | 3 | 4 | 6 | 5 | 6 | | 3 | 4 | 6 | 5 | 6 | | 16 | 9 | 0 | 7 | 0 | |
| | | 25 | 1 | 4 | 5 | 6 | 6 | 6 | | 4 | 5 | 6 | 6 | 6 | | 17 | 9 | 0 | 8 | 0 | D |
| 11 | H | 4 | 1 | 5 | 6 | 6 | 6 | 6 | | 5 | 6 | 6 | 6 | 6 | | 18 | 9 | 0 | 9 | 0 | H, h1 |
| 12 | V | 15 | | | | | | | | | | | | | | 19 | 9 | 0 | 9 | 1 | |
| | | 16 | 1 | 6 | 6 | 6 | 6 | 6 | | 5 | 6 | 6 | 6 | 6 | | 20 | 9 | 1 | 9 | 1 | |
| 13 | h6 | 31 | 1 | | | | | | | | | | | | | 21 | 9 | 1 | 10 | 1 | h6 |

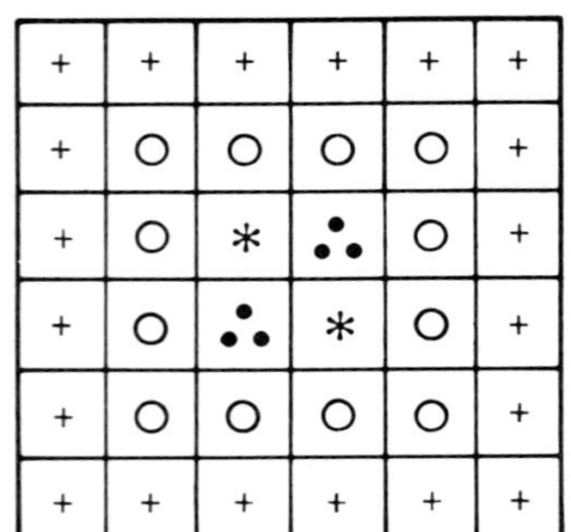

Table 5

Subset Description Table

| Type of Subset | Number of Subsets | Size of Subset | Weight of Subset |
|---|---|---|---|
| 2 x 2 | 10 | 4 | 0.02 |
| line | 14 | 6 | 0.03 |
| mid-lines | 1 | 8 | 0.04 |
| 3 x 3 | 4 | 9 | 0.045 |
| 4 ring | 1 | 12 | 0.06 |
| 6 ring | 1 | 20 | 0.1 |

Table 6

Complexity of the Examples Under $K_2$

| Example Number | 1 | 2 | 3 | 4 | 5 |
|---|---|---|---|---|---|
| Complexity $K_2$ | 1.00 | 2.74 | 3.19 | 3.66 | 2.14 |

Table 7

Pattern Domains Which Subdivide Evenly by P(2,2) and P(3,3)

| Side Length | Total Number of Points | Number of Major Subblocks | Number of Points in Major Subblocks | Subblock Side Length |
|---|---|---|---|---|
| 6 | 36 | 9 | 4 | 2 |
| 6 | 36 | 4 | 9 | 3 |
| 12 | 144 | 4 | 36 | 6 |
| 18 | 324 | 9 | 36 | 6 |
| 36 | 1,296 | 9 | 144 | 12 |
| 36 | 1,296 | 4 | 324 | 18 |
| 72 | 5,184 | 4 | 1,296 | 36 |
| 108 | 11,664 | 9 | 1,296 | 36 |
| 216 | 46,656 | 4 | 11,664 | 108 |
| 216 | 46,656 | 9 | 5,184 | 72 |
| 432 | 186,624 | 4 | 46,656 | 216 |
| 648 | 419,904 | 9 | 46,656 | 216 |

## Table 8

Search of P(12, 12) From Figure 10

No zone searched unless more than 40% of the total intensity sought found therein

Zone search terminated when 80% of the total intensity sought is located by cell

| Quadrant Search | | | | Four/Nine/Four Search | | | |
|---|---|---|---|---|---|---|---|
| Total Number of Queries | Zone Searched | Total Intensity Located by Cell | Area Intensity | Total Number of Queries | Zone Searched | Total Intensity Located by Cell | Area Intensity |
| 1 | a | | 4.9 | 1 | a | | 4.9 |
| 2 | a.a | | 0.3 | 2 | a.1 | | 0 |
| 3 | a.b | | 0.3 | 6 | a.5 | | 1.3 |
| 4 | a.c | | 1.3 | 9 | a.8 | | 1.0 |
| 5 | a.d.1 | | | 10 | a.9.a | 0.3 | |
| 8 | a.d.4 | 0.7 | | 12 | a.9.c(→d) | 2.3 | |
| 9 | a.d.5 | 1.0 | | 13 | b | 2.3 | 0.7 |
| 10 | a.d.6 | 1.3 | | 14 | c | 2.3 | 1.0 |
| 11 | a.d.7 | 1.3 | | 15 | d.1 | 2.3 | 3.7 |
| 12 | a.d.8 | 2.0 | | 16 | d.1.a | 3.3 | |
| 13 | a.d.9 | 3.0 | | 17 | d.1.b | 4.3 | |
| 14 | a.c.3 | 3.7 | | 18 | d.1.c(→d) | 6.0 | |
| 15 | a.c.6 | 4.0 | (>80% a) | 19 | d.2 | 6.0 | 1.3 |
| 16 | b | 4.0 | 0.7 | 20 | d.2.a | 6.3 | |
| 17 | c | 4.0 | 1.0 | 21 | d.2.b | 6.3 | |
| 18 | d.a | 4.0 | 5.6(≡ d) | 22 | d.2.c | 7.3 | (>80% d) |
| 19 | d.a.1 | 5.0 | | 23 | a.5.a | 7.6 | |
| 20 | d.a.2 | 6.0 | | 24 | a.5.b | 7.9 | |
| 21 | d.a.3 | 6.3 | | 25 | a.5.c | 8.6 | |
| 22 | d.a.4 | 7.0 | | 26 | a.8.a | 8.9 | |
| 23 | d.a.5 | 8.0 | | 27 | a.8.b | 9.6 | |
| 24 | d.a.6 | 9.0 | | 28 | c.3.a | 9.9 | (>80% P(12, 12)) |
| 25 | d.a.7 | 9.0 | | | | | |
| 26 | d.a.8(→9) | 9.6 | | | | | |
| 27 | c.b.3 | 10.3 | (>80% P(12, 12)) | | | | |

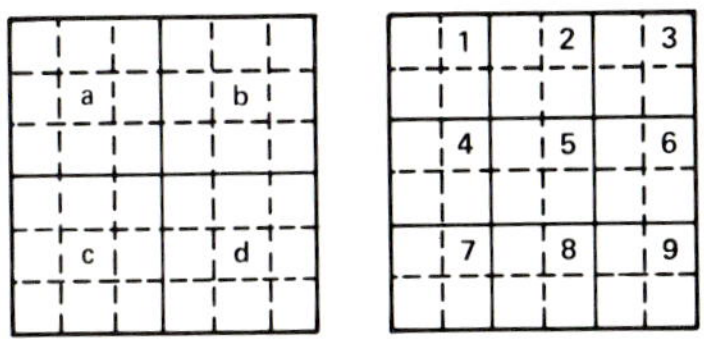


1.a

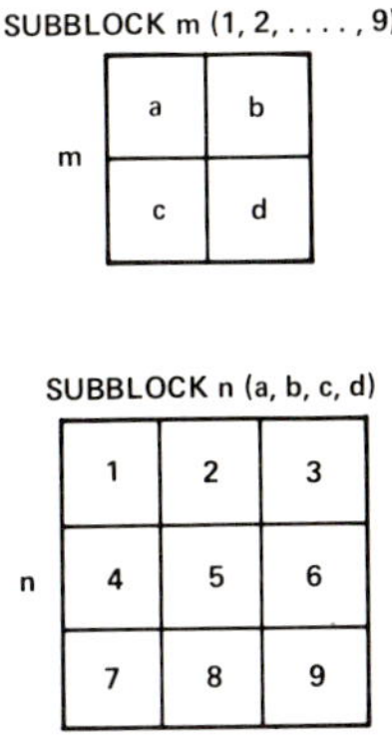


CELL LOCATION BY LETTER-NUMBER PAIR DENOTES "SUBBLOCK" FOLLOWED BY "CELL OF SUBBLOCK": THE 3RD ROW 4TH COLUMN CELL CAN BE WRITTEN b7 OR 5b.

1.b

MATRIX NOTATION AND ONE LINEAR LIST NOTATION FOR P(6, 6)

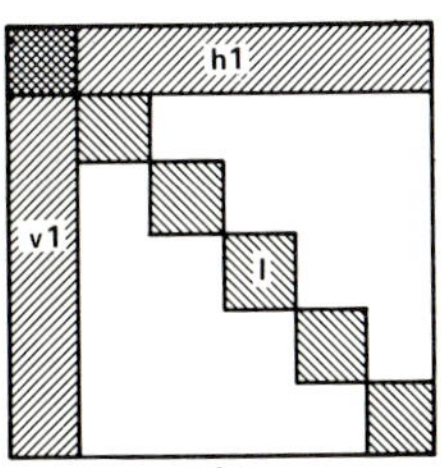


1.c

COLUMN j

| ROW i | 1 | 2 | 3 | 4 | 5 | 6 |
|---|---|---|---|---|---|---|
| 1 | 1 | 2 | 3 | 4 | 5 | 6 |
| 2 | 7 | 8 | 9 | 10 | 11 | 12 |
| 3 | 13 | 14 | 15 | 16 | 17 | 18 |
| 4 | 19 | 20 | 21 | 22 | 23 | 24 |
| 5 | 25 | 26 | 27 | 28 | 29 | 30 |
| 6 | 31 | 32 | 33 | 34 | 35 | 36 |

1.d

Figure 1. Sublock Decompositions of "Patterns"

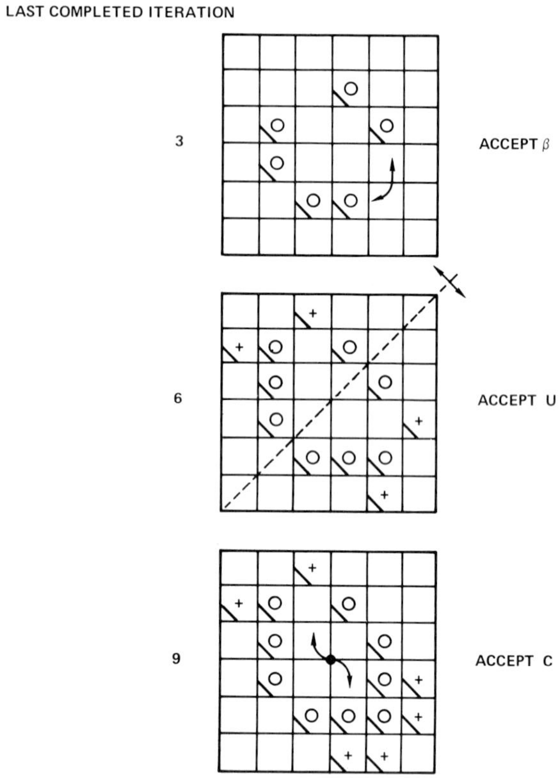


Figure 2. Example 5 Queries By Search Iteration

LAST COMPLETED ITERATION

9

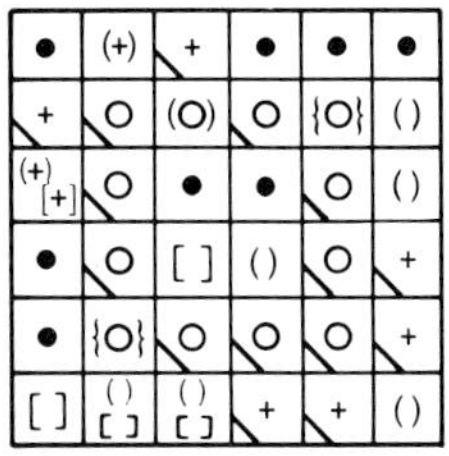

SYMBOL INDUCED BY

( ) U

[ ] C

{ } $\beta$

EIGHT FURTHER QUERIES NEEDED (NO ADDED HYPOTHESES) – QUERIES SHOWN BY DOT; INDUCED SYMBOLS ONLY FROM QUERIES.

12

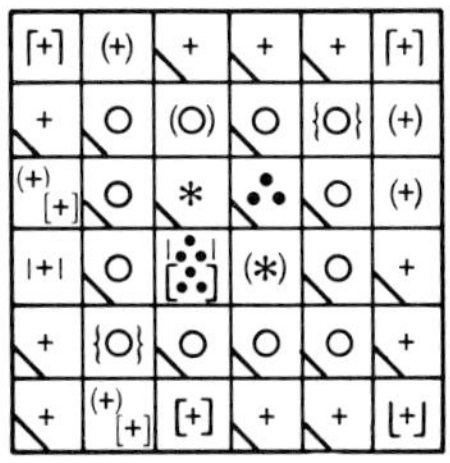

SYMBOL INDUCED BY

| | D

⌈ ⌉ h1

⌊ ⌋ h6

FINAL HYPOTHESIS
[RESULT M (6, 6)] : 21 QUERIES*, NO ERRORS.

*USE OF ADDED HYPOTHESIS D, H, h1, & h6 REDUCED ADDED QUERIES FROM EIGHT (AFTER ITERATION 9) TO SIX.

Figure 3. Example 5 Hypotheses By Search Iteration

EXAMPLE (1) (2) (3)

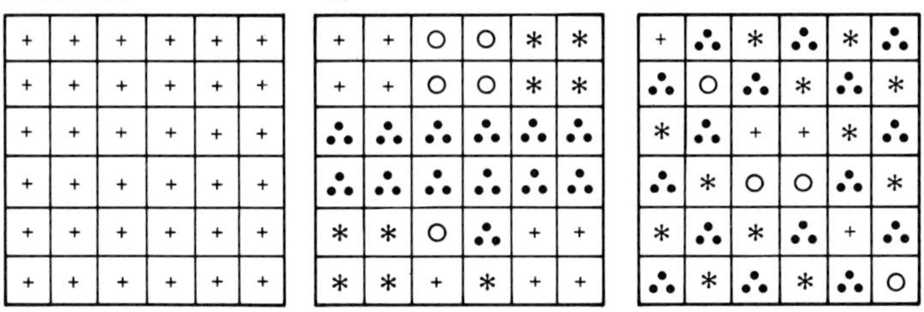

$K_1 = 1$ $K_1 = \frac{1}{13}(8+4+9+4) = \frac{25}{13}$ $K_1 = \frac{1}{13}(4.3 + 7.2 + 2.4) = \frac{34}{13}$

Figure 4. Three Patterns and Complexity $K_1$

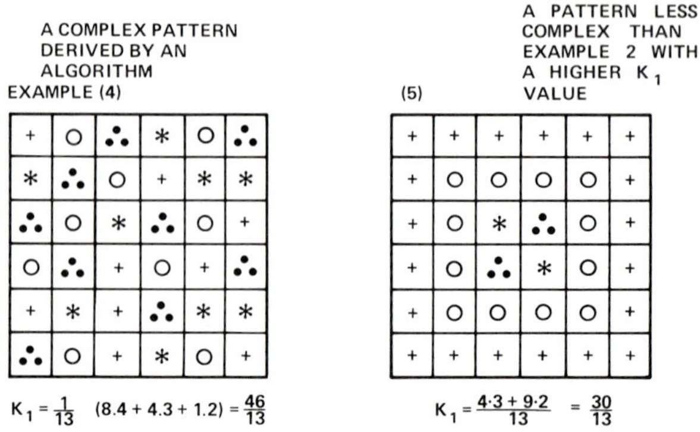


Figure 5. Ring Patterns and Complexity $K_1$

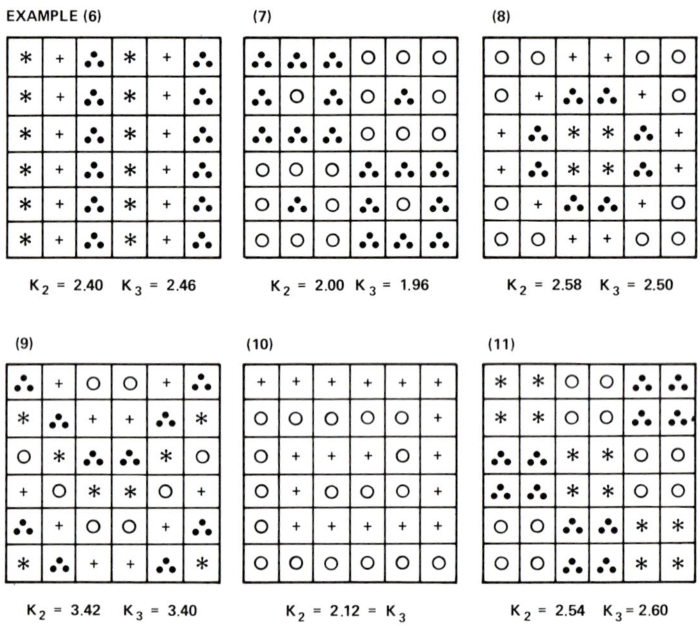


Figure 6. Six Patterns with Their $K_2$ and $K_3$ Values

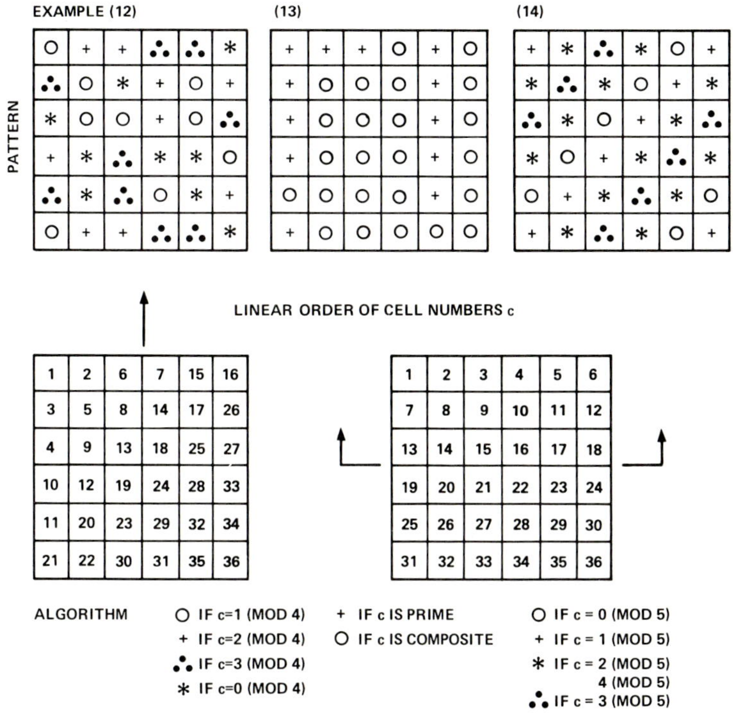


Figure 7. Algorithmic Patterns

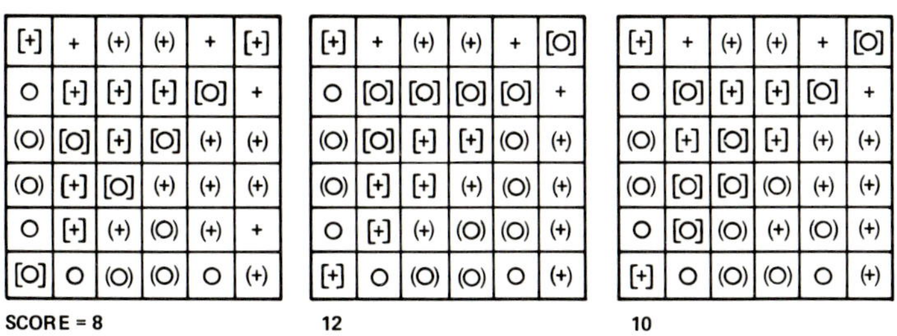


**Figure 8.** Eight-Query Simulated Hypotheses for Example 10
Key: [ ] Randomly Generated; ( ) Induced By U or 4-Cell Chain

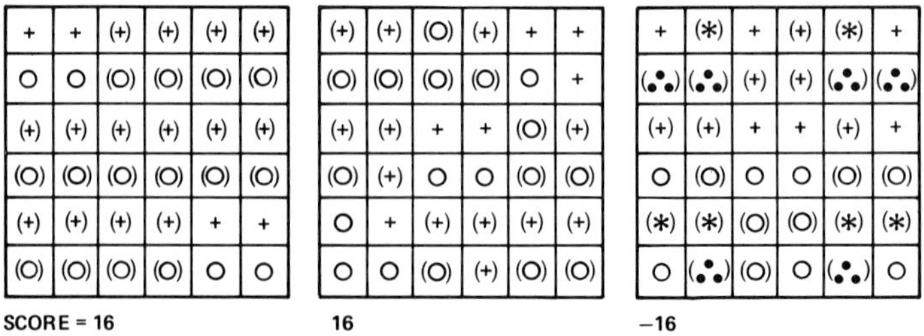


Figure 9. Two Symbol and Four Symbol Hypotheses for Example 10.

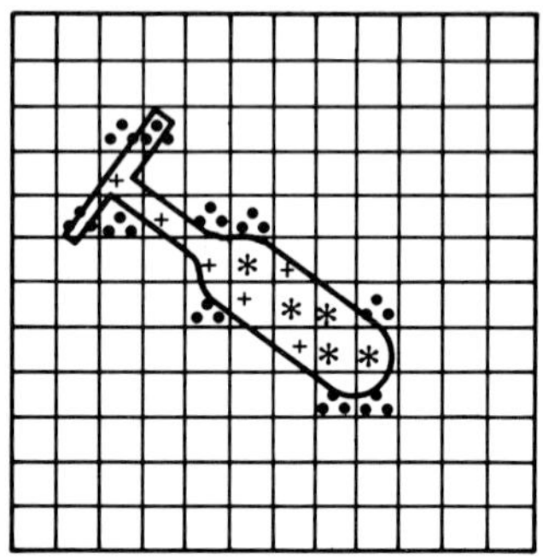

KEY: GRID SIZE 12 X 12 = 144 CELLS

| GREY LEVEL† | SYMBOL |
|---|---|
| 1.0 | ✱ |
| 0.7 | + |
| 0.3 | ∴ |
| 0.0 | blank (0 IN "PATTERNS") |

† SIGNAL INTENSITY

TOTAL OF 12.2 UNITS OVER THE ENTIRE ARRAY

| 6 x 6 SUBBLOCK | NUMBER OF UNITS |
|---|---|
| a | 4.9 |
| b | 0.7 |
| c | 1.0 |
| d | 5.6 |

Figure 10. Manmade Object and Corresponding Pattern

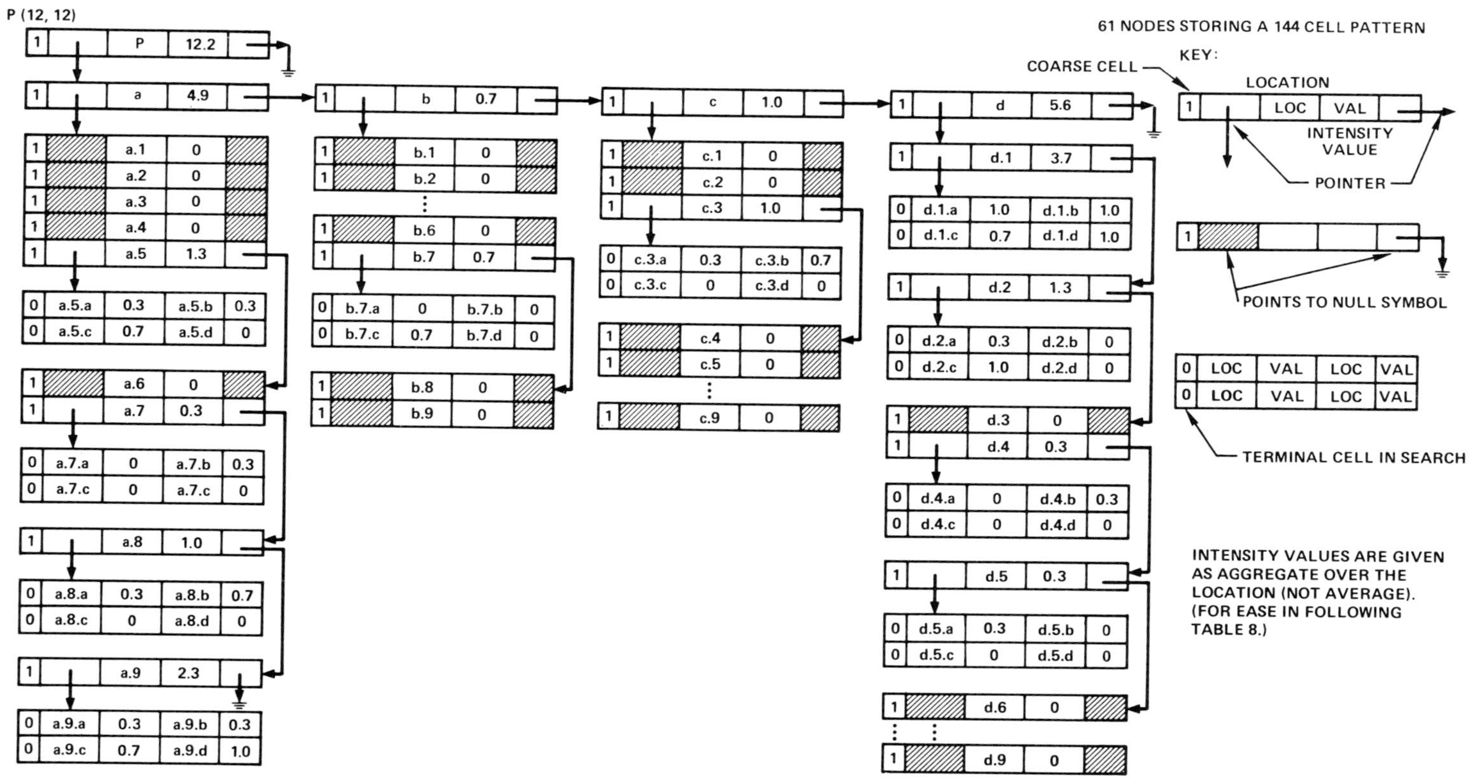
P (12, 12)
61 NODES STORING A 144 CELL PATTERN
KEY:
COARSE CELL
LOCATION
LOC
VAL
INTENSITY VALUE
POINTER
POINTS TO NULL SYMBOL
TERMINAL CELL IN SEARCH
INTENSITY VALUES ARE GIVEN AS AGGREGATE OVER THE LOCATION (NOT AVERAGE). (FOR EASE IN FOLLOWING TABLE 8.)

Figure 11

# NECESSARY CONDITIONS FOR A LOCAL OPTIMUM WITHOUT PRIOR CONSTRAINT QUALIFICATION

G. V. Reklaitis and D. J. Wilde

School of Chemical Engineering
Purdue University
Department of Chemical Engineering
Stanford University

Summary

Constructive necessary conditions for constrained optima which are independent, not only of rank restrictions on the Jacobian of the binding constraints, but also of the Qualified Kuhn-Tucker Theorem are derived. The linear necessary conditions involve determining only $L+1$ multipliers, where L is the difference between the row dimension and the rank of the Jacobian matrix. In addition, necessary conditions second order only in the constraints are presented to complement the linear conditions. The discriminating quantities for both results, constrained first and second derivatives, are directly computable by elementary matrix manipulations. Examples illustrating the tests are included.

## 1 Introduction

In previous publications [21,22,23], a differential form of the Kuhn-Tucker necessary conditions [14] for a constrained local minimum of a general nonconvex nonlinear programming problem was derived together with explicit closed form expressions for the essential constrained first derivatives in terms of which the conditions were formulated. However, the criteria required nonsingularity of a Jacobian matrix of first partial derivatives of the binding constraints. This paper extends the differential approach by providing necessary conditions which require neither rank restrictions on the Jacobian nor any other such constraint qualification.

From a practical point of view, the differential theory is of interest in providing a common framework for a class of mathematical programming algorithms which are based upon Simplex-like operations [23]. Algorithms of this type generally require nondegeneracy-nonsingularity assumptions (see the GRG Algorithm [2] or the Convex Simplex Method [24], for example). Application of the extensions to the differential theory presented here permits these assumptions to be dropped [18].

As pointed out in [1,6], the qualified Kuhn-Tucker theorem is a special case of the fundamental theorem due to Fritz John [9]. John's necessary conditions, originally formulated for the inequality constrained optimization problem, have recently been extended to the mixed equality-inequality constrained problem by Mangasarian and Fromovitz [15]. In either case, the John's type conditions are distinguished from the Kuhn-Tucker result in requiring the introduction of a multiplier for the terms corresponding to the objective or criterion function of the program. In principle but not in detail, Kuhn and Tucker showed that if their constraint qualification holds, then this multiplier can be taken as positive; in particular, it can be normalized to unity. Since this pioneering work, research concerned with optimality tests has taken two directions: first, to weaken the constraint qualifications, that is, to obtain weaker sufficient conditions under which the "Fritz John" multiplier would be positive (notably [1,6]); and, second, to obtain sufficient conditions which, although stronger, would be convenient to apply. The research effort in the former direction has recently been capped by the qualification of Gould and Tolle [7] which is the weakest in the sense of being both necessary and sufficient for the positivity of the Fritz John multiplier. Efforts in the practical direction of obtaining usable qualifications however, is still ongoing. Numerous more restrictive and specialized hypotheses [1,3,6,11,12,13,20] have been advanced because direct verification of the weaker but more abstract constraint qualifications (including that of Gould and Tolle) is difficult - even in situations involving constraint sets in only three dimensional euclidean space.

This paper proposes to obviate the need for formal

qualifications by reintroducing the objective function multiplier. However, the analysis and results presented here differ significantly from the previous work in several respects. First, the constraints satisfied as equalities at a given point are partitioned according to whether the gradient of a given constraint at the point being tested does or does not belong to a selected maximal set of linearly independent constraint gradients at the point. (These are respectively called basic and nonbasic constraints.) This is an interesting contrast to classifying constraints in accordance with linearity-nonlinearity [1] for stating a "regularity condition", or convex-nonconvexity [4] for deriving still another sufficient condition for the constraint qualification. Partitioned this way and further augmented by the introduction of slack variables into the constraints, the program is locally reduced by applying the Implicit Function Theorem to the basic constraints to eliminate some of the variables. This gives the objective function and the nonbasic constraints expressed as explicit functions of the non-negative slack variables and the remaining independent variables. The transformation, described in Section 3, is easily effected since it involves only differentiation and elementary matrix manipulation. The Linear Necessary Conditions, derived (Section 4) in terms of this transformed system, require existence of semi-positive multipliers only for the nonbasic constraints and the objective function. Theoretically, this again is an interesting contrast to requiring semipositive multipliers for all constraints and the objective function [9] or to requiring semipositive multipliers for all nonlinear constraints and the objective function, [1]. Computationally, since in practical situations the rank of the matrix of constraint gradients is frequently close to the number of tight constraints, this means that commonly only three or fewer multipliers will need to be determined. Finally, it is interesting that because of the particular structure of the transformed problem, the positivity or vanishing of the objective function multiplier can be determined by inspection, since a certain easily computable vector of constrained derivatives must satisfy a "complimentary slackness" relation with the multiplier. If this constrained derivative vector is not zero, the differential formulation further permits construction of sharper semi-quadratic necessary conditions, second order in the constraints but linear in the objective function. These conditions, constructed in Section 5 using the pseudo inverse of a linear transformation [8,17], are constructive in that failure of the criteria will generate a direction for improvement.

## 2 Statement of the Problem and Notation

The problem under consideration is to identify a local miminum of a continuously differentiable function y of the N-vector argument $\underset{\sim}{x}$ subject to K inequality constraints in the form of a continuously differentiable mapping $\underset{\sim}{f}$ from $R^N$ to $R^K$; that is,

$$\text{Minimize:} \quad y(\underset{\sim}{x})$$

$$\text{Subject to:} \quad \underset{\sim}{f}(\underset{\sim}{x}) = (f_1, f_2, \ldots, f_k)^T (\underset{\sim}{x}) \geqq \underset{\sim}{0} \qquad \text{(A)}$$

Although we require that y and f be continuously differentiable on $\underset{\sim}{X}$, the closed feasible reqion defined by the common domain of the constraint functions, strictly speaking, the analysis requires this only over some open neighborhood about a particular point $\underset{\sim}{x}^o$ under consideration. To obtain semi-quadratic necessary conditions, we shall in addition require that the constraint functions be twice continuously differentiable. It should be emphasized that although we have stated the above nonlinear program as an inequality constrained problem, we have done so mainly to reduce notational complexity. The formulation using slack variables to be introduced in the next section readily permits the inclusion of equality constraints. The required modification of the main results will be summarized as needed.

Throughout the paper our notation is chosen to conform to that introduced in [23]. In particular, the vector $\underset{\sim}{x}$ and any of its partitions are column vectors; the derivative of a real valued function $g(\underset{\sim}{x})$ at a point $\underset{\sim}{x}^o$ is a row vector denoted by

$$\left( \frac{\partial g}{\partial \underset{\sim}{x}} \right)^o \quad ;$$

the Jacobian matrix of partial derivatives of a vector valued function $\underset{\sim}{g}$ of several vector arguments $\underset{\sim}{x}$, $\underset{\sim}{y}$, $\underset{\sim}{z}$ at a point $(x^o, y^o, z^o)$ is denoted by $J_g(x^o, y^o, z^o)$; $I_M$ denotes the identity matrix of order M; $|a|$, the absolute value of the scalar a; and $||x||$, the conventional Euclidean norm. In addition, C' (C") denotes the class of functions once (twice) continuously differentiable over a domain compatible with context. Finally, in considering vector inequalities, it is understood that a vector $x = (x_1, \ldots, x_i, \ldots, x_n)$ is

nonnegative $(x \geqq 0)$ if each $x_i \geq 0$

semipositive $(x \geq 0)$ if $x \geqq 0$ and not all $x_i = 0$

positive $(x > 0)$ if each $x_i > 0$

## 3 Preliminary Considerations

To facilitate handling the inequality constraints of (A), we introduce non-negative slack variables $f_k$ for each constraint function $f_k(x)$, $k = 1, \ldots, K$ and $f = (f_1, \ldots, f_k)$. With this the program becomes to find $(x, f)$ such as to,

$$\text{minimize} \quad y(x)$$

$$\text{subject to} \quad h_k(x, f) \equiv f_k(x) - f_k = 0, \; k = 1, \ldots, K \qquad f \geqq 0$$

In the subsequent development, let $x^o$ be a point at which exactly T constraints are tight, i.e., $f_k(x) = 0$. Clearly, the corresponding slacks $f_k = 0$. For convenience, let these be indexed $k = 1, \ldots, T$ and observe that $f_k(x) > 0$, $k = T+1, \ldots, K$ implies there is a sufficiently small neighborhood of $x^o$ within which the constraints loose at $x^o$ remain loose. Hence, for the differential, i.e. local, analysis of this paper these constraints may be disregarded. With this proviso, consider the first order Taylor expansion of h, the vector of tight constraints for $(x, f)$ within some proper neighborhood of the point $(x^o, f^o)$,

$$0 = \partial h = h(x, f) - h(x^o, f^o) = J_h \, \partial x - \partial f + o[||\partial||^2]$$

$$\text{where}, \quad \partial x = x - x^o$$

or equivalently,

$$\partial \underset{\sim}{f} = \underset{\sim}{J}_h \, \partial \underset{\sim}{x} + O[\|\partial \underset{\sim}{x}\|^2] \tag{1}$$

where $\underset{\sim}{J}_h$ is the Jacobian matrix of $\underset{\sim}{h}$ with respect to $\underset{\sim}{x}$ -- the T x N matrix whose elements are given by,

$$\underset{\sim}{J}_h = \underset{\sim}{J}_h(\underset{\sim}{x}^o) = \begin{pmatrix} \frac{\partial f_1}{\partial x_1} & \frac{\partial f_1}{\partial x_2} & \cdots & \frac{\partial f_1}{\partial x_N} \\ \frac{\partial f_2}{\partial x_1} & \frac{\partial f_2}{\partial x_2} & \cdots & \frac{\partial f_2}{\partial x_N} \\ \vdots & \vdots & & \vdots \\ \frac{\partial f_T}{\partial x_1} & \frac{\partial f_T}{\partial x_2} & \cdots & \frac{\partial f_T}{\partial x_N} \end{pmatrix}$$

with partial derivatives evaluated at $\underset{\sim}{x}^o$.

By appropriate rearrangement of rows and columns, partition the Jacobian matrix so that,

$$\underset{\sim}{J}_h(\underset{\sim}{x}^o) = \begin{pmatrix} \bar{\underset{\sim}{J}} & \bar{\underset{\sim}{C}} \\ \hat{\underset{\sim}{J}} & \hat{\underset{\sim}{C}} \end{pmatrix}$$

where $\bar{\underset{\sim}{J}}$ is a nonsingular submatrix of $\underset{\sim}{J}_h(\underset{\sim}{x}^o)$ -- not necessarily unique -- of maximal rank M, with $1 \leq M = \text{rank}(\underset{\sim}{J}_h) \leq \text{MIN}(T,N)$. Corresponding to this choice define the following labelling:

<u>STATE</u> variables ($s_m$; m = 1, ... ,M): all $x_i$ such that the column corresponding to $x_i$ appears in the submatrix

$$\begin{pmatrix} \bar{\underset{\sim}{J}} \\ \hat{\underset{\sim}{J}} \end{pmatrix}$$

<u>DECISION</u> variables ($d_r$; r = 1, ... ,R(=N-M)): the remaining N-M independent variables $x_i$.

<u>BASIC</u> constraints ($\bar{f}_k(\underset{\sim}{x})$; k = 1, ... ,M): all tight

constraints such that the row corresponding to $\bar{f}_k$ appears in the submatrix $(\bar{\underset{\sim}{J}},\bar{\underset{\sim}{C}})$.

NON-BASIC constraints ($\hat{f}_\ell(x)$; $\ell = 1, \ldots, L(= T-M)$): the remaining T-M tight constraints.

NOTE: Since with each constraint function is associated a distinct slack variable, the above labelling of the tight constraints can be unambiguously assigned to their slacks as well.

In terms of this labelling, all $(\underset{\sim}{x},\underset{\sim}{f}) = (\underset{\sim}{s},\underset{\sim}{d},\bar{\underset{\sim}{f}},\hat{\underset{\sim}{f}})$ satisfying the constraints of (A) within a neighborhood of $(\underset{\sim}{x}^o,\underset{\sim}{f}^o)$ must satisfy the inequality system.

$$\partial\bar{\underset{\sim}{f}} = \bar{\underset{\sim}{J}}\,\partial\underset{\sim}{s} + \bar{\underset{\sim}{C}}\,\partial\underset{\sim}{d} + O[\|(\partial\underset{\sim}{s},\partial\underset{\sim}{d})\|^2] \geqq 0 \tag{2a}$$

$$\partial\hat{\underset{\sim}{f}} = \hat{\underset{\sim}{J}}\,\partial\underset{\sim}{s} + \hat{\underset{\sim}{C}}\,\partial\underset{\sim}{d} + O[\|(\partial\underset{\sim}{s},\partial\underset{\sim}{d})\|^2] \geqq 0 \tag{2b}$$

If we perform a similar linear expansion for the objective function $y(\underset{\sim}{x})$ about $(\underset{\sim}{x}^o,\underset{\sim}{f}^o)$,

$$\partial y \equiv y(\underset{\sim}{x}) - y(\underset{\sim}{x}^o) = \left(\frac{\partial y}{\partial \underset{\sim}{s}}\right)^o \partial\underset{\sim}{s} + \left(\frac{\partial y}{\partial \underset{\sim}{d}}\right)^o \partial\underset{\sim}{d} + O[\|(\partial\underset{\sim}{s},\partial\underset{\sim}{d})\|^2] \tag{3}$$

then it is possible to state the conditions that this differential system must satisfy in order that $\underset{\sim}{x}^o$ be a local minimum for program (A).

Theorem 1: (Linear Necessary Conditions)

If $\underset{\sim}{x}^*$ is a local minimum of (A), then there exists a rearrangement and partitioning of the Jacobian and a

$$\underset{\sim}{\lambda} = \begin{pmatrix} \underset{\sim}{\lambda} \\ \lambda_o \end{pmatrix} \geq \underset{\sim}{0}$$

such that,

$$\text{(i)} \quad \left[ \left(\frac{\partial y}{\partial \underset{\sim}{d}}\right)^* - \left(\frac{\partial y}{\partial \underset{\sim}{s}}\right)^* \bar{\underset{\sim}{J}}^{-1} \bar{\underset{\sim}{C}} \right]^T \lambda_o = \underset{\sim}{0}$$

$$\text{(ii)} \quad \left[ \left( \frac{\partial y}{\partial \underset{\sim}{s}} \right)^* \underset{\sim}{\bar{J}}^{-1} \right]^T \lambda_o - \left[ \underset{\sim}{\hat{J}} \, \underset{\sim}{\bar{J}}^{-1} \right]^T \underset{\sim}{\hat{\lambda}} \geqq \underset{\sim}{0}$$

Moreover, if the constraint functions are twice continuously differentiable, then the expansions for the basic and nonbasic constraints can be carried out to second order and corresponding semi-quadratic necessary conditions can be stated. However, the discussion of these conditions as well as the proof of the above result will be deferred to later sections. Instead, we first consider several preliminary constructions and lemmas.

Lemma - 1 The system of equations (2a) and (3) may be transformed into the system

$$\partial y \equiv \left[ \left( \frac{\partial y}{\partial \underset{\sim}{d}} \right)^* - \left( \frac{\partial y}{\partial \underset{\sim}{s}} \right)^* \underset{\sim}{\bar{J}}^{-1} \underset{\sim}{\bar{C}} \right] \partial \underset{\sim}{d} + \left[ \left( \frac{\partial y}{\partial \underset{\sim}{s}} \right)^* \underset{\sim}{\bar{J}}^{-1} \right] \partial \underset{\sim}{\bar{f}} + O[\|(\partial \underset{\sim}{d}, \partial \underset{\sim}{\bar{f}})\|^2] \tag{4a}$$

$\partial \underset{\sim}{\bar{f}} \geqq 0$. Moreover,

$$\partial \underline{s} = -\underset{\sim}{\bar{J}}^{-1} \bar{C} \, \partial \underline{d} + \underset{\sim}{\bar{J}}^{-1} \partial \underset{\sim}{\bar{f}} + O[\|(\partial \underline{d}, \partial \underset{\sim}{\bar{f}})\|^2] \tag{4b}$$

The proof follows using the Implicit Function Theorem, the nonsingularity of $\underset{\sim}{\bar{J}}$, and the continuous differentiability of the program functions. The details may be found in [21].

Lemma - 2 The system of equations (2a) and (2b) can be transformed into the system,

$$\partial \underset{\sim}{\hat{f}} = (\underset{\sim}{\hat{C}} - \underset{\sim}{\hat{J}} \, \underset{\sim}{\bar{J}}^{-1} \underset{\sim}{\bar{C}}) \, \partial \underset{\sim}{d} + (\underset{\sim}{\hat{J}} \, \underset{\sim}{\bar{J}}^{-1}) \, \partial \underset{\sim}{\bar{f}} + O[\|(\partial \underset{\sim}{d}, \partial \underset{\sim}{\bar{f}})\|^2] \geqq 0$$

$$\partial \underset{\sim}{\bar{f}} \geqq 0$$

The construction is analogous to that of Lemma 1 with the vector valued function $\underset{\sim}{\hat{h}}(\underset{\sim}{s}, \underset{\sim}{d}, \underset{\sim}{\hat{f}}) \equiv \underset{\sim}{\hat{f}}(\underset{\sim}{s}, \underset{\sim}{d}) - \underset{\sim}{\hat{f}}$ replacing y.

The coefficients occurring in the previous lemmas

suggest the following abbreviations:

$$\left(\frac{\delta \hat{\underset{\sim}{f}}}{\delta \bar{\underset{\sim}{f}}}\right)^{o} \equiv \hat{\underset{\sim}{J}}\, \bar{\underset{\sim}{J}}^{-1} \; ; \quad \left(\frac{\delta y}{\delta \bar{\underset{\sim}{f}}}\right)^{o} \equiv \left(\frac{\partial y}{\partial \underset{\sim}{s}}\right)^{o} \bar{\underset{\sim}{J}}^{-1} \; :$$

respectively, the matrix (vector) of <u>constrained</u> <u>derivatives</u> of $\hat{f}(\underset{\sim}{x})$ $(y(\underset{\sim}{x}))$ with respect to the <u>basic</u> slacks, evaluated at $\underset{\sim}{x}^{o}$.

$$\left(\frac{\delta \hat{\underset{\sim}{f}}}{\delta \underset{\sim}{d}}\right)^{o} \equiv \hat{\underset{\sim}{C}} - \hat{\underset{\sim}{J}}\, \bar{\underset{\sim}{J}}^{-1} \bar{\underset{\sim}{C}} \; ; \quad \left(\frac{\delta y}{\delta \underset{\sim}{d}}\right)^{o} \equiv \left(\frac{\partial y}{\partial \underset{\sim}{d}}\right) - \left(\frac{\partial y}{\partial \underset{\sim}{s}}\right)^{o} \bar{\underset{\sim}{J}}^{-1} \bar{\underset{\sim}{C}} :$$

the matrix (vector) of <u>constrained</u> <u>derivatives</u> of $\hat{f}(\underset{\sim}{x})$ $(y(\underset{\sim}{x}))$ with respect to the <u>decision</u> variables, evaluated at $\underset{\sim}{x}^{o}$.

The result of Lemma 2 may be simplified by means of the following fact,

<u>Lemma -3</u> Let $\underset{\sim}{P}$ be any T x N matrix of rank M < min(T,N) and consider the partition

$$\underset{\sim}{P} = \begin{pmatrix} \underset{\sim}{H} & \underset{\sim}{J} \\ \underset{\sim}{K} & \underset{\sim}{L} \end{pmatrix} ,$$

where $\underset{\sim}{H}$ is a non-singular submatrix of maximal rank M. Then, $\underset{\sim}{L} = \underset{\sim}{K}\, \underset{\sim}{H}^{-1} \underset{\sim}{J}$.

<u>Proof</u>: By construction, the columns of

$$\begin{pmatrix} \underset{\sim}{H} \\ \underset{\sim}{K} \end{pmatrix}$$

are linearly independent and hence the columns of

$$\begin{pmatrix} \underset{\sim}{J} \\ \underset{\sim}{L} \end{pmatrix}$$

can be expressed as non-trivial linear combinations of the columns of

$$\begin{pmatrix} \tilde{H} \\ \tilde{K} \end{pmatrix} .$$

Thus, there exists a matrix $\tilde{B}$ such that

$$\begin{pmatrix} \tilde{J} \\ \tilde{L} \end{pmatrix} = \begin{pmatrix} \tilde{H} \\ \tilde{K} \end{pmatrix} \tilde{B} .$$

Equivalently, $\tilde{J} = \tilde{H}\tilde{B}$ and $\tilde{L} = \tilde{K}\tilde{B}$. Thus since $\tilde{H}$ is non-singular, $\tilde{H}^{-1}$ exists. Hence, $\tilde{B} = \tilde{H}^{-1}\tilde{J}$ and $\tilde{L} = \tilde{K}\tilde{B} = \tilde{K}\tilde{H}^{-1}\tilde{J}$.

Lemma -4

$$\left( \frac{\delta \hat{\tilde{f}}}{\delta \bar{\tilde{d}}} \right)^{o} \equiv \tilde{0}$$

Proof: Since

$$\tilde{J}_h = \begin{pmatrix} \bar{\tilde{J}} & \bar{\tilde{C}} \\ \hat{\tilde{J}} & \hat{\tilde{C}} \end{pmatrix} ,$$

by Lemma 3, $\hat{\tilde{C}} = \hat{\tilde{J}}\,\bar{\tilde{J}}^{-1}\,\bar{\tilde{C}}$. But by definition

$$\left( \frac{\delta \hat{\tilde{f}}}{\delta \bar{\tilde{d}}} \right)^{o} = \hat{\tilde{C}} - \hat{\tilde{J}}\,\bar{\tilde{J}}^{-1}\,\bar{\tilde{C}} = \tilde{0} .$$

Let P be as in Lemma -3 and consider the partition $\tilde{P} = (\tilde{A},\tilde{B})$ where rank (A) = M and the columns of A are linearly independent. Let $\alpha_j(i)$ be a row, $1 \leq j \leq T$ of $\tilde{P}$ such that

$$\alpha_j(i) = (\delta_{1i}, \delta_{2i}, \ldots, \delta_{Ni}) \text{ where } \delta_{ik} = \begin{cases} 0 & \text{if } i \neq k \\ 1 & \text{if } i \neq k \end{cases}$$

Lemma -5 The i-th column of $\tilde{P}$ must be contained in $\tilde{A}$.

The proof follows by assuming the contrary and obtaining the contradiction that there exists a vector $\tilde{\beta}$ such that $1 = \tilde{0} \cdot \tilde{\beta}$ .

The lemma is useful for understanding how the special

structure adopted in the paper permits us to bypass the issue if degeneracy [21]. As a consequence of the lemma, if $h_k(\underset{\sim}{x},\underset{\sim}{f}) = x_i - f_k$ for some k and i (in general, if $h_k(\underset{\sim}{x},\underset{\sim}{f}) = a\underset{\sim}{x}_i + b - f_k$) then $x_i$ must be a state variable and therefore degenerate, i.e., $x_i^o = 0$. However, whether $f_k$ is basic or non-basic, the restriction $f_k \geq 0$ always insures $x_i$ is feasible, so that no special consideration need be given to the degenerate state. Finally, since only state variables may be degenerate, the decision variables are always free to undergo both positive and negative perturbations, subject only to explicit constraints.

## 4 First Order Necessary Conditions

Having the above constructions, we can now state and prove the conditions that the differential system,

$$\partial y = \left(\frac{\delta y}{\delta \underset{\sim}{d}}\right)^o \partial \underset{\sim}{d} + \left(\frac{\delta y}{\delta \underset{\sim}{\bar{f}}}\right)^o \partial \underset{\sim}{\bar{f}} + O[\|(\partial \underset{\sim}{d}, \partial \underset{\sim}{\bar{f}})\|^2]$$

$$\partial \underset{\sim}{\hat{f}} = \left(\frac{\delta \underset{\sim}{\hat{f}}}{\delta \underset{\sim}{\bar{f}}}\right)^o \partial \underset{\sim}{\bar{f}} + O[\|(\partial \underset{\sim}{d}, \partial \underset{\sim}{\bar{f}})\|^2] \geqq \underset{\sim}{0} \tag{5}$$

$$\partial \underset{\sim}{\bar{f}} \geqq \underset{\sim}{0} \tag{6}$$

must satisfy in order that $\underset{\sim}{x}^o$ be a local minimum for program (A). Note that for the discussion of this section, $y(\cdot)$ and $\underset{\sim}{f}(\cdot)$ need only be once continuously differentiable.

Lemma -6 If there exists $(\underset{\sim}{d},\underset{\sim}{\bar{f}})$ within some neighborhood of $(\underset{\sim}{d}^*,\underset{\sim}{\bar{f}}^*)$ such that

$$\left(\frac{\delta \underset{\sim}{\hat{f}}}{\delta \underset{\sim}{\bar{f}}}\right)^* \partial \underset{\sim}{\bar{f}} > \underset{\sim}{0};\ \partial \underset{\sim}{\bar{f}} \geq \underset{\sim}{0};\ \text{and}\ \left(\frac{\delta y}{\delta \underset{\sim}{d}}\right)^* \partial \underset{\sim}{d} + \left(\frac{\delta y}{\delta \bar{f}}\right)^* \partial \underset{\sim}{\bar{f}} < 0,$$

then $\underset{\sim}{x}^*$ is not a local minimum of (A).

Proof: Suppose $\underset{\sim}{x}^*$ is a local minimum and let $(\underset{\sim}{d},\underset{\sim}{\bar{f}})$ be such that the hypothesis is satisfied. Let

$$\partial \underset{\sim}{d}' = t(\underset{\sim}{d} - \underset{\sim}{d}^*) = t\, \partial \underset{\sim}{d}$$

$$\partial\bar{f}' = t(\bar{f} - \bar{f}^*) = t\,\partial\bar{f}$$

where $0 < t \leq 1$. Then,

$$\partial\hat{f} = \left(\frac{\delta\hat{f}}{\delta\bar{f}}\right)^* (\partial\bar{f}') + 0(\|(\partial d', \partial\bar{f}')\|^2) = t\left(\frac{\delta\hat{f}}{\delta\bar{f}}\right)^* \partial\bar{f} +$$

$$t^2\, 0(\|(\partial d, \partial f)\|^2)$$

Similarly,

$$\partial y = t\left(\frac{\delta y}{\delta d}\right)^* \partial d + t\left(\frac{\delta y}{\delta\bar{f}}\right)^* \partial\bar{f} + t^2\, 0(\|(\partial d, \partial\bar{f})\|^2)$$

For each $\ell = 1, \ldots, L$ and for the objective function y, choose t sufficiently small such that

$$\left|\left(\frac{\delta\hat{f}_\ell}{\delta\bar{f}}\right)^* \partial\bar{f}\right| t > t^2 \left| 0_\ell(\|\partial d, \partial\bar{f})\|^2)\right| \qquad \ell = 1, \ldots, L$$

$$\left|\left(\frac{\delta y}{\delta d}\right)^* \partial d \quad \left(\frac{\delta y}{\delta\bar{f}}\right)^* \partial\bar{f}\right| \quad t > t^2 \left| 0(\|(\partial d, \partial\bar{f})\|^2)\right|$$

Let $\tau > 0$ be the minimum of these t's. Then, using the hypothesis of the lemma, for $d'' = d^* + \tau\,\partial d$, $\bar{f}'' = \bar{f}^* + \tau\,\partial\bar{f}$

$$\partial\hat{f} = \left(\frac{\delta\hat{f}}{\delta\bar{f}}\right)^* \partial\bar{f}'' + 0(\|(\partial d'', \partial\bar{f}'')\|^2 \geqq$$

$$\tau\left(\frac{\delta\bar{f}}{\delta\bar{d}}\right)^* \partial\bar{f} - \tau^2 \left| 0(\|(\partial d, \partial\bar{f}\|^2)\right| > 0.$$

Similarly, $\partial y = y(d'', \bar{f}'') - y(d^*, \bar{f}^*) =$

$$\left(\frac{\delta y}{\delta d}\right)^* \partial d'' + \left(\frac{\delta y}{\delta\bar{f}}\right)^* \partial\bar{f}'' + 0(\|(\partial d'', \partial\bar{f}'')\|^2) \leq$$

$$\tau\left(\left(\frac{\delta y}{\delta d}\right)^{*}\partial d+\left(\frac{\delta y}{\delta \bar{f}}\right)^{*}\partial \bar{f}\right)+\tau^{2}\left|O(\|(\partial d,\partial \bar{f})\|^{2})\right|<0,$$

since by hypothesis the linear terms are negative. Moreover, $\partial \bar{f} \geq 0$. But this contradicts the assumption that $x^*$ is a local minimum of (A).

Theorem 1 follows immediately from the above lemma and the following result due to T. S. Motzkin [16].

Lemma -7 (Transposition Theorem)

Let $A$, $B$, and $C$ be matrices; $z$ $\omega$, $\delta$, and $\lambda$ be column vectors conformable respecively to $A$, $B$, $C$; $A^T$, $C^T$, and $B^T$. Exactly one of the following is true:

(i) There exists $z$ such that $Az \geqq 0$, $Cz = 0$, and $Bz < 0$ or

(ii) There exist $\omega$, $\delta$, and $\lambda$ with $\omega \geqq 0$ and $\lambda \geq 0$ such that

$$A^T\omega + C^T\delta = B^T\lambda \quad .$$

Theorem 1 (Linear Necessary Conditions)

If $x^*$ is a local minimum of (A), then there exists a rearrangement and partitioning of the Jacobian and a

$$\lambda = \begin{pmatrix} \lambda \\ \lambda_o \end{pmatrix} \geq 0$$

such that

(i) $$\left[\left(\frac{\delta y}{\delta d}\right)^{*}\right]^{T}\lambda_o = 0$$

(ii) $$\left[\left(\frac{\delta y}{\delta \bar{f}}\right)^{*}\right]^{T}\lambda_o - \left[\left(\frac{\delta \hat{f}}{\delta \bar{f}}\right)^{*}\right]^{T}\hat{\lambda} \geqq 0$$

Proof: Consider the following restatement of the hypothesis of Lemma 6; there exists $(d,\bar{f})$ such that

$$(\underset{\sim}{I}_M, \underset{\sim}{0}) \begin{pmatrix} \partial \underset{\sim}{\bar{f}} \\ \partial \underset{\sim}{d} \end{pmatrix} \geqq \underset{\sim}{0} \text{ and } \begin{pmatrix} -\left(\dfrac{\delta \underset{\sim}{\hat{f}}}{\delta \underset{\sim}{\bar{f}}}\right)^* & \underset{\sim}{0} \\ \left(\dfrac{\delta y}{\delta \underset{\sim}{\bar{f}}}\right)^* & \left(\dfrac{\delta y}{\delta \underset{\sim}{d}}\right)^* \end{pmatrix}^* \begin{pmatrix} \partial \underset{\sim}{\bar{f}} \\ \partial \underset{\sim}{d} \end{pmatrix} < \underset{\sim}{0} .$$

This is exactly in the form of (i) of Lemma -7 with $\underset{\sim}{C}$ vacuous. If $\underset{\sim}{x}^*$ is a local minimum of (A), them, by Lemma -6 the above is not true and hence by (ii) of Lemma -7, there exist $\underset{\sim}{\omega} \geqq \underset{\sim}{0}$ and $\underset{\sim}{\lambda} \geq \underset{\sim}{0}$ such that

$$\begin{pmatrix} \underset{\sim}{I}_M \\ \underset{\sim}{0} \end{pmatrix} \underset{\sim}{\omega} = \begin{pmatrix} -\left[\left(\dfrac{\delta \underset{\sim}{\hat{f}}}{\delta \underset{\sim}{\bar{f}}}\right)^*\right]^T & \left[\left(\dfrac{\delta y}{\delta \underset{\sim}{\bar{f}}}\right)^*\right]^T \\ \underset{\sim}{0} & \left[\left(\dfrac{\delta y}{\delta \underset{\sim}{d}}\right)^*\right]^T \end{pmatrix} \underset{\sim}{\lambda}$$

Let $\underset{\sim}{\lambda} = \begin{pmatrix} \underset{\sim}{\lambda} \\ \lambda_o \end{pmatrix}$, where $\lambda_o$ is a scalar, then

$$\underset{\sim}{\omega} = -\left[\left(\frac{\delta \underset{\sim}{\hat{f}}}{\delta \underset{\sim}{\bar{f}}}\right)^*\right]^T \underset{\sim}{\lambda}^T + \left[\left(\frac{\delta y}{\delta \underset{\sim}{\bar{f}}}\right)^*\right]^T \lambda_o \geqq \underset{\sim}{0}$$

and

$$\left[\left(\frac{\delta y}{\delta \underset{\sim}{d}}\right)^*\right]^T \lambda_o = \underset{\sim}{0}$$

which is the desired result.

Theorem 1 is akin to Fritz John's theorem [9] in the introduction of a multiplier $\lambda_o$ for the terms corresponding to the objective function. In fact, the proof strategy parallels that presented in [1] for a particular form of the Fritz John Theorem. However, the result differs in:

(1) partitioning the constraints according to the rank of the Jacobian;

(2) requiring the determination of semipositive multipliers only for non-basic constraints; and

(3) separating the conditions on the multipliers to obtain a complementary slackness condition on the "Fritz John" multiplier $\lambda_o$.

These features significantly simplify the application of the condition in practice where only one or perhaps two non-basic constraints may be anticipated.

The following corollary discusses the non-singular case considered by several investigators [4,5,10,21,22].

Corollary -1: If $\underset{\sim}{J}_h(\underset{\sim}{x}^*)$ is of maximal rank $M = T \leq N$, then if $\underset{\sim}{x}^*$ is a local minimum of (A), then

$$\left(\frac{\delta y}{\delta \bar{\underset{\sim}{f}}}\right)^* \geqq \underset{\sim}{0} \text{ and } \left(\frac{\delta y}{\delta \underset{\sim}{d}}\right)^* = \underset{\sim}{0} \ .$$

Proof: If $\underset{\sim}{J}_h(\underset{\sim}{x}^*)$ is of rank equal to the number of tight constraints, then there are no nonbasic constraints and slacks. Hence, Theorem 1 reduces to: there exists $\lambda_o > 0$ such that

$$\left[\left(\frac{\delta y}{\delta \bar{\underset{\sim}{f}}}\right)^*\right]^T \lambda_o \geqq \underset{\sim}{0} \quad \text{and} \quad \left[\left(\frac{\delta y}{\delta \underset{\sim}{d}}\right)^*\right]^T \lambda_o = \underset{\sim}{0}.$$

Thus,

$$\left(\frac{\delta y}{\delta \bar{\underset{\sim}{f}}}\right)^* \geqq \underset{\sim}{0} \quad \text{and} \quad \left(\frac{\delta y}{\delta \underset{\sim}{d}}\right)^* = \underset{\sim}{0}$$

In this case, as has been shown previously [21,22],

$$\left(\frac{\delta y}{\delta \bar{\underset{\sim}{f}}}\right)^*$$

is the dual variable (multiplier) vector obtained by solving the Kuhn-Tucker conditions [14], and

$$\left(\frac{\delta y}{\delta \underset{\sim}{d}}\right)^*$$

is the gradient of the Lagrangian function.

The second corollary verifies the fact that for linearly constrained problems, $\lambda_o$ will always be positive.

Corollary -2: (Linear Constraints)

If the constraints tight at $x^*$ are linear, then the fact that $\underset{\sim}{x}^*$ is a local minimum of (A) implies that

$$\left(\frac{\delta y}{\delta \underset{\sim}{d}}\right)^* = \underset{\sim}{0}$$

and there exists $\underset{\sim}{\hat{\lambda}} \geqq \underset{\sim}{0}$ such that,

$$\left[\left(\frac{\delta y}{\delta \underset{\sim}{\bar{f}}}\right)^*\right]^T - \left[\left(\frac{\delta \underset{\sim}{\hat{f}}}{\delta \underset{\sim}{\bar{f}}}\right)^*\right]^T \underset{\sim}{\hat{\lambda}} \geqq \underset{\sim}{0} .$$

Proof: Since the linearization of the constraints is now exact, it suffices to apply Lemma 7 to the system,

$$\begin{pmatrix} \underset{\sim}{I}_M & \underset{\sim}{0} \\ \left(\dfrac{\delta \underset{\sim}{\hat{f}}}{\delta \underset{\sim}{\bar{f}}}\right)^* & \underset{\sim}{0} \end{pmatrix} \begin{pmatrix} \partial \underset{\sim}{\bar{f}} \\ \partial \underset{\sim}{d} \end{pmatrix} \geqq \underset{\sim}{0} \quad \text{and} \quad \left[\left(\frac{\delta y}{\delta \underset{\sim}{\bar{f}}}\right)^*, \left(\frac{\delta y}{\delta \underset{\sim}{d}}\right)^*\right] \begin{pmatrix} \partial \underset{\sim}{\bar{f}} \\ \partial \underset{\sim}{d} \end{pmatrix} < 0$$

Hence, if $x^*$ is a local minimum of (A), there must exist $\lambda_o > 0$ and

$$\underset{\sim}{\omega} = \begin{pmatrix} \underset{\sim}{\omega}_M \\ \underset{\sim}{\omega}_{T-M} \end{pmatrix} \geqq \underset{\sim}{0}$$

such that

$$\begin{pmatrix} \underset{\sim}{I}_M & \left[\left(\dfrac{\delta \underset{\sim}{\hat{f}}}{\delta \underset{\sim}{\bar{f}}}\right)^*\right]^T \\ \underset{\sim}{0} & \underset{\sim}{0} \end{pmatrix} \begin{pmatrix} \underset{\sim}{\omega}_M \\ \underset{\sim}{\omega}_{T-M} \end{pmatrix} = \begin{pmatrix} \left[\left(\dfrac{\delta y}{\delta \underset{\sim}{\bar{f}}}\right)^*\right]^T \\ \left[\left(\dfrac{\delta y}{\delta \underset{\sim}{d}}\right)^*\right]^T \end{pmatrix} \lambda_o$$

or equivalently,

$$\frac{1}{\lambda_o}\underset{\sim}{\omega}_M = \left[\left(\frac{\delta y}{\delta \underset{\sim}{\bar{f}}}\right)^*\right]^T - \left[\left(\frac{\delta \underset{\sim}{\hat{f}}}{\delta \underset{\sim}{\bar{f}}}\right)^*\right]^T \frac{1}{\lambda_o}\underset{\sim}{\omega}_{T-M} \geq \underset{\sim}{0};$$

$$\left(\frac{\delta y}{\delta \underset{\sim}{d}}\right)^* = \underset{\sim}{0}$$

Finally, let $\underset{\sim}{\hat{\lambda}} \equiv \frac{1}{\lambda_o}\underset{\sim}{\omega}_{T-M} \geq \underset{\sim}{0}$ .

Although Theorem 1 was derived for the inequality constrained problem, it can readily be modified to accomodate equality constraints. Suppose, for example, that $f_p(\underset{\sim}{x})$ is such an equality constraint for problem (A) and suppose further that, at the point $\underset{\sim}{x}^*$ in question, $f_p(\underset{\sim}{x}^*) = 0$. Following the partitioning introduced above, either $f_p$ will be basic or it will be nonbasic. In the former instance, $f_p = \bar{f}_{k'}$ for some k' and in this case define

$$\underset{\sim}{\bar{\bar{f}}} = (\bar{f}_1, \ldots, \bar{f}_{k'-1}, \bar{f}_{k'\ 1}, \ldots, \bar{f}_M)^T;$$

in the second instance, $f_p = \hat{f}_{\ell'}$ for some $\ell'$. The following corollary summarizes the required modification of Theorem 1.

<u>Corollary -3</u>: (Equality Constraints)

If $\underset{\sim}{x}^*$ is a local minimum of (A) and $f_p(\underset{\sim}{x})$ with $f_p(\underset{\sim}{x}^*) = 0$ is an equality constraint of (A), then there exists $\lambda_o \geq 0$ such that

(i) $\left(\frac{\delta y^*}{\delta \underset{\sim}{d}}\right)^T \lambda_o = 0$ and

(ii) if $f_p = \bar{f}_{k'}$ for some k', $1 \leq k' \leq M$ then there exists

$$\underset{\sim}{\lambda} = \begin{pmatrix} \lambda \\ \lambda_o \end{pmatrix} \geq 0$$

such that,

$$\left(\frac{\delta y^*}{\delta \bar{\bar{f}}}\right)^T \lambda_o - \left(\frac{\delta \hat{\tilde{f}}^*}{\delta \bar{\bar{f}}}\right)^T \tilde{\lambda} \geq 0 \quad \text{or}$$

(iii) if $f_p = \hat{f}_{\ell'}$ for some $\ell'$, $1 \leq \ell' \leq L$, then there exists $\lambda^{\ell'} \neq 0$ with $\lambda_i \geqq 0$, $i = 0,1, \ldots ,L$, $i \neq \ell'$ and $\lambda_{\ell'}$ unrestricted, such that

$$\left(\frac{\delta y^*}{\delta \bar{\tilde{f}}}\right)^T \lambda_o - \left(\frac{\delta \tilde{f}^*}{\delta \bar{\tilde{f}}}\right)^T \tilde{\lambda} \geqq \tilde{0} \; .$$

Proof: If $f_p$ is basic, then $\partial \bar{f}_{k'} \equiv 0$ in eq. (4) - (6). Hence, the terms

$$\left(\frac{\delta y^*}{\delta \bar{f}_{k'}}\right), \left(\frac{\delta \hat{\tilde{f}}^*}{\delta \bar{f}_{k'}}\right)$$

are eliminated from the linearization. The balance of the proof is unchanged. If $f_p$ is nonbasic, then $\partial \hat{f}_{\ell'} \equiv 0$ in eq. (5). Thus, Lemma 6 must be modified to include

$$\left(\frac{\delta \hat{f}^*_{\ell'}}{\delta \bar{\tilde{f}}}\right) \partial \bar{\tilde{f}} = 0$$

and Lemma 7 is applied with $\tilde{C} \equiv \left(\frac{\delta \hat{f}^*_{\ell'}}{\delta \bar{\tilde{f}}}\right)$.

If $J_h(x^*)$ is restricted to be of maximal rank, then the above corollary reduces to the classical Lagrange Multiplier result.

Corollary -4: (Lagrange)

Let $\tilde{J}_h(\tilde{x}^*)$ be of maximal rank $M = T \leq N$ and the constraints be equalities. If $\tilde{x}^*$ is a local minimum of (A), then

$$\left(\frac{\delta y}{\delta \tilde{d}}\right)^* = \tilde{0} \; .$$

Remark: Theorem 1 and its corollaries not only function as optimality criteria but also can be used in a constructive sense to generate directions for improvement in the case of singular or degenerate points. Theorem 1 together with Lemma 6 is especially adaptable for incorporation within the class of mathematical programming algorithms based on Simplex-like operations or an implicit function type formulation [2,23,24]. Such algorithms typically require the Jacobian to have maximal row rank and use a construction similar to that of Lemma 1. In particular, the constrained decision and slack derivatives of y are employed to choose useful perturbations in the independent decision and slack variables; then these perturbations are substituted into eq. (4b) to calculate the subsequent changes in the state or dependent variables. The algorithms terminate when Corollary 1 is satisfied or when a point is reached at which Jacobian does not have maximal row rank (a singular or degenerate point). As Lemma 6 shows, if the point is singular then the perturbations in the independent variables are not truly independent but rather must be chosen to satisfy equations (5) and (6). The details required to implement these computations will be reported separately [18].

## 5 Semi Quadratic Necessary Conditions

Since in the expansion for the non-basic constraint, eq. (5), the linear terms corresponding to the decision variables vanish, while

$$\left( \frac{\delta y}{\delta \underset{\sim}{d}} \right)^*,$$

in general, does not, a more exclusive sequel to Theorem 1 can be obtained by considering terms of second order in the decisions. The constrained derivative formulation readily permits computation of just these terms, without requiring second order information about the objective function. If the tight constraints are twice continuously differentiable, then a corollary to the Implicit Function Theorem insures that the solution functions are also in C'' so that the expansion given in Lemma 2 can be carried out to second order. An explicit expression for the second order terms can be obtained using the Chain Rule. The results of this computation may be found in the Appendix.

For present purposes existence is sufficient. We define,

$$Z^o_\ell \equiv \left( \frac{\delta^2 \hat{f}_\ell}{\delta \bar{f}_k \delta d_r} \right) :$$

the $N^2$ symmetric matrix of <u>constrained second derivatives</u> of $\hat{f}_\ell(\underset{\sim}{d},\underset{\sim}{\bar{f}})$, $1 \leq \ell \leq L$, evaluated at $(\underset{\sim}{d}^o,\underset{\sim}{\bar{f}}^o)$ with general term as written.

Introducing the second order term, eq. (5) becomes, for $1 \leq \ell \leq L$,

$$\partial \hat{f}_\ell = \left( \frac{\delta \hat{f}_\ell}{\delta \underset{\sim}{\bar{f}}} \right)^o \partial \underset{\sim}{\bar{f}} + \tfrac{1}{2}(\partial \underset{\sim}{d}^T, \partial \underset{\sim}{\bar{f}}^T)\ \underset{\sim}{\hat{Z}}_\ell\ (\partial \underset{\sim}{d}, \partial \underset{\sim}{\bar{f}}) +$$

$$O[\|(\partial \underset{\sim}{d}, \partial \underset{\sim}{\bar{f}})\|^3] = \left( \frac{\delta \hat{f}_\ell}{\delta \underset{\sim}{\bar{f}}} \right) \partial f + \tfrac{1}{2}(\partial \underset{\sim}{d}^T)\ \underset{\sim}{\hat{Z}}_{dd\ell}(\partial \underset{\sim}{\bar{f}}) +$$

$$(\partial \underset{\sim}{d}^T)\ \underset{\sim}{\hat{Z}}_{df\ell}(\partial \underset{\sim}{\bar{f}}) + \tfrac{1}{2}(\partial \underset{\sim}{\bar{f}}^T)\ \underset{\sim}{\hat{Z}}_{ff\ell}(\partial \underset{\sim}{\bar{f}}) + O[\|(\partial \underset{\sim}{d}, \partial \underset{\sim}{f})\|^3] \qquad (7)$$

In the subsequent discussion use will be made of the following well known fact stated in terms of the <u>pseudo-inverse</u> of a matrix, a generalization of the conventional inverse [8,17].

<u>Lemma -8</u>: The system of equations $\underset{\sim}{A}\underset{\sim}{x} = \underset{\sim}{b}$ has a solution if and only if $\underset{\sim}{A}\underset{\sim}{A}^\dagger \underset{\sim}{b} = \underset{\sim}{b}$, in which case the general solution is $\underset{\sim}{x} = \underset{\sim}{A}^\dagger \underset{\sim}{b} + (\underset{\sim}{I} - \underset{\sim}{A}^\dagger \underset{\sim}{A})\underset{\sim}{z}$ where z is an arbitrary vector.

<u>Lemma -9</u>: If there exists $\underset{\sim}{d}$ such that

$$(\partial \underset{\sim}{d})^T\ \underset{\sim}{\hat{Z}}^*_{dd\ell}(\partial \underset{\sim}{d}) > 0,\ \ell = 1, \ldots, L, \text{ and } \left( \frac{\delta y}{\delta \underset{\sim}{d}} \right)^* \partial \underset{\sim}{d} < 0,$$

then $\underset{\sim}{x}^*$ is not a local minimum of (A).

<u>Proof:</u> Set $\partial \underset{\sim}{\bar{f}} = \underset{\sim}{0}$ in equations (4) and (7). The result follows by an argument analogous to that of Lemma -6.

Suppose not all components of

$$\left(\frac{\delta y}{\delta \underset{\sim}{d}}\right)^{*}$$

are zero, and let

$$\underset{\sim}{B} \equiv \underset{\sim}{I}_{N-M} - \frac{\left(\frac{\delta y}{\delta \underset{\sim}{d}}\right)^{*T} \left(\frac{\delta y}{\delta \underset{\sim}{d}}\right)^{*}}{\left\|\left(\frac{\delta y}{\delta \underset{\sim}{d}}\right)^{*}\right\|^{2}} \; ; \; \underset{\sim}{\beta} \equiv \frac{\left(\frac{\delta y}{\delta \underset{\sim}{d}}\right)^{*T}}{\left\|\left(\frac{\delta y}{\delta \underset{\sim}{d}}\right)^{*}\right\|^{2}} \tag{8}$$

Theorem -2 If there exists a vector $\underset{\sim}{z}$ and an $\epsilon > 0$ such that

$$(\underset{\sim}{B}\underset{\sim}{z} - \underset{\sim}{\beta}\epsilon)^{T} \, \hat{\underset{\sim}{Z}}_{dd\ell} \, (\underset{\sim}{B}\underset{\sim}{z} - \underset{\sim}{\beta}\epsilon) > 0 \text{ for all } \ell = 1, \ldots, L$$

then $\underset{\sim}{x}^{o}$ is not a local minimum.

Proof: The hypothesis of Lemma -9 is equivalent to the statement that there exists $\partial\underset{\sim}{d} = \underset{\sim}{d} - \underset{\sim}{d}^{*}$ and $\epsilon = \epsilon(\underset{\sim}{d}) > 0$ such that

$$\partial y = \left(\frac{\delta y}{\delta \underset{\sim}{d}}\right)^{*} \partial\underset{\sim}{d} = -\epsilon \quad \text{and} \quad (\partial\underset{\sim}{d})^{T} \, \hat{\underset{\sim}{Z}}_{dd\ell} (\partial\underset{\sim}{d}) > 0, \tag{9}$$

$\ell = 1, \ldots, L$. By Lemma 8, the solution to

$$\left(\frac{\delta y}{\delta \underset{\sim}{d}}\right)^{*} \partial\underset{\sim}{d} = -\epsilon$$

is given by

$$\partial\underset{\sim}{d} = \left(\frac{\delta y}{\delta \underset{\sim}{d}}\right)^{*\dagger} (-\epsilon) + \left(I - \left(\frac{\delta y}{\delta \underset{\sim}{d}}\right)^{*\dagger} \left(\frac{\delta y}{\delta \underset{\sim}{d}}\right)^{*}\right) \underset{\sim}{z}$$

for arbitrary $\underset{\sim}{z}$, provided that

$$\left(\frac{\delta y}{\delta \underset{\sim}{d}}\right)^{*} \left(\frac{\delta y^{*}}{\delta \underset{\sim}{d}}\right)^{\dagger} (-\epsilon) = -\epsilon$$

or, in this case,

$$\left(\frac{\delta y^*}{\delta \underset{\sim}{d}}\right)\left(\frac{\delta y^*}{\delta \underset{\sim}{d}}\right)^{\dagger} = 1.$$

However,

$$\left(\frac{\delta y^*}{\delta \underset{\sim}{d}}\right)^{\dagger} = \left(\frac{\delta y^*}{\delta \underset{\sim}{d}}\right)^{T}\left[\left(\frac{\delta y^*}{\delta \underset{\sim}{d}}\right)\left(\frac{\delta y^*}{\delta \underset{\sim}{d}}\right)^{T}\right]^{-1} = \left(\frac{\delta y^*}{\delta \underset{\sim}{d}}\right)^{T}\left\|\frac{\delta y^*}{\delta \underset{\sim}{d}}\right\|^{-2}$$

(see [8]). Hence,

$$\left(\frac{\delta y^*}{\delta \underset{\sim}{d}}\right)\left(\frac{\delta y^*}{\delta \underset{\sim}{d}}\right)^{\dagger} = 1$$

and so the solution

$$\partial \underset{\sim}{d} = -\left(\frac{\delta y^*}{\delta \underset{\sim}{d}}\right)^{T}\left\|\frac{\delta y^*}{\delta \underset{\sim}{d}}\right\|^{-2}\epsilon + \left[\underset{\sim}{I}_{N-M} - \frac{\left(\frac{\delta y^*}{\delta \underset{\sim}{d}}\right)^{T}\left(\frac{\delta y^*}{\delta \underset{\sim}{d}}\right)}{\left\|\frac{\delta y^*}{\delta \underset{\sim}{d}}\right\|^{2}}\right]\underset{\sim}{z}$$

is valid for all $\underset{\sim}{z}$. Using definitions (8), set

$$\partial \underset{\sim}{d} = -\underset{\sim}{\beta}\epsilon + \underset{\approx}{B}\underset{\sim}{z} . \tag{10}$$

Then by substitution,

$$(\partial \underset{\sim}{d})^{T}\,\hat{\underset{\sim}{Z}}_{dd\ell}(\partial \underset{\sim}{d}) = (\underset{\approx}{B}\underset{\sim}{z} - \underset{\sim}{\beta}\epsilon)^{T}\,\hat{\underset{\sim}{Z}}_{dd\ell}(\underset{\approx}{B}\underset{\sim}{z} - \underset{\sim}{\beta}\epsilon).$$

Clearly, there exists $\tilde{d}$ and $\epsilon > 0$ such that (9) holds if and only if there exists $\underset{\sim}{z}$ and $\epsilon > 0$ such that

$$(\underset{\approx}{B}\underset{\sim}{z} - \underset{\sim}{\beta}\epsilon)^{T}\,\hat{\underset{\sim}{Z}}_{dd\ell}(\underset{\approx}{B}\underset{\sim}{z} - \underset{\sim}{\beta}\epsilon) > 0 \quad \text{for all } \ell = 1, \ldots, L \tag{11}$$

the conclusion follows directly from Lemma 9.

<u>Corollary -5</u>: If

$$\left(\frac{\delta y}{\delta \underset{\sim}{d}}\right)^{*} \neq 0, \text{ then if } \left(\frac{\delta y^*}{\delta \underset{\sim}{d}}\right)\hat{\underset{\sim}{Z}}_{dd\ell}\left(\frac{\delta y^*}{\delta \underset{\sim}{d}}\right)^{T} > 0$$

for all $\ell = 1, \ldots, L$, then $\underset{\sim}{x}^*$ is not a local minimum of (A).

Proof: In this case $\underset{\sim}{z} = \underset{\sim}{0}$ satisfies Theorem 2 for all $\epsilon > 0$.

Corollary - 6: If

$$\left(\frac{\delta y}{\delta \underset{\sim}{d}}\right)^* \neq \underset{\sim}{0},$$

then if there exists $\underset{\sim}{z}$ such that $\underset{\sim}{z}^T \underset{\sim}{B}\, \hat{\underset{\sim}{Z}}_{dd\ell}\, \underset{\sim}{Bz} > 0$ for all $\ell = 1, \ldots, L$, then $\underset{\sim}{x}^*$ is not a local minimum of (A).

Proof: Suppose $\underset{\sim}{z}^o$ satisfies the hypothesis.

Let $\alpha_\ell = \frac{1}{2}(\underset{\sim}{z}^o)^T \underset{\sim}{B}\, \hat{\underset{\sim}{Z}}_{dd\ell}\, \underset{\sim}{Bz}^o > 0$ and define $\alpha = \min\{\alpha_\ell : \ell = 1, \ldots, L\}$. Furthermore, let

$$M_\ell = \tfrac{1}{2}\underset{\sim}{\beta}^T \hat{\underset{\sim}{Z}}_{dd\ell}\underset{\sim}{\beta}\,; \qquad \gamma_\ell = (\underset{\sim}{z}^o)^T \underset{\sim}{B}\, \hat{\underset{\sim}{Z}}_{dd\ell}\, \underset{\sim}{\beta} \quad .$$

Choose

$$0 < \epsilon < \min\left\{1, \frac{\alpha}{|\gamma_\ell|}, \frac{(|\gamma_\ell|^2 + 4\alpha|M_\ell|)^{\frac{1}{2}} - |\gamma_\ell|}{2|M_\ell|} : \ell = 1, \ldots, L\right\} \tag{12}$$

Then, expanding eq. (11), for each $\ell$

$$\partial\hat{f}_\ell = \tfrac{1}{2}(\underset{\sim}{z}^o)^T \underset{\sim}{B}^T \hat{\underset{\sim}{Z}}_{dd\ell}\, \underset{\sim}{Bz}^o + \frac{\epsilon^2}{2} \underset{\sim}{\beta}^T \hat{\underset{\sim}{Z}}_{dd\ell}\, \underset{\sim}{\beta} - \epsilon(\underset{\sim}{z}^o)^T \underset{\sim}{B}^T \hat{\underset{\sim}{Z}}_{dd\ell}\, \underset{\sim}{\beta}$$

by symmetry, $\underset{\sim}{B}^T = \underset{\sim}{B}$, therefore for the above choice of $\epsilon > 0$

$$= \alpha_\ell + \epsilon^2 M_\ell - \epsilon\gamma_\ell \geq \alpha - \epsilon^2 |M_\ell| - \epsilon\, |\gamma_\ell| > 0$$

Since the given choice of $\epsilon$ satisfies Theorem 2, the proof is complete.

Throughout the preceding discussion it was assumed that rank $\underset{\sim}{J}_h \geq 1$. The contrary special case may now be covered. Observe that rank $\underset{\sim}{J}_h = 0$ implies that $\underset{\sim}{J}_h \equiv \underset{\sim}{0}$ and, thus, the labelling conventions require that all $\underset{\sim}{x}_i$ be decisions and all $f_k$ nonbasic. Since the linear parts vanish, a second order analysis must be undertaken. Hence, the following corollary can be stated,

Corollary -7: If $\underset{\sim}{J}_h^* = \underset{\sim}{0}$

$$\left(\frac{\partial y}{\partial \underset{\sim}{x}}\right)^* \equiv \left(\frac{\delta y}{\delta \underset{\sim}{d}}\right)^* \neq \underset{\sim}{0},$$

and the hypothesis of Theorem 2 is satisfied, then $\underset{\sim}{x}^*$ is not a local minimum of (A).

The following result discusses the singular case, for which Lagrange's Method is invalid.

Theorem -3: (Equality Constraints)

If

$$\left(\frac{\delta y}{\delta \underset{\sim}{d}}\right)^* \neq \underset{\sim}{0}$$

and the tight constraints are equality constraints, then, if there exists $\underset{\sim}{z}$ and $\epsilon > 0$ such that

$$(\underset{\sim}{B}\underset{\sim}{z} - \underset{\sim}{\beta}\epsilon)^T \, \hat{\underset{\sim}{Z}}_{dd\ell}(\underset{\sim}{B}\underset{\sim}{z} - \underset{\sim}{\beta}\epsilon) = 0, \quad \ell = 1, \ldots, L$$

$\underset{\sim}{x}^*$ is not a local minimum of (A).

Proof: If the constraints are equalities, then for every feasible point both $\bar{\underset{\sim}{f}}$ and $\hat{\underset{\sim}{f}}$ must be zero. Hence Lemma -9 holds with

$$(\partial \underset{\sim}{d})^T \, \hat{\underset{\sim}{Z}}_{dd\ell}(\partial \underset{\sim}{d}) = 0 .$$

With this change, the proof of Theorem 2 follows through. Note, moreover, that Corollary -5 also holds with equality replacing strict inequality; however, Corollary -6 may not be so adapted.

Remark 1: The fact that the linear conditions alone are inadequate to characterize a singular point of an equality constrained problem was observed in [15]. The authors proved that such a point always satisfies the generalized Fritz John conditions. The differential formulation indicates very simply why this must be the case.

Remark 2: Observe that Corollaries -5 and -6 immediately yield feasible directions for improvement should the hypothesis be satisfied. Equations (10) and (12) may be directly used to compute the next step.

Remark 3: It should be pointed out that the results of this section can be generalized by considering variations in both decisions and slacks together. The constructions using the pseudo-inverse are similar, but, because they involve interactions between non-negative and unrestricted variables, the resulting conditions are considerably more difficult to apply. These calculations were not undertaken here, because more convenient, entirely second order conditions can adequately cover this gap in the theory.

## 6 Examples

The following examples illustrate the computations required by the necessary conditions presented in this paper.

(1) Consider the problem to:

$$\text{minimize} \quad Y(\underset{\sim}{x}) = \exp(-x_2) + x_3^2$$

$$\text{subject to} \quad f_1(\underset{\sim}{x}) = x_1^2 + x_2 \geq 0$$

$$f_2(\underset{\sim}{x}) = -x_2 - x_3^2 \geq 0$$

at the point $\underset{\sim}{x}^* = (0,0,0)$.

The derivatives at the point are

$$\underset{\sim}{J}_h{}^* = \begin{pmatrix} 0 & 1 & 0 \\ 0 & -1 & 0 \end{pmatrix} ; \quad \frac{\partial y^*}{\partial \underset{\sim}{x}} = (0,-1,0)$$

rank $\underset{\sim}{J}_h{}^* = 1$ and the vector $(0,0,a)^T$ violates the requirements of the Kuhn-Tucker Qualification.

Choose the partition

$$\begin{pmatrix} \bar{\underset{\sim}{J}} & \bar{\underset{\sim}{C}} \\ \hat{\underset{\sim}{J}} & \hat{\underset{\sim}{C}} \end{pmatrix} = \left(\begin{array}{c:cc} 1 & 0 & 0 \\ \hline -1 & 0 & 0 \end{array}\right) ; \left(\frac{\partial y^*}{\partial \underset{\sim}{s}}\right), \left(\frac{\partial y^*}{\partial \underset{\sim}{d}}\right) = (-1 \,\vdots\, 0 \;\; 0)$$

Computing the pertinent constrained derivatives gives

$$\left(\frac{\delta \hat{f}}{\delta \bar{\underset{\sim}{f}}}\right)^* = (-1)(1) = -1; \left(\frac{\delta y}{\delta \bar{\underset{\sim}{f}}}\right)^* = (-1)(1) = -1;$$

$$\left(\frac{\delta y}{\delta \underset{\sim}{d}}\right)^{*} = (0,0)$$

The necessary conditions of Theorem 1 require that there exist $\underset{\sim}{\hat{\lambda}}$ and $\lambda_o$ non-negative but not all zero such that

$$\begin{pmatrix} 0 \\ 0 \end{pmatrix} \lambda_o = 0 \qquad \text{and} \qquad (1,-1) \begin{pmatrix} \hat{\lambda} \\ \lambda_o \end{pmatrix} \geq 0$$

Any choice $\hat{\lambda} \geq \lambda_o$ 0 satisfies these relations so that $\underset{\sim}{x}^*$ passes the necessary conditions for a local minimum.

(2) Suppose for the above example we alter the objective function to $y(\underset{\sim}{x}) = \exp(-x_2) - x_3$. Then for the same partition

$$\left(\frac{\delta y}{\delta \underset{\sim}{d}}\right)^{*} = (0,-1)$$

and the necessary conditions are satisfied for $\lambda_o = 0$, $\hat{\lambda} >$ 0. Since

$$\left(\frac{\delta y}{\delta \underset{\sim}{d}}\right)^{*} \neq \underset{\sim}{0} \ ,$$

the second order conditions are applicable. To do this we compute,

$$\underset{\sim}{\hat{Z}}_{dd1} = \underset{\sim}{\hat{H}}_{dd1} - \left(\frac{\delta \hat{f}_1}{\delta \bar{f}}\right) \underset{\sim}{\bar{H}}_{dd1} = \begin{pmatrix} 0 & 0 \\ 0 & -2 \end{pmatrix} - (-1) \begin{pmatrix} 2 & 0 \\ 0 & 0 \end{pmatrix} =$$

$$\begin{pmatrix} 2 & 0 \\ 0 & -2 \end{pmatrix} \ ; \qquad \frac{1}{\left\|\frac{\delta y}{\delta \underset{\sim}{d}}\right\|^2} \left(\frac{\delta y}{\delta \underset{\sim}{d}}\right)^{T} = \begin{pmatrix} 0 \\ -1 \end{pmatrix} ;$$

$$\underset{\sim}{B} = \left( \underset{\sim}{I}_{N-M} - \frac{\left(\frac{\delta y}{\delta \underset{\sim}{d}}\right)^{T} \left(\frac{\delta y}{\delta \underset{\sim}{d}}\right)}{\left\|\frac{\delta y}{\delta \underset{\sim}{d}}\right\|^2} \right) = \begin{pmatrix} 1 & 0 \\ 0 & 1 \end{pmatrix} - \begin{pmatrix} 0 \\ -1 \end{pmatrix} (0 \ -1)$$

$$= \begin{pmatrix} 1 & 0 \\ 0 & 0 \end{pmatrix}$$

Applying Corollary -5,

$$\left(\frac{\delta y}{\delta \underset{\sim}{d}}\right) \hat{\underset{\sim}{Z}}_{dd1} \left(\frac{\delta y}{\delta \underset{\sim}{d}}\right)^T = (0 \quad -1) \begin{pmatrix} 2 & 0 \\ 0 & -2 \end{pmatrix} \begin{pmatrix} 0 \\ -1 \end{pmatrix} = -2 < 0$$

the condition of positivity is not satisfied; we continue with Corollary 6, i.e., does there exist $\underset{\sim}{z}$ such that $\underset{\sim}{z}^T \underset{\sim}{B} \hat{\underset{\sim}{Z}}_{dd1} \underset{\sim}{B}z > 0$?

$$(z_1, z_2) \begin{pmatrix} 1 & 0 \\ 0 & 0 \end{pmatrix} \begin{pmatrix} 2 & 0 \\ 0 & -2 \end{pmatrix} \begin{pmatrix} 1 & 0 \\ 0 & 0 \end{pmatrix} \begin{pmatrix} z_1 \\ z_2 \end{pmatrix} = 2 z_1^2 > 0$$

if $z_1 \neq 0$. Therefore, x* is not a local minimum. A direction for improvement can be obtained using equations (11) and (12)

$$\begin{pmatrix} \Delta x_1 \\ \Delta x_3 \end{pmatrix} = \begin{pmatrix} 0 \\ 1 \end{pmatrix} \varepsilon + \begin{pmatrix} z_1 \\ 0 \end{pmatrix} \quad \text{for } 0 < \varepsilon < \min \{1, |z_1|\}$$

and $\bar{f}_1 = 0$. For example, with the choice $z_1 = \frac{1}{4}$ and $\varepsilon = 1/5$, $x_1$ becomes $\frac{1}{4}$ and $x_3 = 1/5$. Hence, $x_2 = -1/16$. Then,

$$Y(\tfrac{1}{4}, -1/16, 1/5) \cong 0.86 < y(\underset{\sim}{x}^*) .$$

Observe that the second order conditions have permitted us to correctly identify a point which could not be identified by existing criteria.

In summary, the present work extends the constrained derivative optimality criteria of previous papers to the non-singular case. Moreover, within the framework of the differential approach, we are able to derive constructive semi-quadratic necessary conditions which follow naturally from the linear analysis and which can treat previously untractable problems.

Acknowledgement. This research was supported by National Science Foundation Grant GK-651.

## References


[1] Abadie, Jean, "On the Kuhn-Tucker Theorem." Report ORC 65-18, June 1965, Operations Research Center, University of California, Berkeley.

[2] Abadie, Jean and J. Carpentier, "Generalization of the Wolfe Reduced Gradient Method to the Case of Nonlinear Constraints," in Optimization, R. Fletcher, ed., Academic Press, New York, 1969.

[3] Arrow, K. J. and A. C. Enthoven, "Quasi-Concave Programming." Econometrica 29 (1961) pp. 779-800.

[4] Arrow, K. J., L. Hurowicz, and H. Uzawa, "Constraint Qualification in Maximization Problems." Naval Research Logistics Quarterly, 8 (1961) pp. 175-191.

[5] Bernholtz, B., "A New Derivation of the Kuhn-Tucker Conditions." Opns. Res. 12 (1964) pp. 295-299.

[6] Cottle, R. W., "A Theorem of Fritz John in Mathematical Programming," The Rand Corp., Santa Monica, Research Mem. RM-3858-PR (Oct. 1963).

[7] Gould, F. J. and J. W. Tolle, "A Necessary and Sufficient Qualification for Constrained Optimization," Institute of Statistics Mimeo Series No. 662, University of North Carolina, Chapel Hill, January 1970.

[8] Greville, T. N. E., "The Pseudo-Inverse of a Rectangular or Singular Matrix and its Application to the Solution of Systems of Linear Equations." Siam Review, 1 (1959) pp. 38-43.

[9] John, Fritz, "Extremum Problems with Inequalities as Subsidiary Conditions," Studies and Essays, Courant Anniversary Volume, Interscience, New York, 1948, pp. 187-204.

[10] King, R. P., "Necessary and Sufficient Conditions for Inequality Constrained Extreme Values," Ind. Engng. Chem. Fundamentals, 3 (1966) pp. 484-489.

[11] Kortanek, K. O. and J. P. Evans, "Pseudo-Concave Programming and Lagrange Regularity," Opns. Res., 15 (1967) pp. 882-891.

[12] Kortanek, K. O. and J. P. Evans, "Asymptotic Lagrange Regularity for Pseudo-Concave Rogramming with Weak Constraint Qualification," Tech. Report No. 24, June 1967, Dept. of Ind. Eng. and Opns. Res., Cornell University.

[13] Kortanek, K. O. and J. P. Evans, "Asymptotic Lagrange Regularity: Infinum not Assumed," Tech. Report No. 38,

Dept. of Opns. Res., Cornell University.

[14] Kuhn, H. W. and A. W. Tucker, "Non-Linear Programming," Proc. Second Berkeley Symposium Math. Stat. Prob., (1951) pp. 481- 492.

[15] Mangasarian, O. and S. Fromovitz, "The Fritz John Necessary Optimality Conditions in the Presence of Equality and Inequality Constraints," J. Math. Anal. Appl. 17, (1967) pp. 37-47.

[16] Motzkin, T. S., "Beitrage zur Theorie der Linearen Gleichungen," (Disertation Basel,1933) cited in Linear Inequalities and Related Systems, H. W. Kuhn and A. W. Tucker, eds., Princeton 1956.

[17] Penrose, R., "A Generalized Inverse for Matrices," Proc. Cambridge Philos. Soc., 51 (1955) pp. 406-413.

[18] Reklaitis, G. V. and D. J. Wilde, under preparation.

[19] Reklaitis, G. V. and D. J. Wilde, "Computationally Compact Sufficient Conditions for Constrained Optima," Opns. Res. 17 (1969) pp. 425-436.

[20] Slater, Morton, "Lagrange Multipliers Revisited: A Contribution to Non-Linear Programming, "Cowles Commisssion Discussion Paper, Math. 403 (1950).

[21] Wilde, Douglas J., "Differential Calculus in Non-Linear Programming," Opns. Res., 10 (1962) pp. 764-773.

[22] Wilde, Douglas J.,"Jacobians in Constrained Non-Linear Optimization," Opns. Res., 13 (1965) pp. 848-855.

[23] Wilde, Douglas J., and C. Beightler, Foundations of Optimization Theory, Prentice-Hall, Englewood Cliffs, 1967.

[24] Zangwill, W., "The Convex Simplex Method," Management Science, 14 (1967) pp. 221-232.

## Appendix: Constrained Second Derivatives

This section summarizes the formulas necessary for computing the constrained second derivatives employed in Section 5. The expressions can readily be derived, assuming twice continuous differentiablility of the constraint functions, either by directly applying the Chain Rule, as in the linear analysis of Lemma 2 (see [21]), or by using an indirect method such as that of [23]. The details of the computation will not be elaborated.

Let $\hat{\underset{\sim}{H}}{}^{o}_{\ell}$ and $\bar{\underset{\sim}{H}}{}^{o}_{\ell}$ be the Hessian matrices of second derivatives of the non-basic and basic constraints, respectively, evaluated at the point $\underset{\sim}{x}^{o}$. Using the constrained derivatives,

$$\left(\frac{\delta \hat{f}_{\ell}}{\delta \bar{f}_k}\right)^{o},$$

for each $\ell = 1, \ldots, L$, form the sum,

$$\hat{\underset{\sim}{Q}}{}^{o}_{\ell} = \hat{\underset{\sim}{H}}{}^{o}_{\ell} - \sum_{k=1}^{M} \left(\frac{\delta \hat{f}_{\ell}}{\delta \bar{f}_k}\right)^{o} \bar{\underset{\sim}{H}}{}^{o}_{k}$$

Corresponding to the partition of x into states $\underset{\sim}{s}$ and decision variables $\underset{\sim}{d}$, partition each quadratic form

$$\hat{\underset{\sim}{Q}}{}^{o}_{\ell} = \begin{pmatrix} \hat{\underset{\sim}{Q}}{}^{o}_{dd\ell} & \hat{\underset{\sim}{Q}}{}^{o}_{ds\ell} \\ \hat{\underset{\sim}{Q}}{}^{o}_{sd\ell} & \hat{\underset{\sim}{Q}}{}^{o}_{ss\ell} \end{pmatrix}$$

Then, each nonbasic constraint can to second order be expanded in terms of the slack variables $\bar{\underset{\sim}{f}}$ and decision variables, $\underset{\sim}{d}$, as follows,

$$\partial \hat{f}_{\ell} = \left(\frac{\delta \hat{f}_{\ell}}{\delta \bar{\underset{\sim}{f}}}\right) \partial \bar{\underset{\sim}{f}} + (\partial d, \partial \bar{f})^{T} \hat{\underset{\sim}{Z}}{}^{o}_{\ell} (\partial \underset{\sim}{d}, \partial \bar{\underset{\sim}{f}}) + 0[\|(\partial \underset{\sim}{d}, \partial \bar{\underset{\sim}{f}})\|^{3}].$$

where $$\hat{\underset{\sim}{Z}}{}^{o}_{\ell} = \begin{pmatrix} \hat{\underset{\sim}{Z}}{}^{o}_{dd\ell} & \hat{\underset{\sim}{Z}}{}^{o}_{d\bar{f}} \\ \hat{\underset{\sim}{Z}}{}^{o}_{fd\ell} & \hat{\underset{\sim}{Z}}{}^{o}_{ff\ell} \end{pmatrix}$$

and $$\hat{\underset{\sim}{Z}}_{dd\ell} = \hat{\underset{\sim}{Q}}_{dd\ell} - \hat{\underset{\sim}{Q}}_{ds\ell} \bar{\underset{\sim}{J}}^{-1} \bar{\underset{\sim}{C}} - (\hat{\underset{\sim}{Q}}_{ds\ell} \bar{\underset{\sim}{J}}^{-1} \bar{\underset{\sim}{C}})^{T} + (\bar{\underset{\sim}{J}}^{-1} \bar{\underset{\sim}{C}})^{T} \hat{\underset{\sim}{Q}}_{ss\ell} (\bar{\underset{\sim}{J}}^{-1} \bar{\underset{\sim}{C}})$$

$$\hat{\underset{\sim}{Z}}_{df\ell} = \hat{\underset{\sim}{Q}}_{ds\ell} \bar{\underset{\sim}{J}}^{-1} - (\bar{\underset{\sim}{J}}^{-1} \bar{\underset{\sim}{C}})^{T} \hat{\underset{\sim}{Q}}_{ss\ell} \bar{\underset{\sim}{J}}^{-1} = (\hat{\underset{\sim}{Z}}_{fd\ell})^{T}$$

$$\hat{\underset{\sim}{Z}}_{ff\ell} = (\bar{\underset{\sim}{J}}^{-1})^{T} \hat{\underset{\sim}{Q}}_{ss\ell} (\bar{\underset{\sim}{J}}^{-1})$$

For purposes of Theorems 2 and 3, only the first of these equations is needed. The remaining one are included at this time only for completeness.

# MATHEMATICAL MODELS FOR STATISTICAL DECISION THEORY[1]

Bernard Harris

Mathematics Research Center
The University of Wisconsin

## 1. Introduction

In the typical problem of statistical inference, the statistician is confronted with the problem of selecting one out of the vast number of possible decision rules. I will refer to this as the "fundamental problem of statistics". In this survey paper, I will try to give some mathematical characterizations of the problem of selecting an optimal decision procedure. This will not constitute a resolution of the "fundamental problem", inasmuch as this is intimately tied up with irresolvable philosophical difficulties. These arise, since in all but a few exceptional problems, there is no single procedure which can be regarded as uniformly dominating all other possible procedures. As a consequence, "reasonable people" have disagreed and will continue to disagree on the specific procedure that should be selected.

Despite all of the above difficulties, a great deal can still be done to formalize statistical decision theory. Thus, in what follows, I will give some characterizations of optimality in statistical inference. The discussion will of necessity be brief and is intended only to provide an introduction to these characterizations. More extensive treatments are in preparation and will appear subsequently.

The objective here will be to provide a mathematical

[1] Supported by the United States Army under Contract No.: DA-31-124-ARO-D-462 and by the Technological University, Eindhoven, Netherlands.

structure in which the details of the original problem are replaced by abstract mathematical statements which retain those features common to large classes of problems. This enables us to isolate those aspects of decision theory which are relevant to the selection of a single decision rule.

For the most part, this paper does not really contain new mathematical results; instead it is largely concerned with the adaptation of known mathematical results to statistical problems.

Specifically, I will describe three ways of providing a mathematical model for statistical decision theory. The first is motivated by comparatively recent ideas of approximation theory. The second is obtained by representing the statistical problem as an optimization problem. The third approach is a discussion of some possible axiomatizations of statistical decision theory.

## 2. Preliminaries

Let $\Theta$ and $\mathcal{A}$ be topological spaces; $\theta$ and $a$ will be used to denote generic elements of these spaces. Let $L(\theta,a)$ be a mapping from $\Theta \times \mathcal{A}$ into the reals. We will assume throughout that $L(\theta,a)$ is uniformly bounded from below; that is, there exists a real number $M \geq 0$, such that $L(\theta,a) \geq -M$ for all $\theta \in \Theta$, $a \in \mathcal{A}$. It is customary to refer to $\Theta$ as the parameter space, $\mathcal{A}$ as the space of actions, and $L$ as the loss function. We also require a probability space $(\mathcal{X},\mathcal{B}_{\mathcal{X}},P_\theta)$, where $\mathcal{X}$ is a space of elements $x$ and $\mathcal{B}_{\mathcal{X}}$ is a $\sigma$-algebra of subsets of $\mathcal{X}$, and $P_\theta$, $\theta \in \Theta$ is one of a family of probability measures on $\mathcal{B}_{\mathcal{X}}$ indexed by $\Theta$.

An experiment is conducted and the random variable $X$ is observed. $X$ is assumed to take values in $\mathcal{X}$ and is distributed by $P_\theta$, where $\theta$ is an element of $\Theta$ whose value is unknown to the statistician.

We can now outline the steps in a statistical problem. The statistician observes the event $X = x$, $x \in \mathcal{X}$; then given the family of measures $P_\theta$, but not the value of $\theta$, he selects $a \in \mathcal{A}$ and is assessed the penalty $L(\theta,a)$. Thus, the objective is to choose $a$ so that $L(\theta,a)$ is "kept small"; ideally, if there exists $a_0$ such that

$$L(\theta,a_0) = \min_{a \in \mathcal{A}} L(\theta,a) \quad ,$$

then one should choose $a_0$. However, it is obvious that in order to be able to choose $a_0$, in general, one would need to know which parameter $\theta \in \Theta$ prevailed. Since X is distributed by $P_\theta$, the event $X = x$ contains information about $\theta$, hence the choice of $a \in \mathcal{A}$ should generally depend on the outcome of the experiment. This, given the outcome of the experiment, a mapping $\delta: \mathcal{X} \to \mathcal{A}$ is chosen. Since X is a random variable, $\delta(X)$ is a random variable, provided that $\delta$ is a measurable mapping. We denote the set of such measurable mappings by $\mathfrak{D}$.

In repetitions of the experiment, X will change; hence, unless $\delta(X)$ is almost surely constant, $L(\theta,\delta(X))$ will vary. Thus, the average loss, $E_\theta\, L(\theta,\delta(X))$ is used as a criterion for choosing $\delta$ rather than the loss in any given experiment. We call $E_\theta\, L(\theta,\delta(X))$ the risk function (of $\delta$) and denote it by $R_\delta(\theta) = R(\theta,\delta)$. Note that our assumptions ($\delta$ measurable and $L(\theta,a)$ uniformily bounded from below) insure that $R_\delta(\theta)$ "exists" for each $\theta$; it may, however, be $+\infty$.

It is desirable to augment $\mathfrak{D}$ by introducing the "mixed" or "randomized" decision procedures $\Phi$, whose elements will be denoted by $\varphi$. To do this, we introduce a $\sigma$-algebra $\mathcal{B}_{\mathfrak{D}}$, and let $\Phi$ be the set of all probability distributions on $\mathcal{B}_{\mathfrak{D}}$. Of necessity $\mathcal{B}_{\mathfrak{D}}$ must include each $\{\delta\}$, $\delta \in \mathfrak{D}$, so that $\mathfrak{D} \subset \Phi$, by using distributions such that $P\{\delta\} = 1$. We will refer to the elements of $\Phi$ as decision rules or decision procedures; when it is necessary to specifically identify an element of $\Phi$ as being in $\mathfrak{D}$, we will refer to it as a <u>pure</u> decision rule.

Clearly, if $R(\theta,\delta)$ is $\mathcal{B}_{\mathfrak{D}}$ measurable for each $\theta$,

$$R_\varphi(\theta) = R(\theta,\varphi) = \int_{\mathfrak{D}} R(\theta,\delta)\, d\varphi(\delta)$$

is well defined. We call $R_\varphi(\theta)$, as defined above, the risk function of $\varphi$. Then, the problem of selecting a decision procedure becomes the problem of selecting $\varphi \in \Phi$ so that $R(\theta,\varphi)$ is kept "small".

Let $S = \{R_\varphi(\theta), \varphi \in \Phi\}$ and let T be the mapping defined by $T: \Phi \to S$, that is $T(\varphi) = R_\varphi(\theta)$. We refer to S as the risk set.

If for $\varphi_1, \varphi_2 \in \Phi$, we have $R(\theta,\varphi_1) = R(\theta,\varphi_2)$ for all $\theta \in \Theta$, then $\varphi_1$ and $\varphi_2$ are said to be equivalent. Thus, the elements of S are equivalence classes of elements of $\Phi$. Consequently, we can replace the problem of selecting $\varphi \in \Phi$ with the equivalent problem of selecting an element $s \in S$. Then we can choose any element of $T^{-1}(s)$ as the decision procedure to be employed.

In order to simplify notation, we will adopt the following conventions in the material that follows. For $s_1, s_2 \in S$, $s_1 < s_2$ means $T^{-1}(s_1) = R(\theta,\varphi_1) \leq R(\theta_2,\varphi_2) = T^{-1}(s_2)$ for all $\theta \in \Theta$ and for $\theta_0 \in \Theta$, $R(\theta_0,\varphi_1) < R(\theta_0,\varphi_2)$; here $\varphi_1$ is any element of $T^{-1}(s_1)$ and $\varphi_2$ is any element of $T^{-1}(s_2)$. Similarly, $s_1 \leq s_2$ means $R(\theta,\varphi_1) \leq R(\theta,\varphi_2)$ for all $\theta \in \Theta$.

In particular, we will say that $s_1 \in S$ is inadmissible (and all $\varphi \in T^{-1}(s_1)$ are inadmissible) if there is an $s_0 \in S$ with $s_0 < s_1$. If there is no such $s_0$, then $s_1$, equivalently any $\varphi \in T^{-1}(s_1)$, will be said to be admissible.

Clearly, if $s_1 < s_2$, then $s_2$ should not be employed by the statistician. Further, if there is an $s_0$ such that $s_0 \leq s$, for all $s \in S$; then $s_0$ is to be chosen and there is no problem of selection. This is, unfortunately, an exceptional situation. In most problems, an ordering of this type is not available among the decision rules being considered; it is quite customary to find oneself with the problem of selecting one of a large set of mutually incomparable rules.

Note that S is of necessity a convex set. To see this, observe that $\Phi$ is a convex set. Then let

$$\varphi_0 = \lambda\varphi_1 + (1 - \lambda)\varphi_2\ , \quad \varphi_1,\varphi_2 \in \Phi, \quad \lambda \in [0,1].$$

Then

$$R_{\varphi_0}(\theta) = \int_{\mathcal{D}} R(\theta,\delta)\ d[(\lambda\varphi_1 + (1-\lambda)\varphi_2)(\delta)]$$

$$= \int_{\mathcal{D}} R(\theta,\delta)[\lambda d\varphi_1(\delta) + \lambda d\varphi_2(\delta)]$$

$$= \lambda R_{\varphi_1}(\theta) + (1-\lambda)R_{\varphi_2}(\theta) .$$

Since T is a linear mapping, we have that if $s_0 = T(\varphi_0)$, then

$$s_0 = \lambda_{s_1} + (1-\lambda)s_2,$$

where

$$s_1 = T(\varphi_1) , \quad s_2 = T(\varphi_2).$$

However, as S has been defined, it is not necessarily the convex hull of $T(\mathcal{D})$, denoted by $co(T(\mathcal{D}))$. To see this, let $\mathcal{D} = \{-\infty < \theta < \infty\}$ and $\Theta = \{-\infty < \theta < \infty\}$ and

$$R_\delta(\theta) = \begin{cases} 0 & \theta < \delta \\ 1 & \theta \geq \delta \end{cases} .$$

That is, $T(\mathcal{D})$ is the set of all degenerate cumulative distribution functions. It is easily seen that $co(T(\mathcal{D}))$ is the set of all cumulative distribution functions with a finite number of jumps. However, S is the set of all cumulative distribution functions on $(-\infty, \infty)$.

In the discussion that follows, the "classical" decision criteria known as the minimax criterion, minimax regret criterion, and Laplace's criterion will be used repeatedly as illustrations. For the sake of completeness, they are defined below. Each of these corresponds to a different interpretation of what might be meant by "keeping $R_\varphi(\theta)$ small".

(1). <u>The minimax criterion.</u> Choose $\varphi_0 \in \Phi$ so that $\sup_\theta R_{\varphi_0}(\theta) \leq \sup_\theta R_\varphi(\theta)$ for all $\varphi \in \Phi$.

(2). <u>The minimax regret criterion.</u> Choose $\varphi_0 \in \Phi$ so that

$$\sup_\theta [R_{\varphi_0}(\theta) - \inf_{\gamma \in \Phi} R_\gamma(\theta)] \leq \sup_\theta [R_\varphi(\theta) - \inf_{\gamma \in \Phi} R_\gamma(\theta)]$$

for all $\varphi \in \Phi$.

(3). Laplace's criterion. Choose $\varphi_0 \in \Phi$ so that

$$\int_{\Theta} R_{\varphi_0}(\theta)\, d\mu(\theta) \leq \int_{\Theta} R_{\varphi}(\theta) d\mu(\theta)$$

for all $\varphi \in \Phi$, where $\mu$ is the uniform measure on $\Theta$. (If $\Theta$ is a compact subset of $E_n$, for example,

$$\int_{\Theta} R_{\varphi}(\theta)\, d\mu(\theta)$$

is well-defined. In more general situations, some modifications to this definition may be required).

In each of the three cases above $s_0$, the "optimal" element of s, is defined by $s_0 = T(\varphi_0)$.

Each of these reflects a different interpretation of "keeping $R_\varphi(\theta)$ small", in ignorance of the value of $\theta$. The minimax criterion guarantees that the largest value of $R_\varphi(\theta)$ is as small as possible. The minimax regret criterion identifies

$$\inf_{\varphi \in \Phi} R_{\varphi}(\theta) \quad ,$$

the lower envelope function, as the smallest loss that you could incur if you knew $\theta$; hence

$$R_{\varphi}(\theta) \ - \inf_{\gamma \in \Phi} R_{\gamma}(\theta)$$

is the additional loss that is incurred by one's ignorance of $\theta$. Then you seek to make the maximum of this difference as small as possible. In Laplace's criterion, the philosophy is one of keeping the "average loss" small rather than the maximum - as is the case with minimax and minimax regret. Average is here identified with the uniform measure on $\Theta$.

Historically, there have been two approaches employed in studying the question of what might be meant by a "best" $s \in S$. They are

(1) Specify a criterion, then deduce its properties and see if they are satisfactory.

(2) List the properties that you would like a decision procedure to possess, then determine the existence and construction of decision rules meeting the required conditions.

In what follows, the third and fourth sections will follow the first approach and the fifth section will use the last approach.

For additional material on the above definitions and concepts, the reader is referred to standard treatises on decision theory, such as D. Blackwell and M. A. Girshick [2], T. S. Ferguson [11], and A. Wald [30].

## 3. Approximation Theory and Statistical Decision Theory

Let $\mathcal{L}$ be a normed linear space of real-valued functions of $\theta$, $\theta \in \Theta$; for $x \in \mathcal{L}$, we denote the norm of x by $\|x\|_{\mathcal{L}}$. When there is no danger of confusion concerning the space under consideration, the subscript $\mathcal{L}$ will be deleted. Let $S_{\mathcal{L}}$ be a convex set in $\mathcal{L}$ and let v be a distinguished point in $\mathcal{L}$.

We say that $s_0 \in S_{\mathcal{L}}$ is a best approximation to v if $\|s_0 - v\| \leq \|s - v\|$ for all $s \in S_{\mathcal{L}}$. The existence and determination of $s_0$ is a well-known problem in approximation theory and there is considerable literature about this topic. A few of the more significant of these results are summarized below. If we add the additional assumption that $v \leq s$ for all $s \in S_{\mathcal{L}}$, then, we will show that the notion of a best approximation to v is a possible interpretation of the concept of optimality in statistical decision theory. Consequently, the theorems of this part of approximation theory frequently have a natural reinterpretation in a statistical context.

For general discussions of properties of convex sets in normed linear spaces, the reader is referred to F. A. Valentine [29] and N. Dunford and J. T. Schwartz, Chapter V, [9]. The theory of best approximations by elements of convex sets is discussed in the book by I. Singer [26], (in particular see Appendix I) and in papers by F. R. Deutsch and P. H. Maserick [8], A. L. Garkavi [12,13], V. N. Burov [3,4], and G. S. Rubinstein [23], to name a few.

We denote a hyperplane H in $\mathcal{L}$ as a set of the form

$$H = \{x \in \mathcal{L}: L(x) = c\}$$

where $L \in \mathcal{L}^*$, the adjoint space of $\mathcal{L}$, $L \neq 0$, and c is a real scalar.

Then, the best approximation in $S_{\mathcal{L}}$ to v can be characterized by the following theorems, which will be stated here without proofs.

<u>Theorem 3.1</u> (I. Singer [26], F. R. Deutsch and P. H. Maserick[8]). Let $S_{\mathcal{L}}$ be a convex set in $\mathcal{L}$, a normed linear space, and let $v \in S_{\mathcal{L}}^c$, the complement of $S_{\mathcal{L}}$. Then, there exists an $s_0 \in S_{\mathcal{L}}$ which is a best approximation to v if and only if there exists a linear functional $L \in \mathcal{L}^*$ with

(1) $\|L\| = 1$,

(2) $L(s_0) = \inf_{s \in S_{\mathcal{L}}} L(s)$,

(3) $L(s_0 - v) = \|s_0 - v\|$ .

Geometrically, this says that a point $s_0$ in $S_{\mathcal{L}}$ is a best approximation to $v \in S_{\mathcal{L}}^c$ if and only if there is a hyperplane H separating v from $S_{\mathcal{L}}$ which supports $S_{\mathcal{L}}$ at $s_0$, and whose distance from v is the distance from v to $s_0$.

Secondly, we have the following characterization of a best approximation.

<u>Theorem 3.2.</u> (A. L. Garkavi [13]). Let $S_{\mathcal{L}}$ be a convex set in the normed linear space $\mathcal{L}$. Then $s_0 \in S_{\mathcal{L}}$ is the best approximation to $v \in S_{\mathcal{L}}^c$ if and only if for each $s \in S_{\mathcal{L}}$ there is a linear functional $L_s$ in $\mathcal{L}^*$ such that

(1) $L_s$ is an extreme point of the closed unit ball in $\mathcal{L}^*$,

(2) $L_s(s - s_0) \geq 0$,

(3) $L_s(s_0 - v) = \|s_0 - v\|$ .

We now turn to the connection between the best approximation problem described above and optimality in statistical decision theory.

Let S be the risk set of a statistical decision problem and let $\mathcal{L}$ be a normed linear space. Let $S_{\mathcal{L}} = S \cap \mathcal{L} = \{s \in S, \|s\|_{\mathcal{L}} < \infty\}$. We say that $s_0 \in S$ is $(v, \mathcal{L})$ optimal if for $v \leq s$, for all $s \in S_{\mathcal{L}}$, $s_0$ is the best approximation to v from $S_{\mathcal{L}}$. If $S_{\mathcal{L}}$ is empty, then we define every $s \in S$ as $(v,\mathcal{L})$ optimal.

It remains to be shown that this is in fact a reasonable definition of optimality for statistical decision problems. We will try to justify this in two steps. First, we will show that minimax, minimax regret, and Laplace's criterion are $(v,\mathcal{L})$ optimal procedures for particular choices of v and $\mathcal{L}$. Second, having established that this is in fact a generalization of these "classical" decision criteria, we will give an intuitive interpretation of the notion of $(v,\mathcal{L})$ optimality as a family of decision criteria. We will make the simplifying assumption that $\Theta$ is compact. Modifications to some definitions will be needed, when this is not the case, and can easily be made. However, these will be omitted here, since they do not serve the immediate purpose of this exposition.

Let $v = v(\theta) = -M$, $\theta \in \Theta$ . Then, if we take $\mathcal{L}$ to be the space of bounded functions $f(\theta)$, $\theta \in \Theta$ with

$$\|f\|_{\mathcal{L}} = \sup_{\theta} |f(\theta)| \quad .$$

Then, the best approximation to v from S, that is, the $(v,\mathcal{L})$ optimal decision rule for this case is the minimax decision procedure. Note that our hypotheses insure $v \leq s$ for all $s \in S$.

Now employ the same choice of v and let $\mathcal{L}$ be the space of $\mu$-integrable functions of $\theta$, where $\mu$ is the uniform measure on $\Theta$. Here the $(v,\mathcal{L})$ optimal decision rule is

Laplace's criterion.

To obtain identification of minimax regret as a $(v,\mathcal{L})$ optimal decision procedure, we define

$$v = v(\theta) = \inf_{\varphi \in \Phi} R_{\varphi}(\theta) ,$$

the lower envelope function. Clearly $v \leq s$, for all $s \in S$. Then the same choice of $\mathcal{L}$ as in the representation of the minimax criterion provides the representation of the minimax regret criterion.

Hence, it is evident that this notion of optimality is a generalization of some of the familiar notions of optimality.

We now give an intuitive interpretation of $(v,\mathcal{L})$ optimality as a statistical optimality criterion. The statistician should interpret v as the "ideal" decision rule, that is, what he would like to be able to accomplish, such as in the case of "perfect information". The distance from v to S reflects his inability to accomplish this ideal as a consequence of uncertainty. Since v in general is not attainable, the suggestion is to choose that element of S which comes as close as possible to the ideal v, hence, "a best approximation to v".

To clarify these ideas, consider the following simple example. We consider a decision theroetic model for the problem of testing a simple hypothesis against a simple alternative. Thus $\Theta = (\theta_1, \theta_2)$ and $\mathcal{A} = (a_1, a_2)$.

Let

$$L(\theta,a) = \begin{array}{c} \\ a_1 \\ a_2 \end{array} \begin{array}{c} \begin{array}{cc} \theta_1 & \theta_2 \end{array} \\ \begin{pmatrix} 0 & 1 \\ 1 & 0 \end{pmatrix} \end{array}$$

Then the risk set is the set $(\alpha_{\varphi}, \beta_{\varphi})$, corresponding to the probabilities of errors of the first and second kind using the decision rule $\varphi \in \Phi$. A typical risk set for such a problem is shown in the figure.

In the figure, it appears natural to set $v = (0,0)$, which corresponds to a "perfect" test, that is, one with size zero and power unity. The different choices of $\mathcal{L}$ correspond to different ways of defining the point in S which is closest to (0,0).

We now return to Theorems 3.1 and 3.2 to reexamine them in the light of statistical decision theory, rather than approximation theory. First note that the only significant distinction in the conversion to the statistical problem is the additional assumption $v \leq s$. This readily leads to the following, which we state as Theorem 3.3.

<u>Theorem 3.3.</u> Let $S_{\mathcal{L}}$ be a convex set in $\mathcal{L}$, a normed linear space and let $v \leq s$, for $s \in S_{\mathcal{L}}$. Then, $s_0$ is a best approximation to v if and only if there exists a <u>positive</u> linear functional $L \in \mathcal{L}^*$ satisfying conditions (1), (2), and (3) of Theorem 3.1. Similarly, $s_0$ is a best approximation to v if and only if for each $s \in S_{\mathcal{L}}$, there is a positive linear functional $L_s$ in $\mathcal{L}^*$ satisfying conditions (1), (2), and (3) of Theorem 3.2.

Some of the more immediate consequences, for example, include the following well-known result in statistical decision theory. If all the risk functions in $S_{\mathcal{L}}$ are continuous and $S_{\mathcal{L}} \neq \phi$, then the space $\mathcal{L}$ determined by the sup norm coincides with $\mathcal{L}_\infty$ space, say if $\Theta \subset E_n$. The adjoint space is $\mathcal{L}_1$ space and the positive linear functionals of norm unity are then the probability distributions on $\Theta$, using the usual Borel $\sigma$-algebra on $\Theta$. Then the linear functional L of Theorem 3.1 is the least favorable distribution (for either minimax or minimax regret, depending on the choice of v from the two alternatives specified earlier). In a less restrictive context, the linear functional L of Therorem 3.1, gives a prior distribution against which $s_0$ is Bayes, provided one adds a few minor regularity conditions. Statistical interpretations of Theorem 3.2 are not as useful, although these can be inferred as well.

## 4. Optimization in Statistical Decision Theory Defined by Minimizing Functionals

In this section, we introduce another method of defining optimization in statistical decision problems.

Let $\mathcal{H}$ be a class of extended real valued functionals h on a linear topological space $\mathcal{M}$ of real valued functions of $\theta$, $\theta \in \Theta$, with the following properties:

(1) $h(0) = 0$, $|h(-\bar{M})| < \infty$, where $-\bar{M}$ denotes the function $f(\theta) \equiv M$, $\theta \in \Theta$.

(2) If $x \leq y$, $h(x) \leq h(y)$.

Then we say that $s_0 \in S \subset \mathcal{M}$ is h-optimal if $h(s_0) \leq h(s)$ for all $s \in S$. If $h(s) = +\infty$ for all $s \in S$, then every $s \in S$, is said to be h-optimal.

Since every norm is a functional, h-optimality reduces to $(v,\mathcal{L})$ optimality in some particular instances. That is, if $h(s) = \|s - v\|_{\mathcal{L}}$, when $\|s - v\|_{\mathcal{L}} < \infty$ and $h(s) = \infty$, otherwise, then this is precisely $(v,\mathcal{L})$ optimality. However, there are many instances of h-optimality which are not expressible in terms of norms, and hence, this is in fact a generalization of $(v,\mathcal{L})$ optimality.

It is instructive to examine the principle of Bayesian inference in the light of the definition of h-optimality. Let $\mu$ be a probability measure (equivalently any measure $\mu$ with $\mu(\Theta) < \infty$) on the Borel sets of $\Theta$. Then $s_0$ is the Bayes decision rule with respect to $\mu$ if

$$\int_{\Theta} R_{\varphi_0}(\theta)\, d\mu(\theta) \leq \int_{\Theta} R_{\varphi}(\theta)\, d\mu(\theta) \tag{4.1}$$

for all $\varphi \in \Phi$ and $s_0 = T(\varphi_0)$. Note that we need the additional assumption that $R_{\varphi}(\theta)$ is measurable with respect to the Borel $\sigma$-algebra on $\Theta$ for every $\varphi \in \Phi$.

Observe that (4.1) is a statement of minimization with respect to a linear functional. Since, as a consequence of the Riesz representation theorem, every positive linear functional has a representation of the form employed in (4.1) for compact $\Theta$, Bayesian inference coincides in this case with h-optimality for linear h.

This observation provides an obvious explanation for the assertion that the solution to a Bayesian problem is obtained more easily than the solution using other criteria. Namely, in the sense employed here, Bayesian problems, are linear problems, that is, minimization with respect to linear functionals. Other optimization principles are generally non-linear in this sense.

We can relate this principle to a growing body of mathematical literature as well by noting that this is precisely the structure of mathematical programming problems. The function h becomes the objective function of the mathematical programmer and the convex set S is the set of feasible points. Hence, Bayesian inference is linear programming from this point of view. However, the set of constraints required to generate S need not necessarily be finite. The general optimization problem is, in general, a problem of non-linear programming in an arbitrary linear space.

Some useful work in this area which can be exploited by statisticians are J. W. Daniel [6,7], K. Kirchgässner and K. Ritter [15], K. Ritter [22], and L. W. Neustadt [19]. In particular, it should be noted that some relationships between mathematical programming and the areas of statistics and probability are quite well-known. An extensive discussion of these and a substantial bibliography may be found in the survey paper by O. Krafft [16].

## 5. Axiomatizations of Optimality in Statistical Decision Theory

Another method of defining optimality in statistical decision problems is to list properties that you would like such a procedure to possess. I will briefly summarize the history of this topic and conclude with the statement of some recent results by E. E. Nordbrock and some of their consequences.

In H. Chernoff [5], a set of eight postulates is exhibited for a finite decision problem ($\Theta$, $\mathcal{A}$ both finite). For these postulates, Laplace's criterion is the only rule which satisfies all eight. Chernoff notes that if an additional postulate were added to the list; a postulate of the "nature duplication" type to be discussed below, then a contradiction would result. Chernoff's results "justifying" Laplace's criterion were extended to more general decision problems by H. Uzawa [27,28].

In J. Milnor [18], a list of ten postulates for a finite decision problem are given. Subsets of these which characterize Laplace's criterion, minimax, and minimax regret are exhibited. It should be noted that minimax regret was proposed by L. J. Savage in [24,25], and is referred to as Savage's criterion by Milnor. Milnor also exhibited a set of eight postulates which are consistent and gave a construction of a rule which satisfies these postulates. His rule is a precursor of the rule used by Atkinson, Church, and Harris [1], which will be discussed in greater detail later. Good [14], proposed a restricted type of minimax rule. The thirteenth chapter of R. D. Luce and H. Raiffa [17] and the paper of R. Radner and J. Marschak [21] provide expository treatment of decision principles.

In Atkinson, Church, and Harris, the following set of postulates were proposed for the finite decision problem $\Theta = \{\theta_1, \ldots, \theta_n\}$ and $\mathcal{D} = \{d_1, d_2, \ldots, d_m\}$; here S is a convex polyhedron in $E_n$ and is the convex hull of the row vectors of the matrix $A = \{a_{ij}\}$, where $a_{ij} = L(d_i, \theta_j)$.

1. The optimal class Q(A) is non-empty.

2. Let $\pi_1$ and $\pi_2$ be permutations acting on $\Theta$ and $\mathcal{D}$ respectively. Then if

$$A' = \{a'_{ij}\} = \{a_{\pi_1^{-1}(i)\,\pi_2^{-1}(j)}\},$$

   Q(A') is the set of points of S obtained by applying $\pi_2$ to the coordinates of points in Q(A).

3. Every element of Q(A) is admissible.

4. Q(A) is a convex subset of S.

5. If

$$A_1 = \lambda A_o + \begin{bmatrix} c_1 c_2 \dots c_n \\ c_1 c_2 \dots c_n \\ \vdots \\ c_1 c_2 \dots c_n \end{bmatrix}, \quad \lambda > 0,$$

then $Q(A_1) = \{\lambda \tilde{x} + \tilde{c}, \tilde{x} \in Q(A_0)\}$, where $\tilde{c} = (c_1, c_2, \dots, c_n)$.

6. Let $co(A_1^T) = co(A_2^T)$, where $A^T$ is the transpose of A and co(A) is the convex hull of the row vectors of A. If $A_1$ is obtained from $A_2$ by deleting j columns from $A_2$, then $Q(A_1)$ is obtained from $Q(A_2)$ by deleting the corresponding j coordinates from each element of $Q(A_2)$.

Remark: This type of postulate is usually called a "nature duplication" postulate.

7. If for two statistical decision problems, $A_1$ and $A_2$, with risk sets $S_1$ and $S_2$, the points of $S_1$ which are both extreme points and admissible coincide with the points of $S_2$ which are both extreme points and admissible, then $Q(A_1) = Q(A_2)$.

8. If $\{A_n\}_{n=1}^{\infty}$ converges to $A_0$ and $x_n \in Q(A_n)$ for every n, then every limit point of $\{x_n\}_{n=1}^{\infty}$ is in $Q(A_0)$. (Here convergence of $\{A_n\}$ is element by element).

In [1], it was exhibited that these postulates are consistent for finite decision problems and a decision rule called "iterated minimax regret" (IMR) was shown to satisfy all eight postulates. IMR is closely related to the rule given by Milnor [18] and it is described below.

The iterated minimax regret principle (IMR) selects any element $s_0 \in S$ as optimal which is obtained by the following process.

Let

$$v_1 = v_1(\theta) = \inf_{\varphi} R_{\varphi}(\theta)$$

and let $\mathcal{L}$ be the normed linear space with, for $s = T(\varphi)$,

$$\|s\| = \|R_{\varphi}(\theta)\| = \sup_{\theta \in \Theta} R_{\varphi}(\theta) .$$

Let

$$z_1 = \inf_{s \in S} \|s - v_1\|$$

and let $\{\epsilon_n\}_{n=1}^{\infty}$ be a sequence of positive numbers with

$$\lim_{n \to \infty} \epsilon_n = 0.$$

Let $Q_1 = S$ and inductively, for $n \geq 1$, define $Q_{n+1} = \{s \in Q_n : \|s - v_n\| \leq z_n + \epsilon_n z_1\}$ where

$$v_n = \inf_{\varphi \in T^{-1}(S_n)} R_{\varphi}(\theta)$$

and

$$z_n = \inf_{s \in S_n} \|s - v_n\| .$$

If $z_1 = \infty$, then all $s \in S$ are said to be optimal. If $z_1 < \infty$, define

$$Q = \bigcap_{n=1}^{\infty} Q_n$$

and choose as $s_0$ any element of Q.

B. Efron [10] extended these results in part to infinite decision problems. Here some modifications in the postulates must be made and these are listed below in the form given by E. E. Nordbrock [20].

2'. If $h: \Theta \to \Theta'$ is a homeomorphism and if $S' = \{R'_{\varphi}(\theta'); \theta' \in \Theta'; R' = R \circ h\}$, then $Q' = Q \circ h$.

5'. If $S' = \lambda S + c$, where $c = c(\theta)$ is a continuous function of $\theta$, then $Q' = \lambda Q + c$.

6'. Let $S' = \{R_\varphi(\theta')\}$ and $S = \{R_\varphi(\theta) = R|\Theta\}$, where $R|\Theta$ means the restriction of the domain of R to $\Theta$ and $\Theta \subset \Theta'$. Then if for every $\theta_0 \in \Theta'$, there is a probability measure

$$\mu_{\theta_0}$$

on $\Theta$ such that for all $\varphi \in \Phi$, we have

$$R_\varphi(\theta_0) = \int_\Theta R_\varphi(\varphi)\, d\mu_{\theta_0}(\theta) ,$$

then $Q'|\Theta = Q$.

7'. If S and S' have a common complete class, then $Q = Q'$.

8'. Define $d(S,S') = \max \{ \sup_{s \in S} \inf_{r \in S'} \|r - s\|,$ $\sup_{r \in S'} \inf_{s \in S} \|r - s\|\}$. Then if $d(S_n,S) \to 0$ as $n \to \infty$ and if $s^{(n)} \in Q_n$ for all n, and if $d(s^{(n)}, s) \to 0$, then $s \in Q$.

Efron showed that these postulates (1,2',3,4,5',6',7', 8') are satisfied by IMR for S a closed bounded convex set in $E_n$. He also claimed that with the exception of postulate 1 and the weakening of the conclusion of postulate 8 to $s \in \bar{Q}$, this holds for closed bounded convex sets in $L_\infty$. However, the following counter example shows that inadmissible decision procedures may result.

Example 5.1. Nordbrock [20]. Let $\Theta = \{1,2, \ldots \}$ and let $\mathcal{D} = \{0,1,2, \ldots \}$. For $\delta = 1,2, \ldots$ , let

$$R_\delta(\theta) = \begin{cases} 0 & \delta = \theta \\ 1 & \delta \neq \theta \end{cases}$$

and let

$$R_0(\theta) = 1 \text{ for all } \theta \in \Theta .$$

Let S be the closed convex hull of $\{R_\delta(\theta)\}$. Then it is easily seen that

$$\bigcap_{n=1}^{\infty} Q_n = Q_1 = S$$

and $s_0 = R_0(\theta)$ is inadmissible.

We now conclude with a statement of Nordbrock's results.

S is said to be weak intrinsically compact (Wald[30]), if for every sequence $\{s_n\} \in S$, there is a subsequence $\{s_{n_k}\}$ and an $s' \in S$ such that

$$\lim \quad \inf R_{n_k}(\theta) \geq R'(\theta) \text{ for all } \theta,$$

where

$$s_{n_k} = R_{n_k}(\theta)$$

and $s' = R'(\theta)$.

Theorem 5.1. IMR satisfies properties 2',4,5',6' generally. Property 1 holds if S is weak intrinsically compact. Property 8' holds if S is closed and properties 3 and 7 hold if S is compact.

## 6. Summary

In this exposition, I have attempted to give some illustrations of the possible directions in which the mathematical foundations of statistical decision theory might be developed. The limitations of this volume preclude the extensive development of these ideas which are necessary in order to determine its possible impact on the subject of theoretical statistics. However, it is hoped that this brief exposition will encourage research workers to further examine the implications of the ideas contained herein. At this stage it is not yet apparent whether the material of sections three and four will in fact produce new basic results in statistics as such. To date, it is possible to identify many familiar statistical results in the writings of functional analysts and further display the correspondence between these two areas. The notion of iterated minimax regret developed in the fifth section is handicapped by its apparent imcomputability, except in rather artificial examples. The principle has not been successfully applied to any concrete statistical problem as yet.

The author is indebted to Professor Herman Chernoff for some discussions concerning his paper [5] and for calling the two papers by H. Uzawa [27,28] to his attention.

References

[1] F. V. Atkinson, J. D. Church, and B. Harris, Decision procedures for finite decision problems under complete ignorance, Ann. Math. Statist. 35 (1964), 1644-1655.

[2] D. Blackwell and M. A. Girshick, Theory of Games and Statistical Decisions, John Wiley, New York, 1954.

[3] V. N. Burov, Approximation with constraints in linear normed spaces, I., Ukrain. Mat. Ž. 15 (1963), 3-12. (in Russian)

[4] V. N. Burov, Approximation with constraints in linear normed spaces, II., Ukrain. Mat. Ž. 15(1963), 135-144. (in Russian)

[5] H. Chernoff, Rational selection of decision functions, Econmetrica 22 (1954), 422-443.

[6] J. W. Daniel, Applications and methods for the minimization of functionals in Non-linear Functional Analysis and Applications, 399-424, Editor L. B. Rall, Academic Press, New York, 1971.

[7] J. W. Daniel, The Approximate Minimization of Funtionals, Prentice-Hall, Englewood Cliffs, N. J. (1971).

[8] F. R. Deutsch and P. H. Maserick, Applications of the Hahn-Banach theorem in approximation theory, SIAM Rev. 9 (1967), 516-530.

[9] N. Dunford and J. T. Schwartz, Linear Operators, I. Interscience Publishers, New York, 1957.

[10] B. Efron, Note on decision procedures for finite decision problems under complete ignorance, Ann. Math. Statist. 36 (1965), 691-697.

[11] T. S. Ferguson, Mathematical Statistics, A Decision Theoretic Approach, Academic Press, New York, 1967.

[12] A. L. Garkavi, On a criterion for an element of best approximation, Sibirsk Mat. Ž. 5 (1964), 472-476. (in Russian)

[13] A. L. Garkavi, Duality theorems for approximation by elements of convex sets, Uspehi Mat. Nauk, 16 (1961), 141-145. (in Russian).

[14] I. J. Good, Rational decisions, J. Royal Statist. Soc. Ser. B, 14 (1952), 107-114.
[15] K. Kirchgässner and K. Ritter, On stationary points of nonlinear maximum problems in Banach spaces, SIAM J. Control 4, (1966), 732-739.
[16] O. Krafft, Programming methods in statistics and probability theory, in Nonlinear Programming, 425-446, Edited by J. B. Rosen, O. L. Mangasarian, and K. Ritter, Academic Press, New York, 1970.
[17] R. D. Luce and H. Raiffa, Games and Decisions, John Wiley, New York, 1957.
[18] J. Milnor, Games against nature, in Decision Processes, edited by R. M. Thrall, C. H. Coombs, and R. L. Davis, John Wiley, New York, 1954.
[19] L. W. Neustadt, Sufficiency conditions and a duality theory for mathematical programming problems in arbitrary linear spaces, in Nonlinear Programming, 323-348, Edited by J. B. Rosen, O. L. Mangasarian, and K. Ritter, Academic Press, New York, 1970.
[20] E. E. Nordbrock, A rational approach to decision problems, MRC Technical Summary Report #1159, Mathematics Research Center, University of Wisconsin, 1971.
[21] R. Radner and J. Marschak, Note on some proposed decision criteria, in Decision Processes, 61-68, Edited by R. M. Thrall, C. H. Coombs, and R. L. Davis, John Wiley, New York, 1954.
[22] K. Ritter, Optimization theory in linear spaces, III. Mathematical programming in partially ordered Banach spaces, Math. Ann. 184 (1970), 133-154.
[23] G. S. Rubinstein, On a method of investigation of convex sets, Dokl. Akad. Nauk. SSSR (N.S.), 102 (1955), 451-454. (in Russian).
[24] L. J. Savage, The theory of statistical decision, J. Amer. Stat. Assoc. 46 (1951), 55-67.
[25] L. J. Savage, The Foundations of Statistics, John Wiley, New York, 1954.
[26] I. Singer, Best approximation in normed linear spaces by elements of linear subspaces (translation of original Romanian book). Publishing House of the Academy of the Socialist Republic of Romania, Bucharest, 1970.
[27] H. Uzawa, A generalization of Laplace criterion for decision problems, Ann. Inst. Statist. Math. 7 (1956), 123-129.
[28] H. Uzawa, Note on the rational selection of decision

functions, Econometrica 25, (1957), 166-174.
[29] F. A. Valentine, Convex Sets, McGraw-Hill Book Company, New York, 1964.
[30] Abraham Wald, Statistical Decision Functions, John Wiley, New York, 1950.

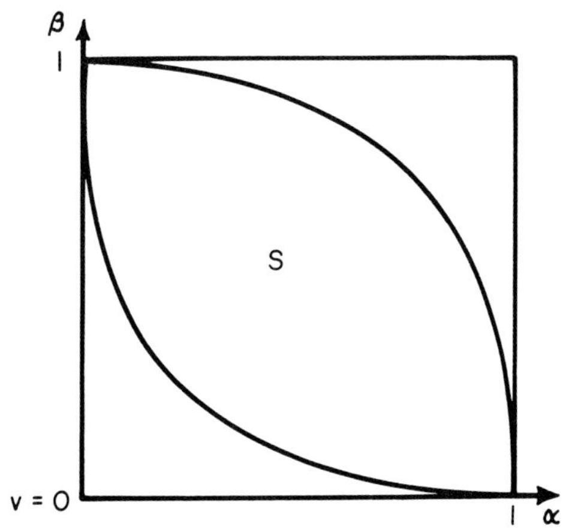


Figure

CHANCE-CONSTRAINED PROGRAMMING:
AN EXTENSION OF STATISTICAL METHOD[1]

A. Charnes,
W. W. Cooper,
M. J. L. Kirby

University of Texas,
Carnegie-Mellon University,
Dalhousie University

## Origin and Concept

Chance-constrained programming was originated in the latter part of 1953 by Charnes, Cooper and Symonds[2] in the context of developing a planning and operations model for Standard Oil of New Jersey for the scheduling of heating oil manufacture, storage and distribution with weather-dependent demand. Initially the situation was modeled as a constrained game against nature since a.) many statistical-decision problems could be reduced to this form (cf. McKinsey [23]), and b.) Charnes [2] had just shown how to reduce 2-person zero-sum constrained games from computationally (and theoretically) impractical fixed-point problems to effectively computable (and analyzable) linear programs.

The constrained game approach, however, was not entirely satisfactory as an extension of statistical method.

[1]This research was partly supported by a grant from the Farah Foundation and by ONR Contracts N00014-67-A-0126-0008 and N00014-67-A-0126-0009 with the Center for Cybernetic Studies, The University of Texas. Reproduction in whole or in part is permitted for any purpose of the United States Government.

[2]Reported by Charnes at the Christmas meeting of the Econometric Society in Washington, D. C. (Also, vide the preface in Symonds [27]).

While the constraints on permissible (mixed) strategies of nature were substantially better than none at all, the remanent "clairvoyant malevolence" of nature implicit in the minimax solution did not seem appropriate to the behavior of weather. Another important consideration was that while the policy of the company as the major supplier of a product tinged with the national interest was to supply all emergent demand, one could determine random combinations of weather whose associated demand would be impossible for the company to meet from its available mechanisms or resources. It was discovered, on further investigation, that several incidents had actually occurred in the past in which the company could not meet the demand. Extraordinary measures outside the company's control had to be taken to alleviate the crises.

An extension of statistical method was therefore sought which (a) did not place nature in the role of a clairvoyant malevolent competitor, (b) would express a "policy" as a rule which was to hold almost all, but not all, the time, and (c) would develop a plan and an operationally implementable procedure for "most", but not all, randomly emergent situations. These challenges chance-constrained programming met through (a) representation of nature more classically via joint probability distributions, (b) representation of a "policy" as a "chance constraint," and (c) posing the problem of chance-constrained programming as determing from a preassigned admissible class a vector of stochastic decision rules which would satisfy the chance constraints and optimize the expected value of a preassigned functional.

It is clear that within these prescriptions chance-constrained programming may be considered a method of tremendous flexibility with innumerable individual possible variations or models. A chance constraint, e.g., the requirement that a given function of the sought vector satisfy a given condition to at least a prescribed level of probability, admits as many interpretations as does the probability operator, e.g., total or conditional, etc. Again, where a system of constraints is desired to hold, chance constraints may be imposed individually on the members of the system ("individual chance constraints") or jointly ("joint chance constraints"). One might employ functionals ranging from minimization of expected costs to maximization of the probability of some event. Additionally, the pre-assigned class of admissible stochastic decision rules, which represent the operational prescriptions of the model, can range from preset parameters to be determined ("zero-order decision rules") to be the most general class of "informationally feasible" measurable functions of sample points of the random variables. In any particular situation to be modeled, of course, the choice is in the hands of the statistician to select the blend of representation of the real features of the situation with an effectively operable and analyzable structure.

To exemplify more concretely the nature of the opportunities and difficulties of chance-constrained programming we shall next formulate a general class of chance-constrained programming problems.

## The Problem of Chance-Constrained Programming

Throughout this paper we will use $b = (b_1, \ldots, b_m)$ to denote a vector of random variables $b_i$, $i = 1, \ldots, m$ with a known joint distribution function,

$$F(t) = F(t_1, \ldots, t_m) = P\{b_i \leq t_i, i = 1, \ldots, m\}.$$

From its definition F is right continuous. The subset of m space which is the range of the random vector b will be denoted $\Omega$. The operators P and E will denote respectively probability and expectation operators; the corresponding integrals are to be evaluated with respect to the Lebesgue-Stieltjes measure (also denoted by F) induced by the function F. The letter $\alpha$ will denote throughout a given fixed real number such that $0 < \alpha \leq 1$.

For any uni-variate distribution function G, the $\alpha$ fractile of G is given by

$$G^{-1}(\alpha) = \inf \{s: G(s) \geq \alpha\}.$$

A <u>decision rule</u> (or <u>rule</u>) $x(t): R^m \to R^n$, is a vector function which is a member of some prescribed nonempty subclass C of the Lebesgue measurable functions. Examples of the class C which we will consider are:

(a) $C_0 = \{x: x(t) = x_0 \text{ for all } t\}$, the class of constant or zero-order rules.

(b) $C_L = \{x : x(t) = Dt + d$, where D is m x n, d is m x 1$\}$, the class of linear rules.

(c) $C_M$ or $C_{PA}$ or $C_{PL}$, the class of measurable or piecewise analytic or linear rules, respectively.

(d) $C_k = \{x$: $x_i(t)$, the i-th component of x, is an arbitrary measurable function of $t_1, \ldots, t_{j_i}$, some $j_i$, and a constant function of $t_j$ for $j > j_i\}$, the class of k-stage or k-order rules.

Note that if $j_i = 0$ for all i then $C_k = C_0$, while if $j_i = m$ for all i then $C_k = C_M$.

For our purposes, the interpretation of a decision rule is that the x(b) represents a decision which will be taken in response to the event b. That is, the "response" to an "observation $\hat{b}$ of the random variable b is to be $\hat{x} = x(\hat{b})$. We have thus defined what ought to be called a nonrandomized rule in that x is uniquely determined once b has been observed.

Within the above framework we can define a fairly general (nonlinear) chance-constrained programming problem as follows:

Select, if one exists, a decision rule x(b) with $x \in C$ which yields an infimum to

$$E\ f(x(b),b)$$

subject to

$$P\ \{g_j(x(b),b) \geq 0\} \geq \alpha_j, \quad j = 1, \ldots, J \tag{1}$$

and

$$x(b) \in K \text{ with probability one (w.p.1).}$$

In (1) the functions $f, g_j$ are assumed to be Borel measurable and to be such that

$$f: R^n \times R^m \to R^1 \qquad \text{and} \qquad g_j: R^n \times R^m \to R^{m_j}.$$

The set K is assumed to be some nonempty subset of $R^n$.

We remark that interesting and useful generalizations of (1) can be obtained by allowing $f$ and $g_j$ to be random functions (e.g.,

$$g_j: R^n \times R^m \times R^{n_j} \to R^{mj}, \text{ so that } g_j = g_j(x,b,u_j),$$

where u is a random vector whose range and sample space are $R^{n_j}$), and by permitting the decision rules to depend upon the $u_j$ as well as b. We shall not treat these cases here.

EXAMPLE 1. Take $C = C_k$ $(k = 2)$ and $j_i = 0$ for $i = 1, \ldots, n_1$ and $j_i = m$ for $i = \bar{n} + 1, \ldots, n$. Then

$$x(t) = [x_1, \ldots, x_{n1}, y_1(t), \ldots, y_{n-n_1}(t)].$$

Let

$$K = \{(x, y(t): Dx \geq d, \ x \geq 0, \ y(t) \geq 0\}$$

and set

$$f(x(b),b) = c_1 x + c_2 y(b)$$

for fixed vectors $c_1$ and $c_2$.

Let $J = 2$ and let $g_1 = Ax + My(b) - b = -g_2$. Let $\alpha_1 = \alpha_2 = 1$.

Then (1) becomes the two-stage linear programming under uncertainly problem:

$$\begin{aligned} &\min c_1 x + Ec_2 y(b) \\ &Dx \geq d, \ x \geq 0 \end{aligned} \tag{2}$$

subject to

$$Ax + My(b) = b, \qquad y(b) \geq 0 \text{ w.p.1.}$$

EXAMPLE 2. Use the same choice of C as in Example 1, and take

$$b = (b^1, b^2) \text{ where } b^1 = (b_1, \ldots, b_m),$$

$$b^2 = (b_{m+1}, \ldots, b_{\bar{m}}) \text{ so that } x(b) = (x, y(b^1)).$$

Choose $K = \{x,y): x \geq 0\}$. Choose $J = \bar{m} + (n - n_1)$. Let $m_j = 1$ for all $j = 1, \ldots, J$. Let

$$g_j = A_{11}^j X - b_j, \; j = 1, \ldots, m;$$

$$g_j = A_{21}^j X + A_{22}^j Y(b^1) - b_j, \; j = m+1, \ldots, \bar{m};$$

$$g_j = Y_{j-\bar{m}}(b^1), \; j = \bar{m}+1, \ldots, \bar{m}+(n-n_1);$$

where the superscript j denotes the j-th row of the matrix.

Let $f = C_1X + C_2Y(b^1)$ for fixed vectors $C_1$ and $C_2$. Problem (1) then becomes

$$\min C_1X + EC_2Y(b^1)$$

subject to

$$P(A_{11}^j X \geq b_j) \geq \alpha_j, \quad j = i, \ldots, m,$$

$$P(A_{21}^j X + A_{22}^j Y(b^1) \geq b_j) \geq \alpha_j, \; j = m+1, \ldots, \bar{m}, n \qquad (3)$$

$$P(Y_{j-\bar{m}}^{(b^1)} \geq 0) \geq \alpha_j, \; j = \bar{m}+1, \ldots, n.$$

This is the two-stage E-model of chance-constrained programming.

If we interpret the P operation in the constraints of (3) to be a total probability operator, then we get the chance constrained model with total probability constraints. If we interpret P to mean that we compute the probability using the conditional distribution of $b^2$ given $b^1$, then we get the chance-constrained model with conditional probability constraints.

EXAMPLE 3. If we use the same functions as in Example 2, but redefine f to be equal to 1 if $c_1x + c_2y(b^1) \geq L_0$ and 0 otherwise, for some prescribed $L_0$, we obtain the P-model objective of chance-constrained programming, namely,

$$\max P\{C_1X + C_2Y(b^1) \geq L_0\}.$$

EXAMPLE 4: If we now define f to be the function

$$[C_2Y(b^1) - E(C_2Y(b^1))]^2,$$

then the objective becomes that of the V-model of chance-constrained programming, namely,

$$\min \text{var } [C_1X + C_2Y(b^1)].$$

In the case of (1) in which J = 1, but in which $m_1$ is not necessarily equal to one,

$$P\ \{g(x(b),b) \geq 0\} \geq \alpha,$$

or

$$P\ \{g_i(x(b),b) \geq 0,\ i = 1, \ldots, m_1\} \geq \alpha,$$

where $g_i$: $R^m \times R^n \to R^1$. This type of constraint has been termed a joint chance constraint in the literature.

For reference, we write here the nonlinear and linear joint chance-constrained problems with decision rule class $C_M$:

$$\min Ef(x(b))$$

$$x(b) \in K \quad \text{w.p.1}$$

subject to

$$P\ \{g_i(x(b)) \geq b_i, \text{ all } i = 1, \ldots, m_1\} \geq \alpha \tag{4a}$$

and

$$x \in C_M.$$

$$\min Ecx(b)$$

$$x(b) \in K \quad \text{w.p.1}$$

subject to

$$P\ \{Ax(b) \geq b\} \geq \alpha \tag{4b}$$

and

$$x \in C_m.$$

Note that (4a) is specialized to the extent that $g_i(x(b),b)$ is written (with slight abuse of notation) as

$$g_i(x(b)) - b_i \quad \text{and} \quad f = f(x(b)),$$

while in (4b), $g_i(x(b),b)$ is $a^i x(b) - b_i$, where $a^i$ is the i-th row of the matrix A.

## Applications, Extensions, and Current Problems

We have noted above the highly desirable aspects of chance-constrained programming in allowing one to formulate plans and procedures for "most", but not all, of the time, where the precise exceptional situations are locked up implicitly in the problem and need not be known a priori. The fact, however, that an optimal class of stochastic decision rules is determined on the <u>viable</u> system sample points (and indeed rigorous specification of the problem requires explicit consideration of the unviable or "rejected" regions) suggest analogies with the Neyman-Pearson theory of testing statistical hypotheses. Thus the notion of "acceptance regions" for chance constraints has recently been formulated by Charnes, Kirby and Raike and applied to derive essential simplifications in chance-constrained programming theory, particularly in the more recondite areas of joint and conditional chance constraints.

The thrust of the chance-constrained method for applications is that it provides a very natural framework for stochastic decision situations which are partially controllable. The analytic form of the structure is also substantially simpler than that in other relevant statistical disciplines. For example, many situations of a queueing nature which suffer (from the standpoint of available theory) from being partially controllable can be formulated in a meaningful and constructively solvable manner by chance-constrained programming. A tremendous class of problems ranging from extension of PERT analyses to cover explicit stochastic features, financial planning over time, military tactical operations, new product introduction in marketing, and engineering design with stochastic considerations, can all be comprehended within this class.

Another important direction of extension has been to so-called "chance-constrained games" and "pseudo-games". These refer to games in which payoff elements may be random variables and mixed strategies may only be partially controllable by the players. As an example of phenomena therein, it turns out in some of the simplest 2-person zero-sum games that the random variable interposition requires for optimal play that each player play optimally for a different "non-random" 2-person zero-sum game. The pseudo-games arise, for example, in situations of price competition among transport carriers and involve even in 2-player situations the loss of minimax solutions which may not exist because the actions of one player restrict the actions of his opponent, e.g., the strategy space is no longer a topological product. The results to date are resident in work and papers of Charnes, Kirby, and Raike, also Charnes, Kirby, Littlechild, and Raike, and farther, Charnes, Cooper, Cassidy and Kirby.

The practitioner's need for constructive solutions leads to the use of linear decision rules, etc., and correspondingly leads to the consideration of classes of distribution functions which combine in simple analytic manner and, thereby, to translation of the stochastic problem into a deterministic one of mathematical programming. As may be anticipated, the two most useful classes to date for approximation of problems and inversion of chance constraints have been mixtures of normal distributions and discrete distributions. Charnes and Woodbury and students, however, are currently developing a class of density functions which are linear combinations of exponentials which permit simple determination of distributions of sums and maxima of non-negative random variables. The problem of approximation of tails of distribution functions with parameters yet to be determined is the central problem for constructive solution of chance-constrained programming problems with higher order decision rules. Some characterizations of optimal decision rules within a specified class, e.g., the optimality of piecewise linear decision rules in the class of piecewise analytic rules for certain models, reduces substantially the scope of the approximations needed.

As in many other developments in probability and statistics, results in chance-constrained programming have influenced substantially results in other fields. For example, the Displaced Ray Theorem (and its method of proof) in the "Solution Theorems and Probabilistic Programming" paper of Charnes, Kirby, and Raike has been the basis for the current complete regularization of general convex programming of Jeroslow and Kortanek in terms of the semi-infinite programming of Charnes and Kortanek and the non-Archimedean Hilbert extension of the real field. All of the "asymptotic solvability" cases of Ben-Israel, Charnes, and Kortanek are resident in "real" solvability in terms of the Hilbert field.

For the guidance of the interested reader, a selection of papers in chance-constrained programming follows.

References

[1] Ben-Israel, A., Charnes, A., and Kirby, M.J. L. "On Stochastic Linear Approximation Problems," Operations Research 18, No. 3, 1970, pp. 555-558.

[2] Charnes, A. "Constrained Games and Linear Programming," Proc. National Academy of Sciences, July 1953, pp. 639-641.

[3] Charnes, A., Cooper, W. W. "Chance-constrained Programming," Management Science, October 1959, pp. 73-79.

[4] Charnes, A., Cooper, W. W. "Deterministic Equivalents for Optimizing and Satisficing under Chance Constraints," Operations Research 11, 1963, pp. 18-39.

[5] Charnes, A., Cooper, W. W., and Kirby, M. J. L. "Optimal Decision Rules in Conditional Probabilistic Programming," Rend. Accad. Sci. Fis. Mat. Nat. Lincei, 45, 1968, pp. 231-235.

[6] Charnes, A., Cooper, W. W. , Learner, D. and Devoe, J., "DEMON, Mark II: Extremal Equations, Solutions and Approximations," Management Science 14, May 1968, pp. 513-524.

[7] Charnes, A., Cooper, W. W., and Thompson, G. L. "Constrained Generalized Medians and Linear Programming under Uncertainty," O.N.R. Research Memorandum No. 41, Northwestern University, Evanston, Ill., July 1961.

[8] Charnes, A., Cooper, W. W., and Thompson, G. L. "Constrained Generalized Medians and Hypermedeans as Deterministic Equivalents for Two-stage Linear Programs under Uncertainty, Management Science 12, 1965, pp. 83-112.

[9] Charnes, A., and Kirby, M. J. L. "Application of Chance-Constrained Programming to Solution of the So-called 'Savings and Loan Association' Type of Problem," RAC-P-12, Research Analysis Corporation, McLean

Virginia, October 1965.
[10] Charnes, A. and Cooper, W. W. "Optimal Decision Rules for the E-model of Chance-Constrained Programming," Cahiers Centre Etudes Recherche Operationnelle 8, 1966, pp. 5-44.
[11] Charnes, A. and Cooper, W. W. "Optimal Decision Rules for the Triangular Model of Chance-Constrained Programming," Cahiers Centre Etudes Recherche Operationnelle, 1971.
[12] Charnes, A. and Cooper, W. W. "Some Special P-models in Chance-Constrained Programming," Management Science 15, 1968, pp. 33-48.
[13] Charnes, A., Cooper, W. W., Littlechild, S., and Raike, W. M. "A Chance-Constrained Model for Transport Pricing and Scheduling Under Competition," Transportation Science 1, No. 2, February 1968.
[14] Charnes, A., Cooper, W. W., and Raike, W. M. "An Acceptance Region Theory for Chance-Constrained Programming," J. Math. Anal. and Applic. 32, No. 1, October 1970.
[15] Charnes, A., Cooper, W. W. and Raike, W. M. "Solution Theorems in Probabilistic Programming: A Linear Programming Approach," J. Math. Anal. Appl. 20, 1967, pp. 565-582.
[16] Charnes, A., Cooper, W. W. and Raike, W. M. "Synthesis and Analysis of Some Transport Price Policies with a Chance-Constrained Demand Capture Model," Proc. Second International Symposium on Railway Cybernetics, Montreal, October 1967, pp. 134-142.
[17] Charnes, A., Cooper, W. W., and Raike, W. M. "Zero-zero Chance-Constrained Games," Theor. Probability Appl. 13, 1968, pp. 663-681.
[18] Charnes, A. and Stedry, A. "A Chance-Constrained Model for Real Time Control in R and D," Management Science 12, 1966, pp. 353-362.
[19] Charnes, A. and Thore, S. "Planning for Liquidity in Savings and Loan Associations," Journal of Finance, December 1966.
[20] Dantzig, G. B. "Linear Programming Under Uncertainty," Management Science 1, 1955, pp. 197-206.
[21] Kirby, M. J. L. "Generalized Inverses and Chance-Constrained Programming," Ph.D. Thesis in Applied Mathematics, Northwestern University, Evanston, Ill., June 1965.

[22] Kirby, M. J. L "The Current State of Chance-Constrained Programming," Proc. Symposium on Mathematical Programming, Princeton University, 1971.

[23] McKinsey, J. C. C. Introduction to the Theory of Games. New York: McGraw-Hill Book Company, 1952.

[24] Miller, B. L. and Wagner, H. M. "Chance-Constrained Programming with Joint Constraints," Operations Research 13, 1965, pp. 930-945.

[25] Raike, W. M. "Dissection Methods for Solutions in Chance-Constrained Programming Problems Under Discrete Distributions," Management Science (to appear).

[26] Raike, W. M. "On Mathematical Programming," Ph.D. Thesis in Applied Mathematica, Northwestern University, Evanston, Ill., June 1967.

[27] Symonds, G. Linear Programming: The Solution of Refinery Problems. Esso Standard Oil Co., 1955, L.C.C. Card No. 55-11950.

[28] Walkup, D. W. and Wets, R. J. B. "Stochastic Programs with Recourse," Document D1-82-0551, Boeing Scientific Research Laboratories, Seattle, Washington, July 1966.

[29] Wets, R. J. B. "Programming Under Uncertainty: The Complete Problem," in Wahrscheinlicheitstheorie verw. Geb. Vol. 4, pp. 316-339, 1966.

[30] Wets, R. J. B. "Programming Under Uncertainty: The Equivalent Convex Program," SIAM J. Appl. Math. 14, 1966, pp. 89-105.

[31] Williams, A. C. "On Stochastic Linear Programming," SIAM J. Appl. Math. 13, 1965, pp. 927-940.

# STOCHASTIC ALLOCATION OF SPARE COMPONENTS

Bennet P. Lientz[1]

University of Southern California

## Abstract

This paper presents a method for stochastically allocating spare components in a multicomponent system. The method can be classified as one in the generalized branch and bound, implicit enumeration category. Other applications are briefly discussed.

## Introduction

This paper extends the results of Lientz [10] to the allocation problem related to spares and repair facilities for investment budgets. Of interest in applications is that the methods herein described are meant to apply to large-scale systems. We will begin with a brief description of implicit enumeration methods to date. In section 2 the algorithm is given. The investment problem given by Demmy[3] is examined in section 3.

The motivation for investigating new procedures lies in the need for developing allocation procedures which 1) handle large-scale problems and 2) allow for operational needs. For small problems we could use a standard enumeration method and an off-the-shelf linear programming code (such as those developed by computer manufacturers prior to unbundling). The need to satisfy operational requirements becomes apparent in large problems where the scheduler, drawing from years of experience, can provide inference as to the general framework of a schedule which would meet operational needs. Frequently, this cannot easily be translated

[1]This work was partially supported by the Air Force Office of Scientific Research, AFSC, under grant no. 71-2025, Project 9749.

into a mathematical form. In statistics this is sometimes reflected in the prior distribution. Is there an analog in allocation? This paper provides one such approach.

Before moving to the algorithm, some examples are useful. Consider the problem of scheduling crews for the Military Airlift Command or for a commercial freight or passenger carrier. An allocation solution would be undesirable if it were to split up crews frequently, to allow insufficient time for turnaround at a congested airport, to permit certain night-day round trips, or to not handle weekend scheduling. To quantify all of these conditions would not only be complex and perhaps unrealistic, but also require too much time to obtain a solution.

Another example is a dispatcher of vehicles from various plant yards to service and install equipment. From experience the dispatcher can weigh classes of routings based upon traffic congestion and accidents. Here the dispatcher seeks a number of good near-optimal solutions to reduce total travel time or mileage.

Most methods developed in the past cannot easily be adapted to the above examples. There are two basic classes of algorithms, algebraic and enumerative. Recent interest has centered on the latter or on a combination of the two. Surveys of existing literature are given in Balinski [1] and Balinski and Spielberg [2]. Two criticisms of some methods are that they only yield single solutions and that this is not done rapidly. This can be fatal in an operational environment where single solutions may not be operational. For example, cutting plane techniques (see Gomory [6] and [7]) are only efficient for the first few cuts and also yield one solution (if, indeed, there is convergence). The group-theoretic method (see Shapiro [11]) depends on a group of high order in a large problem. Also, it is not too successful if more than one pass is required. The most recent papers have been away from true mathematical optimality and toward near-optimality.

## 2. General Procedure

Let us consider the general problem of allocating components in parallel to the existing components of a structure. This would be the case with spares and repair facilities as well as with integrated circuit design. Suppose there are n components which are of k distinct categories. Components of category i are located in $L_i$ different places in the system. We seek to place components in the system so as to maximize the reliability of the system subject to constraints on weight, volume, cost, etc.

A complete allocation is a matrix A( k x n) where the i-th row vector, $A_i$, is the i-th class allocation of A. The (i,j)th entry of the matrix A is the number of components of type i allocated to position j (interpreted as the position of the j-th component of the original system). The allocation is viewed as a matrix rather than as a single vector because several allocation methods consider component types as stages (e.g. dynamic programming).

The general problem is

$$\max_{A} \ R(A) \tag{2.1}$$

subject to

$$\sum_{i} c_{im} \sum_{j} a_{ij} \leq C_m \ , \quad 1 \leq m \leq r \tag{2.2}$$

and $a_{ij}$ is a nonegative integer. Here r is the number of constraints, $c_{im}$ is the unit cost of the i-th class with respect to the m-th constraint, and R(A) is the system reliability with the allocation given by A.

We remark here that an alternative formulation is to minimize cost subject to reliability, weight, volume, and other constraints. That this is not done is in part due to the mathematical form of the problem. In the above version all constraints are linear, but the objective function is in a complex form for general series-parallel structure. Thus, most work in the past has been done in the context of all-in-series structures (see Lientz [10] for a literature summary). For the general all-in-series structure, the problem is

$$\max_{A} \prod_{i=1}^{n} [1 - (1 - p_i)^{a_i+1}] \tag{2.3}$$

subject to

$$\sum c_{im} a_i \leq C_m \quad , \quad 1 \leq m \leq r \ , \tag{2.4}$$

and $\{a_i\}$ are nonegative integers. Now define $d_{ij}$, $b_m$, and $x_{ij}$ by

$$d_{ij} = \ell n \{[1 - (1 - p_i)^{j+1}]/[1 - (1 - p_i)^j]\} \tag{2.5}$$

$$b_m = C_m - \sum_i c_{im} \tag{2.6}$$

$$x_{ij} = 0 \text{ or } 1. \tag{2.7}$$

Letting $a_i = \Sigma x_{ij}$ we obtain the following revised form of the problem.

$$\max_A \sum_i \sum_j d_{ij} x_{ij} \tag{2.8}$$

subject to

$$\sum_i \sum_j c_{im} x_{ij} \leq b_m \tag{2.9}$$

and
$$x_{ij} = 0 \text{ implies } x_{iv} = 0 \text{ for } v > j \text{ for all } i. \tag{2.10}$$

These new variables have a direct relationship to the initial problem. $d_{ij}$ is a measure of the unit gain by adding an additional component of type i to the j already installed; $x_{ij}$ is the indicator function of whether the j-th unit of type i is to be added. Thus, if the j-th unit is not added, the (j + 1)st unit will not be added. This is represented by the constraint (2.10).

The problem given by (2.8) - (2.10) will now be considered. Since the algorithm presented here is one of implicit enumeration there are four elements that must be considered. The first is that of enumeration. Since efficient methods already exist, we will use one of these, namely that in Glover [5].

The next two elements are the rules for eliminating branches of the allocation on the basis of optimality and feasibility. Note that (2.8) can be made into finite summations by defining $LI_i = \min_m [b_m/c_{im}]$ so that (2.8) becomes

$$\max_A \sum_{i=1}^{n} \sum_{j=1}^{LI_i} d_{ij} x_{ij}$$

Set $LI = \sum_i LI_i$. An allocation is then a member of

$$\{0,1\}^{\{1, \ldots ,LI\}} .$$

Let us now suppose that k components have been selected for placement in the system. There is a family of allocations, $B_k$, which follows this placement in the system as well as a set of indices, $I_k$, of the (i,j) where $(i,j) \in I_k$ if the allocation $B_k$ has determined the value of $x_{ij}$. Let the accrued value of the objective function be $PO(k)$ and of the m-th constraint by $P_m(k)$. If we now select the next variable and its value ($(i^*,j^*)$ and $x_{i^*j^*}$, respectively), then we have corresponding values for an allocation $a \in B_k$ given by

$$PR_m(k,(i^*,j^*),x_{i^*j^*})(a) = \sum_i \sum_j{}_{(i,j) \in I - I_k - (i^*,j^*)} c_{im}\psi_{(i,j)}(a) \tag{2.11}$$

and

$$PRO(k,(i^*,j^*),x_{i^*j^*})(a) = \sum_i \sum_j{}_{(i,j) \in I - I_k - (i^*,j^*)} d_{ij}\psi_{(i,j)}(a). \tag{2.12}$$

$\psi_{(i,j)}(a) = 1$ if a assigns the j-th unit of type i and 0 otherwise.

Next, the subproblem based on the partial allocation can be given by

$$\max_a PRO\ (k,(i^*,j^*),x_{i^*j^*})(a) + d_{i^*J^*}\ x_{i^*j^*}$$

subject to

$$PR_m(.)(a) \leq b_m - P_m(k) - c_{i^*m}\ x_{i^*j^*}\ , \quad \text{for } 1 \leq m \leq r.$$

We seek to determine if $B_k$ has any "good" elements. With the use of a cathode ray tube terminal, the results of the partial allocation could be displayed to the operator. With a large number of components summaries of modules could be given. Suppose the operator is allowed to use his experience to weigh certain allocations over others. This is equivalent to imposing a probability distribution $P_k$ on $B_k$ prior to fathoming the branch. Graves and Whinston [8] used the uniform distribution for the quadratic assignment problem. This general approach resembles a Bayesian approach in statistics. Data here is construed to be additional information gained from exploring sub-branches.

More formally, suppose $(B_k, P_k)$ is a probability space (we omit $\sigma$ - fields since we are dealing with finite sets and we wish to weigh each allocation). There are two possible approaches for handling data. The first is to add one variable to the partial solution and obtain the set of (k + 1) partial allocations emanating from $B_k$. Doing this yields information on $B_k$ and could be used to obtain an updated probability measure $P_k^*$ for $B_k$. Then for a particular $B_{k+1}$ we could take the relative probability measure of $P_{k+1}$ of $P_k^*$. However, this procedure involves exploring each sub-branch.

The second approach is to reverse the procedure. That is, we first take the projective mapping of $P_k$ onto the most desirable $B_{k+1}$ to obtain $P_{k+1}$. After examining $B_{k+1}$ we could obtain a revised measure $P_{k+1}^*$ for $B_{k+1}$. Of course, if we did this for each $B_{k+1}$, then we could compose a new measure $P_k^*$ for $B_k$.

With the measure $P_k$ the projection quantities given in (2.11) and (2.12) become sums of random variables. Because of (2.10) the summations inside i are not independent. However, in many applications such as reliability the outside summands can be taken to be independent. Also, we could construct $P_k$ as being homeomorphic to a product probability measure where the product is taken over the type of component. If we have added a small number of components (so that k is small) so that the number of allocations in $B_k$ is large, then we estimate these random variables using a form of the central limit theorem. Doing this would give estimates of the form

$$\widetilde{PR}_m(\cdot) = E(PR_m(\cdot)) + c_{m\alpha}\sigma(PR_m(\cdot))$$

and

$$\widetilde{PRO}(\cdot) = E(PRO(\cdot)) - c_\alpha\,\sigma(PRO(\cdot))$$

These estimates of $PR_m(\cdot)$ and $PRO(\cdot)$ respectively satisfy

$$P_k(PR_m(\cdot) \geq \widetilde{PR}_m(\cdot)) = \alpha \quad , \quad 1 \leq m \leq r \tag{2.13}$$

and

$$P_k(PRO(\cdot) \leq \widetilde{PRO}(\cdot)) = \alpha$$

and give the values of $\{c_{m\alpha}\}$ and $c_\alpha$. These estimates insure that the probabilities of making the worst type of error (missing good allocations) small with respect to $P_k$.

The above discussion gives the procedure for truncation by feasibility and optimality. The selection of the next variable can be done by using the approach of Kettelle [9]. This is to select the next variable on the basis of largest ratio in gain and setting this variable equal to unity.

## 3. Spare Allocation

In this section the problem presented by Demmy [3] is considered. The goal is to allocate spares and repair facilities so as to optimize end item availability. The model assumes that all failed units are replaced or repaired, that the arrival rate of failed units are a Poisson process, and that repair and resupply times are i.i.d. In addition, independence is assumed for the time between demands among items. The following notation from [3] will be used.

$V$ - investment limitation

$C^I(I)$ - cost needed to have an average inspection time of $I$,

$C_j^R(r_j)$ - cost needed to have an average active repair time of $r_j$ for failures of type $j$,

$C_j^S(s_j)$ - cost of $s_j^j$ spares of unit $j$,

$\lambda_o$ - mean arrival rate of failed units,

$D$ - average time to obtain replacement unit,

$b$ - percent of failed items which may be repaired,

$p_j$ - probability that repair of a unit will require the removal of unit $j$ given that repair is possible,

$p_o$ - probability that unit may be repaired without requiring the replacement of the unit,

$w_j$ - average time delay caused by the lack of spares for unit j.

The mean time in the system, $T_o$, is then $I + b \sum_{j=0} p_j(w_j + r_j)$. Here n is the number of subunits.

With the above notation and assumptions the problem is

$$\min_{s_o} \min_{\underline{S},\underline{R},I} \sum_{k=s_o+1}^{\infty} (k - s_o) e^{-\lambda_o T_o}$$

subject to

$$\sum_{j=0}^{n} C_j^S(s_j) + \sum_{j=0}^{n} C_j^R(r_j) + C^I(I) = V.$$

By properties of monotoneity and separability it can be shown that it is sufficient to solve

$$\min_{s_o} \min_{\underline{S},\underline{R},I} \left\{ I + b \sum_{j=0}^{n} p_j(w_j + r_j) + D(1 - b) \right\}$$

as the objective function. The problem can then be considered in stages for a specific investment level. There are 2n + 3 stages with n for $\underline{S}$, n for $\underline{R}$, 1 for I, 1 for $r_o$, and 1 for $s_o$. Allocations are then paths through the stages. A method to reduce computation and storage from the dynamic programming approach is to use dynamic programming to evaluate branches, but to reduce the number of branches which must be considered by imposing a probability distribution on the families of partial allocations. The distribution could be treated as a product of individual probabilities associated with the investment level for each stage. The distribution of the set of all allocations would be a product space of dimension 2n + 3. After the k-th stage is examined, the dimensionality of the set of partial allocations is 2n + 3 - k.

In this application assumptions were made on the behavior of arrivals and on independence of variables such as repair and reservice times. These could be informally relaxed in the selection of the distribution. Such factors as past performance on repair times for certain numbers of failed components, losses due to pilferage of spares at certain locations, and interrelationships between service times for units which some of the same repair equipment could be taken into account.

## 4. Remarks

It is of some interest to examine whether the method outlined in the previous sections is applicable to other problems. In a routing problem of equipment, vehicles or messages, there frequently is little time to apply an out-of-kilter algorithm for optimal flow. Rather what is needed is a method which gives rapid, mear-optimal solutions. Usually, such problems require the access of a distance file for a grid system which is stored in disc. A number of tasks or messages are given which must be routed through the grid system. Goals could be the minimization of total distance or the minimization of delay times. A partial routing would reflect a partial allocation. The dispatcher could apply his operational experience to eliminate certain paths or collection of paths due to time-of-day, accident reports, traffic patterns and similar factors in a sector of the grid. The application of a probability distribution to families of allocations allows the dispatcher to insert these factors. Implementation would be similar to that described previously. However, the computations could be more rapid if the set of options available to the dispatcher were limited. This could be done by permitting only a specified level of discretization. As an example, if there were m possible allocations that could be given positive weight, then the weights would only allowed to be in units of 1/(2m). This will be explored in more detail in a later paper.

## References

[1] Balinski, M.L. (1965). Integer programming: methods, uses, and computation, Mgmt. Sci. , 12, 253-313.

[2] Balinski, M.L. and K. Spielberg (1969). Methods for integer programming: algebraic, combinatorial, and enumerative. Progress in Operations Research, vol. III, J.S. Aronofsky (ed.), John Wiley and Sons, New York.

[3] Demmy, W. Steven (1970). Allocation of spares and repair resources to a multi-component system. Air Force Logistics Command Report, 70-17.

[4] Ghare, P.M. and R.E. Taylor (1969). Optimal redundancy

for reliability in series systems, Opns. Res., 17, 838-847.

[5] Glover, F. (1965). A multiphase-dual algorithm for the zero-one integer programming problem, Opns. Res., 13, 879-919.

[6] Gomory, R.E. (1958). An algorithm for integer solutions to linear problems, Recent Advances in Mathematical Programming, Graves and Wolfe, (eds.), McGraw-Hill, New York.

[7] Gomory, R.E.(1960). All-integer programming algorithm, Industrial Scheduling, Muth and Thompson (eds.), Prentice-Hall, Englewood Cliffs, New Jersey.

[8] Graves, G. and A. Whinston (1970). An algorithm for the quadratic assignment problem, Mgmt. Sci., 17, 453-471.

[9] Kettelle, J. D. (1962). Least-cost allocations of reliability investment, Opns. Res., 10, 249-265.

[10] Lientz, B. P. (1970). Allocation of components to maximize reliability - I, submitted for publication.

[11] Shapiro, J. F. (1968). Group theoretic algorithms for the integer programming problem II: extension to a general algorithm, Opns. Res., 16, 928-947.

# OUTLIER PRONENESS OF PHENOMENA AND OF RELATED DISTRIBUTIONS

Jerzy Neyman and Elizabeth L. Scott*

Statistical Laboratory
University of California, Berkekey

## 1. Introduction

Let $S_n$ stand for a sample of $n \geq 3$ observations and let $x_i$ be the $i$th of them in the order of increasing value. We shall be concerned with the case where one or the other of the extremes, that is, $x_1$ or $x_n$, (or both), differs considerably from the remaining observations, suggesting the possibility of gross errors of measurement or of observation.

A recent paper by Grubbs [1], with many references, summarizes the practical methods for detecting "outlier observations," with the possibility of eliminating them from the analysis of the sample $S_n$. A fundamental paper by Ferguson [2] is given to performance of several such methods and to search for those that have preassigned properties of optimality. In both papers it is assumed that, basically, the observations follow a normal distribution.

The purpose of the present paper is to bring out the distinction between cases where the tendency to suspect and to eliminate outlier observations may be justifiable and those in which it is not. For this purpose we introduce three concepts: the concept of a "k-outlier", the concept of families of distributions that are "outlier-resistant" and the concept of families that are "outlier-prone". If a substantial previous experience in a particular domain of study appears sufficient for the statistician to act on the

---

*With partial support of U. S. Army Research Office (Durham) Contract DA-ARO-D-31-124-G816.

assumption that the observable variables follow an outlier-resistant distribution, then his efforts to seek out and, possibly, to eliminate the outliers are justified. However, as will be exemplified in section 6, domains of study exist in which large outliers are a rule, rather than an exception, and in which substantial samples of observations can be fitted only by some outlier-prone distribution. In cases of this kind, after careful checking for miscopying, etc., the elimination of a sample member, merely because its value deviates considerably from those of the others, cannot be justified.

## 2. Basic Concepts

Definition 1. For a positive number k we shall say that $x_n \in S_n$ is a k-outlier on the right if its value exceeds that of $x_{n-1}$ by more than $k(x_{n-1} - x_1)$.

The definition of a k-outlier on the left is analogous. Because of symmetry, the discussions that follow will be concerned only with k-outliers on the right.

From now on, we shall assume that samples $S_n$ represent independent observations on some random variable X which has its distribution function belonging to a specified family indexed by a possibly vector parameter $\varphi \in \Theta$. Generally, the distribution functions considered will be absolutely continuous. They will be denoted by capital letters $F(x|\varphi)$ or $G(x|\varphi)$, etc. The corresponding probability densities will be denoted by lower case letters, such as $f(x|\varphi)$, $g(x|\varphi)$ and so on.

If $x_n$ is a k-outlier, then

$$x_n > x_{n-1} + k(x_{n-1} - x_1) \tag{1}$$

and

$$x_{n-1} < (kx_1 + x_n)/(k+1). \tag{2}$$

Let $P(k,n|F)$ denote the probability that a sample $S_n$ of observations from a distribution F will contain a k-outlier. Clearly

$$P(k,n|F) = n(n-1)\int_{-\infty}^{\infty} dF(x) \int_{x}^{\infty} \left\{F\left(\frac{kx+y}{k+1}\right) - F(x)\right\}^{n-2} dF(y) \tag{3}$$

The probability $P(k,n|F)$ has the following obvious properties:

(i) Whatever be $n > 1$, we have $P(0,n|F) = 1$.

(ii) For a fixed $n > 2$, the probability $P(k,n|F)$ is a decreasing function of $k$ and tends to zero as $k \to \infty$.

Let $\mathfrak{F}$ be a family of distributions and let $\Pi(k,n|\mathfrak{F})$ stand for the least upper bound of probabilites $P(k,n|F)$ for $F \in \mathfrak{F}$.

Definition 2. *If $\Pi(k,n|\mathfrak{F}) < 1$ then we shall say that the family $\mathfrak{F}$ is $(k,n)$-outlier-resistant. Otherwise, that is, if $\Pi(k,n|\mathfrak{F}) = 1$, we shall say that the family $\mathfrak{F}$ is $(k,n)$ -outlier-prone.*

Obviously, if family $\mathfrak{F}$ is $(k,n)$-outlier-prone for a particular $k > 0$, it will also be outlier-prone for all smaller values of $k$. Of particular interest are families of distributions that are outlier-prone for all $k > 0$, no matter how large, and for all $n > 2$.

Definition 3. *If a family of distributions $\mathfrak{F}$ is $(k,n)$-outlier-prone for all $k > 0$ and all $n > 2$, we shall say that $\mathfrak{F}$ is outlier-prone completely.*

## 3. Two Useful Lemmas

Lemma 1. *If a family $\mathfrak{F}$ is composed of distributions $F(x/\sigma)$ that differ only in the scale parameter $\sigma > 0$, then the probability $P(k,n|\sigma) = P(k,n|1)$ and for any $k > 0$ and $n > 2$, the family $\mathfrak{F}$ is outlier-resistant.*

Lemma 2. *If a family $\mathfrak{F}$ is composed of distributions $F(x - \xi)$ that differ only in their location parameter, $\xi$, then $P(k,n|\xi) = P(k,n|0)$ and for any $k > 0$ and $n > 2$, the family $\mathfrak{F}$ is outlier-resistant.*

Both lemmas are easily verified by considering formula (3) and by performing the necessary calculations. The two lemmas imply that the family $N(\xi,\sigma^2)$ of normal distributions (with mean $\xi$ and variance $\sigma^2$) is outlier-resistant.

Also, perhaps unexpectedly, the family $C(\xi,\sigma)$ of Cauchy distributions (centered at $\xi$ and having scale parameter $\sigma$) is outlier resistant. The two lemmas are occasionally useful in dealing with more complicated families $\mathfrak{F}$ of distributions which, among others, depend on either scale or on location parameters. The studies of outlier properties of such families can be limited to subfamilies with standard values for location and scale.

## 4. Outlier Properties of the Family of Gamma Distributions

Because of the two lemmas in the preceding section, the study of the outlier properties of the family of Gamma distributions is reduced to the standard subfamily indexed by the shape parameter only. The probability density to consider is, say,

$$g(x|\alpha) = \frac{1}{\Gamma(\alpha)} x^{\alpha-1} e^{-x}, \quad \text{for } x > 0$$

$$= \quad 0, \text{ otherwise,} \tag{4}$$

with $\alpha > 0$. The corresponding distribution function will be denoted by $G(x|\alpha)$.

<u>Theorem 1. The family of Gamma distributions is outlier-prone completely.</u>

In order to prove Theorem 1, we notice that, because $x_1 > 0$.

$$P(k,n|\alpha) = P\{X_n > X_{n-1} + k(X_{n-1} - X_1)\}$$

$$> P\{X_n > (k+1)X_{n-1}\} = Q(k,n|\alpha)$$

(say). Thus, in order to prove Theorem 1, it is sufficient to prove the stronger assertion, namely that, as $\alpha \to 0$, $Q(k,n|\alpha) \to 1$. The formula for $Q(k,n|\alpha)$ is a little simpler than that for $P(k,n|\alpha)$, namely

$$Q(k,n|\alpha) = n\int_0^\infty G^{n-1}\left(x/(k+1)|\alpha\right) g(x|\alpha)\,dx. \tag{5}$$

Multiplying and dividing the integrand in (5) by the derivative of $G(x/(k+1)|\alpha)$, that is by $g(x/(k+1)|\alpha)/(k+1)$, we obtain,

$$Q(k,n|\alpha) = \int_0^\infty \frac{d}{dx} G\left(\frac{x}{k+1}\middle|\alpha\right) R(x)\, dx \tag{6}$$

with

$$R(x) = (k+1)^\alpha \exp\left(-\frac{kx}{k+1}\right). \tag{7}$$

Integrating (6) by parts and noticing that the integrated part vanishes at both ends of the interval of integration, we obtain

$$Q(k,n|\alpha) = (k+1)^\alpha \int_0^\infty G^n\left(\frac{x}{k+1}\middle|\alpha\right) \exp\left\{-\frac{kx}{k+1}\right\} \frac{kdx}{k+1}$$

$$= (k+1)^\alpha \int_0^\infty G^n\left(\frac{x}{k}\middle|\alpha\right) \exp\{-x\}\, dx. \tag{8}$$

Now we notice that for $0 < \alpha < 1$, $a > 0$ and $t > a$ we have

$$1 - G(t|\alpha) = \frac{1}{\Gamma(\alpha)} \int_t^\infty x^{\alpha-1} e^{-x}\, dx < \frac{e^{-a}}{a^{1-\alpha}\,\Gamma(\alpha)}. \tag{9}$$

Because $\Gamma(\alpha) \to \infty$ as $\alpha \to 0$, the expression in the integrand in (8) tends to unity uniformly for $x/k$ exceeding any arbitrary small positive a. It then easily follows that, as $\alpha \to 0$, the probability $Q(k,n|\alpha) \to 1$. This completes the proof of Theorem 1.

## 5. Outlier Properties of the Family of Lognormal Distributions

For $\xi$, $\sigma > 0$, we say that a positive random variable X follows a lognormal distribution if the distribution of log X is normal with mean log $\xi$ and variance $\sigma^2$. Here, then, the variable

$$U = (\log X - \log \xi)/\sigma = \frac{1}{\sigma} \log (X/\xi) \tag{10}$$

has a normal distribution. It is seen that $\xi$ plays the role of a scale parameter. Thus, in investigating the outlier properties of the lognormal distributions, we may restrict ourselves to the case $\xi = 1$.

The lognormal distribution function, depending upon a single parameter $\sigma > 0$, will be denoted by $L(x|\sigma)$. The corresponding probability density will be

$$\ell(x|\sigma) = (\sigma x)^{-1}(2\pi)^{-\frac{1}{2}} \exp\left\{-\frac{1}{2\sigma^2}\log^2 x\right\}, \tag{11}$$

for $x > 0$, and $\ell(x|\sigma) = 0$ elsewhere.

Theorem 2. The family of lognormal distributions is outlier-prone completely.

The proof of this theorem is based on the same device as that of Theorem 1. As formerly, let $Q(k,n|\sigma)$ denote the probability that the largest of the n sample members will exceed the next largest by more than a factor k+1. Then the probability $P(k,n|\sigma)$ of a k-outlier in a sample of n will satisfy

$$1 > P(k,n|\sigma) > Q(k,n|\sigma), \tag{12}$$

where

$$Q(k,n|\sigma) = n\int_0^\infty L^{n-1}\left(x/(k+1)|\sigma\right)\ell(x|\sigma)\,dx. \tag{13}$$

Using the relationship (10) between the standard normal distribution (with zero mean and unit variance) customarily denoted by $\Phi(x)$, with its derivative $\varphi(x)$, the integral (13) is easily transformed into

$$Q(k,n|\sigma) = n\int_{-\infty}^\infty \Phi^{n-1}\left[u - \frac{1}{\sigma}\log(k+1)\right]\varphi(u)\,du. \tag{14}$$

It is seen that, as $\sigma \to \infty$ while k and n remain fixed, the value of (14) is increased and tends to a limit, namely

$$\lim_{\sigma\to\infty} Q(k,n|\sigma) = n\int_{-\infty}^\infty \Phi^{n-1}(u)\,\varphi(u)\,du = 1, \tag{15}$$

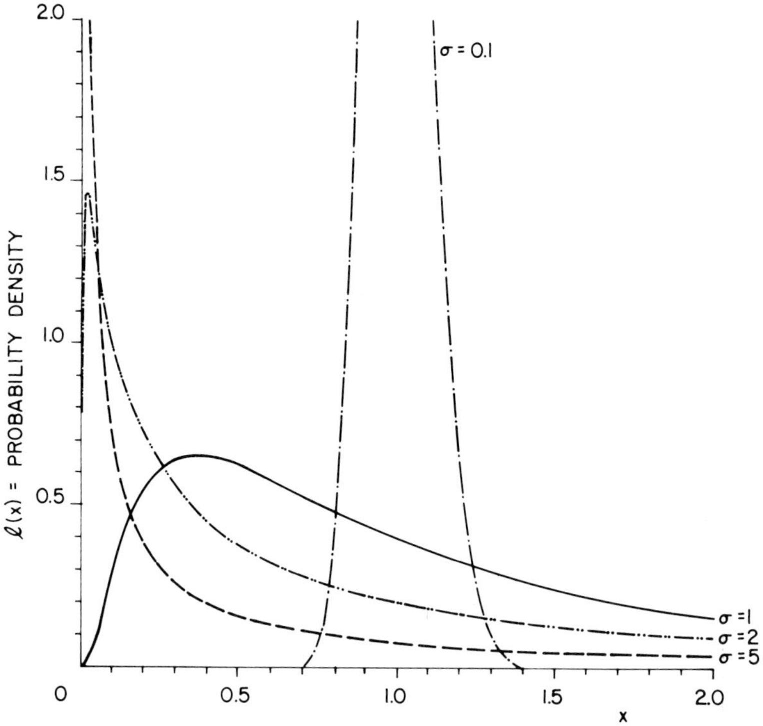


Fig. 1. Lognormal probability density function for various values of the parameter $\sigma$

which proves Theorem 2.

We are indebted to Miss Ellen Sherman for the calculation of Table 1, illustrating the convergence of $Q(k,n|\sigma)$ to unity as the value of $\sigma$ is increased.

The Gamma density curves are very familiar to all statisticians. When the shape parameter is less than unity, the density is infinite at zero and then decreases steadily. The lognormal density is less frequently used. It is unimodal, vanishes at $x = 0$, and reaches its maximum at $x = \exp(-\sigma^2)$ which is less than unity. If $\sigma$ is increased, the mode converges rapidly to zero. Already with $\sigma = 2$ (see Figure 1), the graph of the lognormal density looks very much as if it were monotonically decreasing over the whole range of positive x. With $\sigma = 5$, the mode is, essentially, at zero. These details explain the fact that, for some observations, the lognormal density appears as a competitor to the Gamma.

Table 1

Probability $Q(k,n = 100|\sigma)$ that the largest of $n = 100$ observations from a lognormal distribution will exceed the next largest by a factor of k+1 or more.

| | k+1 | | | k+1 | |
|---|---|---|---|---|---|
| $\sigma$ | 1.10 | 1.30 | $\sigma$ | 1.50 | 2.00 |
| 1 | .784 | .504 | 1 | .338 | .144 |
| 2 | .887 | .715 | 3 | .707 | .548 |
| 3 | .923 | .801 | 5 | .814 | .701 |
| 4 | .942 | .847 | 7 | .864 | .777 |
| 5 | .953 | .876 | 9 | .892 | .822 |
| 6 | .961 | .895 | 11 | .911 | .852 |

## 6. A Category of Phenomena with Frequent Outlier Observations

In statistical practice, when applying a test it is important to see that the theory underlying the test is not in conflict with the phenomenon studied. The customary requirements on the test of a hypothesis H are two: (i) if

H is true, the use of the test should ensure the maintenance of the desired level of significance; (ii) if H is false and some contemplated alternative hypothesis $H_1$ is true, the power with regard to $H_1$ should be high. The first requirement (i) is satisfied by all the nonparametric tests. As to condition (ii), it is obvious that, if a parametric family of distributions provides a satisfactory approximation to the actual distribution of the observable variables, then an optimal test deduced assuming the particular family cannot be worse than any nonparametric test. Our own experience indicates [3] that, in spite of their only asymptotic optimality, the optimal $C(\alpha)$ tests have a substantially higher power than some nonparametric tests in frequent use.

When approaching a category of phenomena it is, then, important to see whether some specific parametric family of distributions can be found which, as a general rule, provides a satisfactory approximation to the actual distribution of the observable variables.

The illustrations that follow are meant to document that, in a particualr domain of study, namely, in experimentation with rain making, the sample distributions of nonzero rainfall per experimental unit (an experimental day or a storm) indicate as a very general rule population distributions that are J-shaped, with long "tails" and frequent "outliers". The steepness of the indicated J-shaped distributions and the frequency of outliers depend upon a number of circumstances and, in particular, upon the area of the "target" over which the precipitation is measured: generally, the smaller the target, the steeper the descent of the indicated distribution and the more frequent are the outliers.

Our own experience indicates that, as a more or less general rule, the Gamma density fits the observations well and the optimal $C(\alpha)$ test deduced for this family [4] is used routinely in our evaluation of rain making experiments. However, we have no intention of insisting on any particular family of parametric distributions, whether Gamma or lognormal or any other. Our intention is to document that, as a general rule, the sample distributions of rainfall are J-shaped and that, frequently, they show substantial outliers. In fact, outlier amounts of rainfall in samples of reasonable size (of the order of 100-300 observations) appear in the very nature of the phenomenon. Thus, if a parametric family is used for deducing optimal statistical procedures, this should be an outlier-prone family. Furthermore, any rejection of the outlier observations amounts to mutilation of the observed samples. The documentation that a phenomenon occurs "as a general rule", requires a reasonably broad range of instances. The histograms exhibited refer to (i) the Swiss experiment Grossversuch III [3], [5], (ii) the Whitetop project in the Missouri-Arkansas region, with several "targets" ranging in area from about 3,000 to 100,000 square miles, [6], (iii) a tight network labelled Little Egypt, consisting of 49 gages spread over an area of 550 square miles [7], (iv) the rainfall at Walnut Gulch, Arizona, [8], observed during the seven year experiment with cloud seeding over the Santa Catalina Mountains, (v) the Israeli experiment [9].

It will be seen that, in spite of the wide range of orographic and climatic conditions, outlier amounts are frequent. The conclusion is that rainfall over a target of moderate area and over a period of time comparable to a day or two, is an outlier-prone phenomenon.

Similar situations occur in other domains of research.

## References


1. Frank E. Grubbs, "Procedures for detecting outlying observations in samples," Technometrics, Vol. 11, (1969), pp. 1-21.
2. T. S. Ferguson, "On the rejection of outliers," Proceedings of the Fourth Berkeley Symposium on Mathematical Statistics and Probability, Vol. 1, (1961) University of California Press, (Berkeley and Los Angeles), pp. 253-287.
3. J. Neyman and E. L. Scott, "Some outstanding problems relating to rain modification," Proceedings of the Fifth Berkeley Symposium on Mathematical Statistics and Probability, Vol. 5, (1967), University of California Press (Berkeley and Los Angeles), pp. 293-326.

4. J. Neyman and E. L. Scott, "Note on techniques of evaluation of single rain stimulation experiments," Proceedings of the Fifth Berkekey Symposium on Mathematical Statistics and Probability, Vol. 5, (1967), University of California Press (Berkeley and Los Angeles), pp. 371-384.
5. J. Neyman, E. L. Scott and M. A. Wells, "Statistics in meteorology," Review Intern. Stat. Institute, Vol. 37, (1969), pp. 119-148.
6. J. L. Lovasich, J. Neyman, E. L. Scott, and J. A. Smith, "Wind directions aloft and effects of seeding on precipitation in the Whitetop trial," Proceedings Natl. Acad. Sci., Vol. 64, (1969), pp. 810-817.
7. S. A. Changnon, Jr., and F. A. Huff, "The effect of natural variability in verification of rain modification experiments," Proceedings of the Fifth Berkeley Symposium on Mathematical Statistics and Probability, Vol. 5, (1967), University of California Press (Berkeley and Los Angeles), pp. 177-198.
8. J. Neyman and H. B. Osborn, "Evidence of widespread effects of cloud seeding at two Arizona experiments," Proceedings Natl. Acad. Sci., Vol. 68, (1971), pp. 649-652.
9. K. R. Gabriel, "The Israeli artificial rainfall stimulation experiment. Statistical evaluation for the period 1961-1965," Proceedings of the Fifth Berkekey Symposium on Mathematical Statistics and Probability, Vol. 5, (1967), University of California Press (Berkeley and Los Angeles), pp. 91-114.

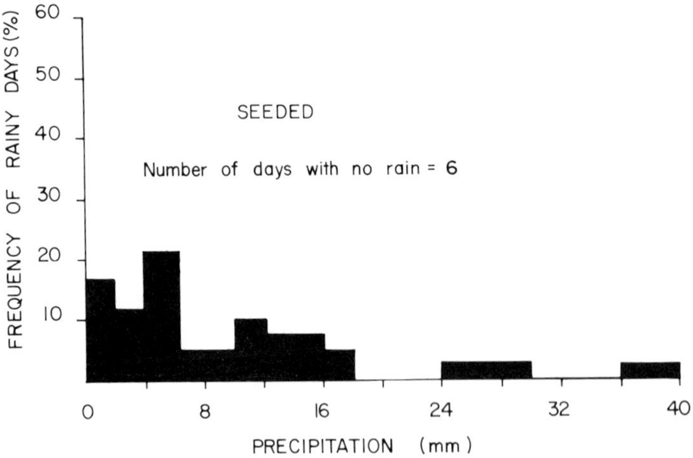


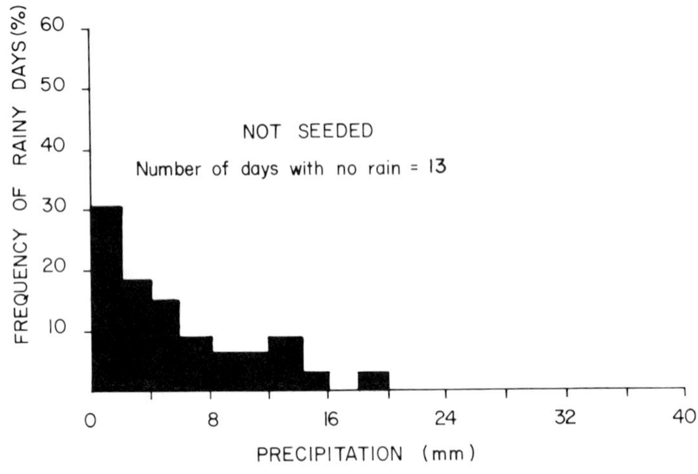


Fig. 2. Empirical frequency distribution of non-zero precipitation per experimental day in Grossversuch III (Ticino, Switzerland) Experiment, summers of 1957-1963. Area near Zurich under conditions of low southerly winds and stability layer over Milan. Total of 48 seeded, 46 not seeded experimental days

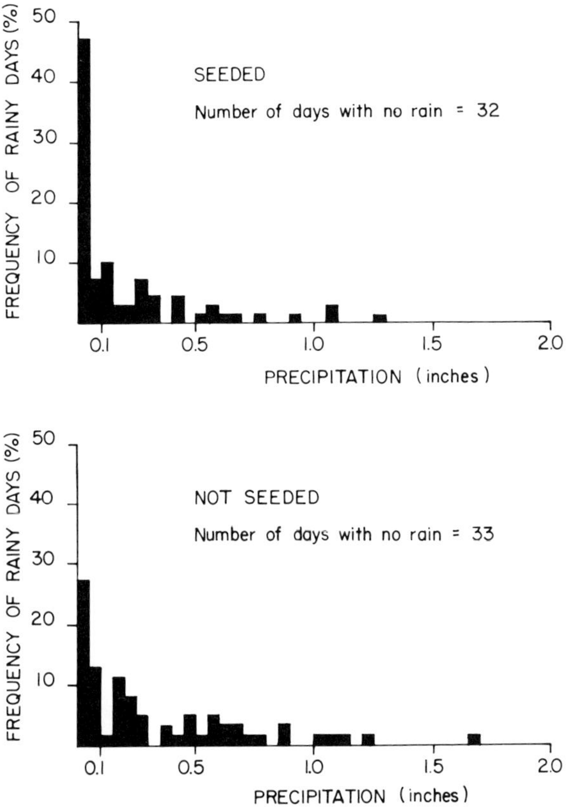


Fig. 3. Empirical frequency distribution of non-zero rainfall per experimental day in Whitetop experiment, summers of 1960-1964. Area: Zone A (radius of 30 miles about target center). Total of 102 seeded, 96 not seeded experimental days

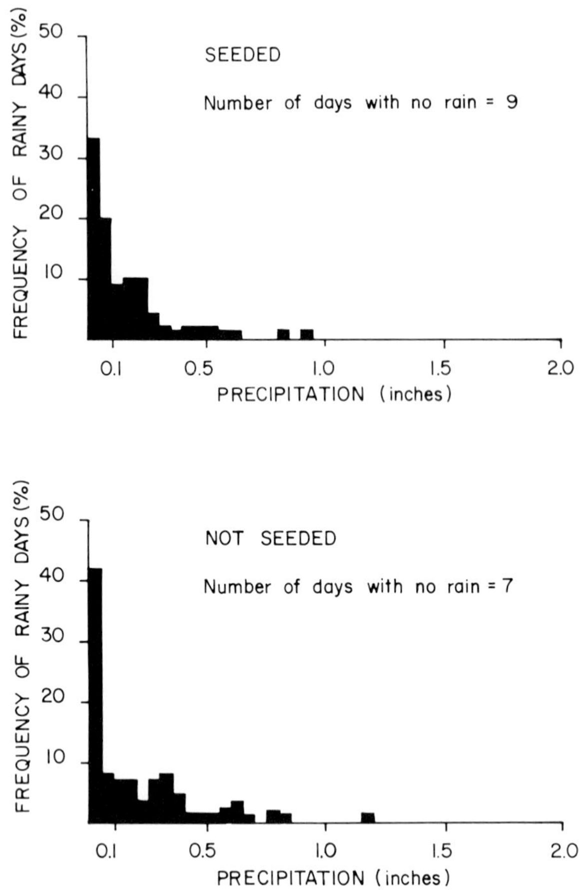


Fig. 4. Empirical frequency distribution of non-zero rainfall per experimental day in Whitetop experiment, summers 1960-1964. Area: Zones A + B + C (circle of radius 90 miles). Total of 102 seeded, 96 not seeded experimental days

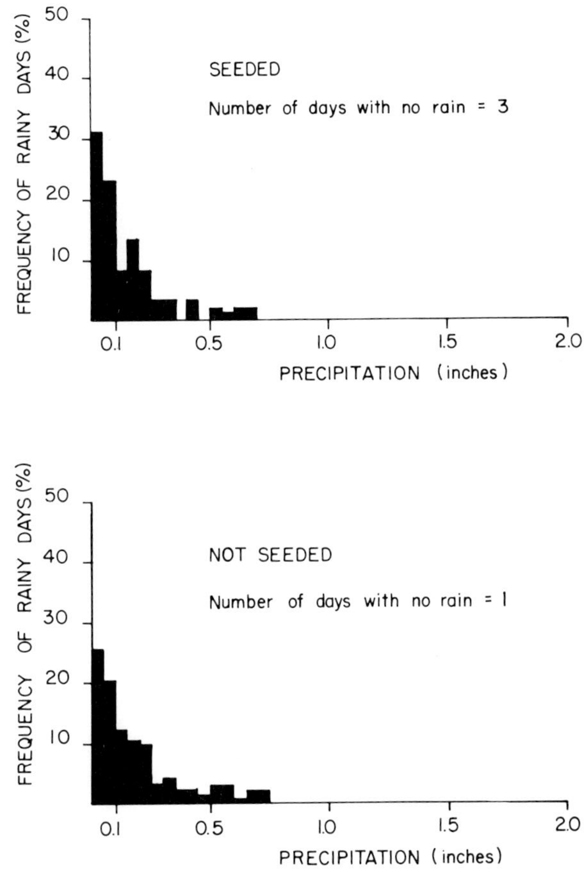


Fig. 5. Empirical frequency distribution of non-zero rainfall per experimental day in Whitetop experiment, summers of 1960-1964. Area: Zones A + B + C + D + E + F (circle of radius 180 miles). Total of 102 seeded, 96 not seeded experimental days

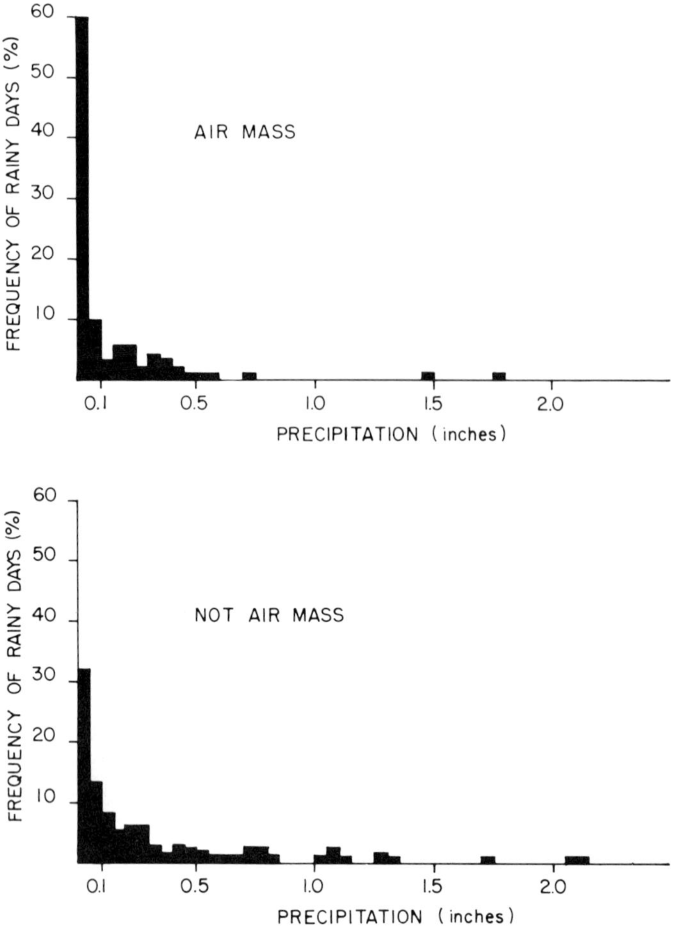


Fig. 6. Empirical frequency of distribution of non-zero rainfall per storm, summer seasons of 1960-1964. Area: Little Egypt Network, Illinois. Total of 116 air mass storms, 115 not air mass storms

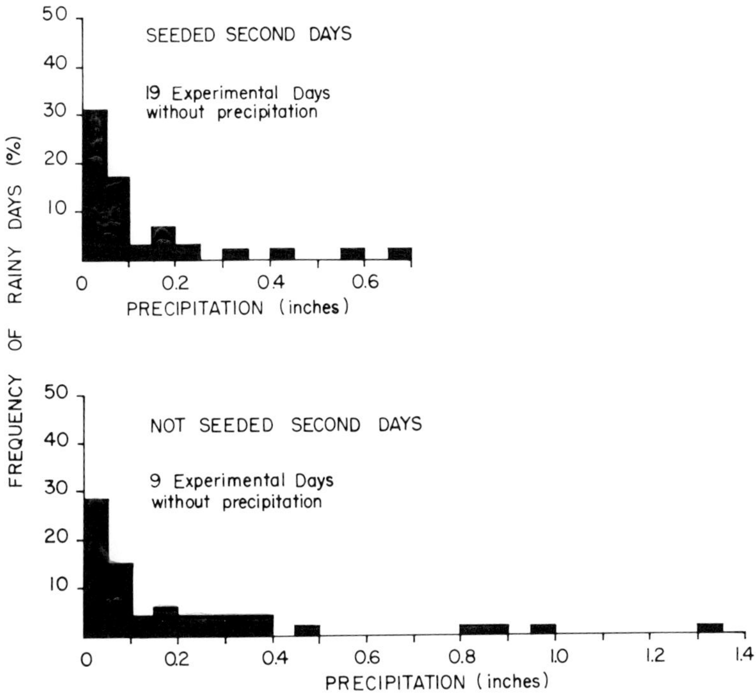


Fig. 7. Empirical frequency distribution of non-zero rainfall per "second" experimental day in Arizona (Santa Catalina Mountains) experiment, summers of 1957-1962, 1964. Area: Walnut Gulch. Total of 59 seeded, 47 not seeded experimental second days

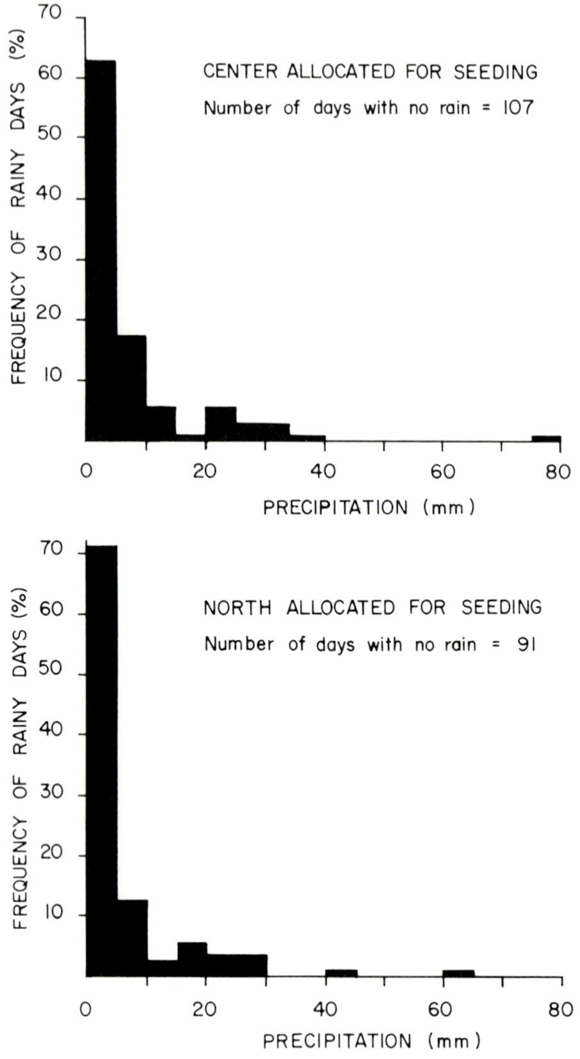


Fig. 8. Empirical frequency distribution of non-zero rainfall per morning-to-morning experimental day in Israeli experiment, 1961, 1963-1965. Area: Center Target. Total of 192 Center Allocated, 180 North Allocated experimental days

# PROBLEM AREAS REQUIRING OPTIMIZING METHODS

Colonel John V. Armitage

Headquarters, United States Air Force

The principal purpose of this paper is to draw your attention to a series of representative Air Force optimization problems for which any improvement in the method of achieving a solution would be welcome. Before detailing these prospective challenges let me very briefly remind you of what you already know very well; that is, the fact that optimization methods have many and varied applications within the Air Force. The reason for expending effort on such methods is different for each area of application.

I find five sets of Air Force problems to which optimization methods have been and are being applied. These are the following:

Problems of design, where the point of a study is the requirement that a system be so designed that some aspect of its performance is optimized; for example, a structure of minimum weight for a given loading condition.

Operational problems, where the point of a study might be to optimize some aspect of the performance of an already designed system; for example, optimal flight of aircraft or optimal rocket trajectories.

Problems of programming under uncertainty, where extensions of the classical, deterministic linear programming approach offer a wide variety of applications. Such studies are usually undertaken for economic reasons.

Problems of optimal time scheduling, where first-order importance must be accorded to the control over a time schedule from the moment a contract is awarded until a system becomes fully operational. In this area, the "Program Evaluation and Review Technique (PERT)" deserves special mention.

Problems in the area of communication and control, smoothing and prediction. These studies are motivated by the hope that a better understanding of a system or process will lead to an improved ability to control it and to make predictions about it. Typical problems are: the reception of radio and television signals in the presence of static and the efficient reproduction of sound and images; the design of control systems, servo-mechanisms and regulators; weather forecasting; and automatic tracking of moving objects.

These five broad areas still represent challenges to the most ingenious among us. I will now turn to a set of four specific Air Force problems. For each of these, the present method of solution will be discussed and the direction in which improvements are desired will be indicated.

The first problem deals with reliability. Specifically, I shall discuss the problem of obtaining a lower confidence bound to the reliability of a system given certain information about the reliability of its components. A Bayesian method, currently used by the Air Force for solving this problem, is outlined in (1). Reference (2) also applies the Bayesian method to reliability problems. The approach of (1) and (2) will be illustrated through the consideration of a hypothetical system consisting of thirty doubly redundant configurations in series, as illustrated in Figure 1.

The function that relates system reliability, $R_s$ to component reliability, $R_{i,j}$ $(j = 1,2)$ is given in this case by:

$$R_s = \prod_{i=1}^{30} [1 - (1 - R_{i,1})(1 - R_{i,2})] \quad (1)$$

Each component is assumed to have been tested 35 times with no failures detected. It is assumed that each test result is an independent random variable drawn from a binomial distribution and that the test described above demonstrates that the applicable binomial distribution is B(35,35,R) where R is unknown. It is assumed that prior to the test nothing is known about a component's reliability; therefore, a uniform distribution on the interval (0,1), expressing that all reliabilities are equally probable, is selected as the component prior distribution. If Bayes' theorem is applied to the above distributions, a Beta distribution, $\beta(R|35,1)$, is obtained for the posterior distribution of the component's reliability. With this as a background, the following procedure is followed to obtain a distribution representing the system reliability. For each component i,j, a random number is drawn from a uniform distribution in the interval (0,1) and then this number is used as a percentage point to the appropriate Beta distribution to obtain samples of the reliability $R_{i,j}$, j = 1,2. These numbers are then combined, using equation (1), to obtain a Monte Carlo replication of the system reliability $R_s$. In the example quoted here, 1000 Monte Carlo replications of the system reliability were made and a histogram, illustrated in Figure 2, was constructed from the results.

As can be seen from the histogram, more than 90 percent of the replications lie to the right of class interval .935, and less than 90 percent lie to the right of .940; hence a 90 percent lower confidence bound, as determined by this procedure, lies between .935 and .940. This appears to be a rather low value in light of the fact that no failures were detected in any of the tests. This low value is due principally to the selection of a uniform distribution as the prior distribution. A method was given in (2) for utilizing engineering estimates and information about analagous components to obtain more realistic prior distributions. However, a method which incorporates subjective information produces non-unique results and loses credibility. A method is needed which minimizes or eliminates the effect of subjective information and which produce lower confidence bounds comparable to those produced by the method described here.

I now turn to a discussion of aircraft vulnerability. Aircraft manufacturers who are interested in increasing the survivability of military aircraft treat this problem in great detail. For example, the vulnerable components of aircraft have been modeled in three dimensions, using arbitrary solids such as spheres, cones, cylinders, etc. This model is then used in Monte Carlo simulations involving various projectiles in order to determine aircraft survivability. Such great detail is ordinarily not incorporated into simulations concerned with the determination of aircraft attrition. Studies of this type utilize information about the vulnerable areas in three principal directions where the size of these areas is a function of the projectile type and velocity. This information is commonly used to obtain a vulnerable area in some arbitrary direction by employing the "shoebox model." In this model the vulnerable area pertaining to a given direction is approximated by the sum of the projections of the three principal areas. In many cases this is not a realistic representation as can be seen by a consideration of the case where the three principal vulnerable areas are equal and these vulnerable areas are projected in the direction of a diagonal through the cube. In this direction, the three-direction cosines are each equal to $\sqrt{3}/3$. Hence, the sum of the three projected areas is equal to $\sqrt{3}$ multiplied by the common vulnerable area; in other words, it is 73 percent larger than the area viewed from one of the principal directions. For aircraft not shaped like a shoebox, this is a rather poor approximation. A somewhat more realistic model can be constructed from the three principal vulnerable areas in the following way:

Let F, S and B represent the vulnerable areas as viewed from the front, side and bottom of the aircraft respectively. Construct a vulnerable ellipsoid with semi-major axes a,b and c satisfying

$$\begin{aligned} F &= \pi b c \\ S &= \pi a c \\ B &= \pi a b \end{aligned} \qquad (2)$$

Now, if a cylinder is constructed tangent to the vulnerable ellipsoid with axis parallel to a principal direction, the area cut out by this cylinder on a plane perpendicular to the axis is the appropriate area F, S or B. It is postulated that this construction will produce an appropriate vulnerable area in any arbitrary direction. If $\alpha$, $\beta$, and $\gamma$ represent the direction cosines in an arbitrary direction, then the equation for calculating the required area is:

$$A = (\alpha^2 F^2 + \beta^2 S^2 + \gamma^2 B^2)^{1/2} \qquad (3)$$

Consider some of the properties of equation (3). First, note that if F = S = B, then this common value may be factored from the right hand side of equation (3) and, since $\alpha^2 + \beta^2 + \gamma^2 = 1$, equation (3) reduces to A = S; that is the vulnerable ellipsoid has become a sphere and the vulnerable area is the same for all directions. Furthermore, equation (3) satisfies the definition of a generalized mean of order two (with the squares of the direction cosines assuming the role of weighting factors). Therefore, area A always lies between the maximum and minimum of F, S and B. These properties are generally more desirable than those exhibited by the "shoebox model" and they produce quite different results in aircraft attrition studies. Probabilities of kill determined from the vulnerable ellipsoid model are about 20 percent lower than those derived from the "shoebox model." However, both models are of a completely arbitrary nature. There is a need to study the actual configurations of vulnerable components of aircraft and to derive empirical formulae, hopefully of a simple nature, which better express the true vulnerability.

I shall now discuss one aspect of aircraft attrition. Combat attrition of aircraft is caused by small arms fire, anti-aircraft artillery fire, surface-to-air missile fire, and air-to-air combat. I will cover the aspect of attrition most difficult to simulate in computer programs, namely attrition due to anti-aircraft artillery fire.

Consider an encounter between an aircraft and a battery of anti-aircraft artillery. Anti-aircraft artillery gun systems (including the crew) must perform three basic functions. They must detect aircraft, they must predict the future positions of aircraft, and they must direct fire toward those predicted positions. Each of these functions is subject to its own errors. An encounter between an aircraft and an anti-aircraft artillery system begins when the aircraft is detected at a random range dependent upon the type of aircraft, the time of day, alertness of the crew, weather conditions, and possibly terrain masking. At the instant of detection, the gun will be oriented in some random direction; therefore, a random time delay will occur between detection and the onset of fire. The gun computer system predicts future aircraft positions based on the assumption that it will follow a straight course during the flight of the projectile. If the aircraft maneuvers during this time, a bias in lead angle occurs. In order to calculate lead angle, it is necessary to measure or to estimate the slant range from gun to aircraft as well as the aircraft's speed, course angle, and climb or dive angle. Errors in measuring or estimating these quantities result in random errors in lead angle. During engagement the aircraft is tracked in both elevation and azimuth and this tracking process is subject to error. There is some evidence to indicate that tracking errors are serially correlated and this correlation cannot be neglected.

When all of the aforementioned errors (plus many not mentioned) are taken into account in making a careful simulation of an encounter between an aircraft and an anti-aircraft artillery gun, the predicted probabilities of kill from the simulation do not compare well with combat attrition. In fact, probabilities of kill derived from a simulation are considerably higher than those experienced in combat. However, present models are useful in making comparisons between various aircraft types. There is a need for improved understanding of the serial correlation which exists between rounds. There is also a need for devising better methods for validating the results of such simulations.

I next discuss fighter commitment policy in air defense. The problem is as follows: suppose that our radar net has detected a wave of enemy bombers approaching the border along a broad front, flying either singly or in groups of two or more. The defensive force consists of one or more types of defensive systems deployed along the attacked border. Associated with each defensive system is a known probability of success, $p_k$, against an enemy bomber. The number of defensive units is assumed to be limited and care must be taken not to saturate the defense through an initial overcommitment of defensive forces. Defensive forces must be committed in such a way that the number of surviving bombers is minimized. In the past, two different commitment policies have been used in computer simulations of such engagements. In one, no information concerning the probability of success of a fighter was assumed, so a constant overcommit ratio was used. Another model committed fighters based upon the nearest integer to the expected number required to destroy a given raid. Computer runs made with the latter model indicate that this commitment policy is somewhat too conservative, particularly for larger raid sizes. A somewhat better policy can be constructed by assuming that most engagements between an attacking group of fighters and a group of enemy bombers occur on the shoot-look-shoot principle. In other words, individual fighters are perfectly assigned to units of the opposing group of bombers. If this is the case then knowing the probability $p_k$ of the fighter's success, the probability of destruction of a given raid can be determined from tables of the cumulative binomial distribution and a fighter commitment policy could be constructed based upon this distribution.

Examination of tables of the binomial distribution shows that the commitment of the expected number of fighters to destroy a given raid becomes an increasingly poor policy as the raid size increases or as the assumed value of $p_k$ decreases. However, even commitment policies based on the binomial distribution become too conservative if the assignment of fighters to bombers is less than perfect. A commitment policy is needed which applies to realistic situations, which does not saturate the defense forces, and which minimizes the number of surviving enemy bombers. This policy should also incorporate a decision rule which is readily, or automatically, applied which selects the best defensive system from those available.

## References


1. "A Bayesian Approach for Assessing the Reliability of Air Force Re-entry Systems", Unpublished, D. V. Mastran Sept., 1966.
2. "A Bayesian Approach to Reliability Assessment", RAND Memorandum RM-5084-NASA, B. L. Fox, Aug., 1966.
3. "Mathematical Optimization Techniques", Edited by R. Bellman, 1963, RAND Report R-396-PR.
4. "PERT - Time System Description Manual", Vol. 1, Sept., 1963.


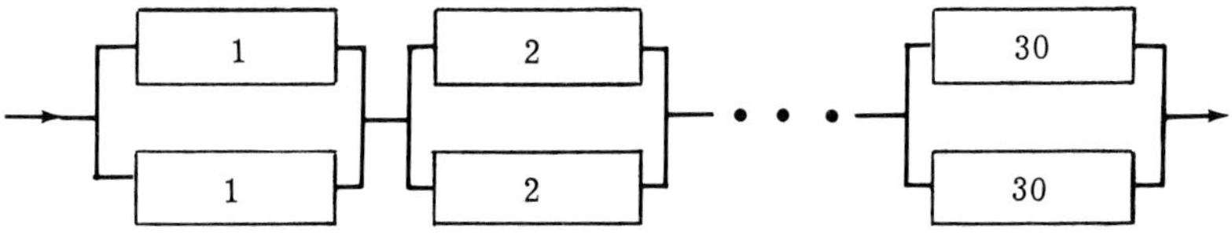


Figure 1.
A system composed of 30 pairs of redundant components in series.

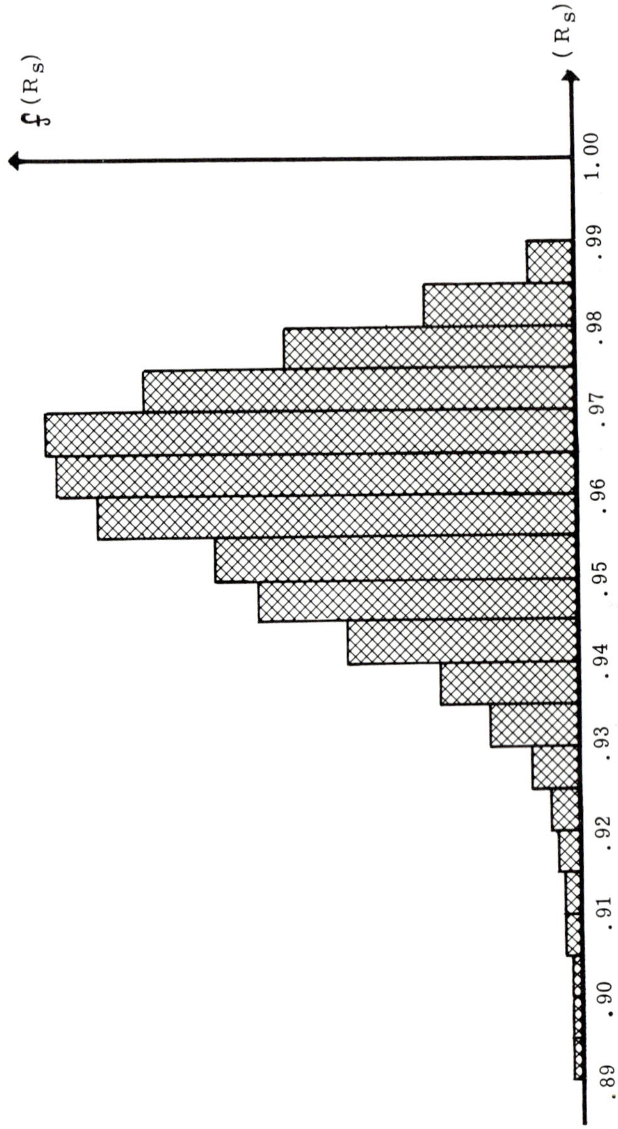


Figure 2
System Histogram

# STOCHASTIC APPROXIMATION

Václav Fabian*

Michigan State University

## 1. Introduction

(1.1) Foreword. This is an attempt to review and discuss basic results on stochastic approximation of a point of minimum of a function of k real variables, with the main emphasis on the speed of the convergence.

The main ideas and results will be explained and described in this introduction, using simplifying assumptions whenever possible. Then in Section 2 some of the results will be reformulated, this time without oversimplification. Section 3 contains proofs.

There are well informed and deep reviews by Schmetterer (1961 and 1969) and it would be futile to add another extensive review now. But perhaps this paper may be useful by its concentration to a narrower topic which can then be discussed in more detail. Also some results are included which were obtained after Schmetterer's (1969) review and the result of Theorem 2.7 on Newton-type stochastic approximation is new.

Thus we are interested in the problem of finding a point $\theta$ of minimum of a function f defined on the k-dimensional Euclidean space $R^k$. What makes stochastic approximation different from, say, methods in numerical analysis, is the restriction on information available. With the assumption that f belongs to a rather general family of functions the only possibility to obtain information is that we can, for any argument x, observe the value f(x) subject to

* Research supported by NSF Grant GP-23480.

a random error; i.e. to obtain an unbiased estimate of f(x).

Let us recall briefly, paying homage, that Robbins and Monro (1951) initiated the area by considering a (simpler) problem of an equation $f(\theta) = 0$ with f defined on R. Kiefer and Wolfowitz (1952) considered our problem of function minimization on R. Blum (1954) generalized these methods to the multidimensional case, i.e. when f is defined on $R^k$, and strengthened the results. A series of papers on the speed and asymptotic distribution followed: Schmetterer (1953,1954a,1954b), Chung (1954), Derman (1956), Dupač (1957), Sacks (1958).

To be more specific let f be a function on $R^k$, let D denote the vector of the first derivatives of f, H a matrix of the second order derivatives. Unless stated otherwise let us assume always that f has an absolute minimum at a point $\theta$ in $R^k$, that H exists and is bounded on $R^k$, H is continuous at $\theta$ and $H(\theta)$ is positive definite and

$$\inf \{\|D(x)\|;\ \|x - \theta\| > \epsilon\} > 0 \text{ for every } \epsilon > 0 . \tag{1}$$

We try to approximate $\theta$ by a sequence $X_n$ satisfying a recursive relation

$$X_{n+1} = X_n - a_n Y_n \tag{2}$$

where $a_n$ are numbers and each $Y_n$ determines the direction of the correction at the n-th step.

(1.2) Numerical analysis methods. It may be useful to look at methods available in numerical analysis, and their properties. In the *gradient methods* one uses (1.1.2) with $Y_n = D(X_n)$ and the $a_n$ are either chosen to minimize $f(X_n - a D(X_n))$ as a function of a (in the so called steepest descent method) or $a_n$ are chosen in another way, usually equal to a suitable small constant (gradient methods with *small steps*).

In both cases we obtain $X_n \to \theta$ provided that, if small steps are used,

$$0 < \delta < a_n < 2\rho^{-1} - \delta \tag{1}$$

where $\delta$ is a positive number and

$$\rho = \sup \{\|H(x)\|;\ x \in R^k,\ f(x) \leqq f(X_1)\}. \tag{2}$$

Moreover, if small steps are used,

$$\|X_{n+1} - \theta\| \leqq \|X_n - \theta\| (q_n + \varepsilon_n) \tag{3}$$

with $\varepsilon_n \to 0$ and, if $\lambda_1, \lambda_k$ are the smallest and the largest eigenvalue of $H(\theta)$,

$$q_n = \max\{1 - a_n\lambda_1, a_n\lambda_k - 1\} \leqq q = \max\{1-\delta\lambda_1, 1-\delta\rho\} < 1. \tag{4}$$

These assertions follow from results in Goldstein (1962) (surprisingly there does not seem to be any reference to an earlier paper.)

Thus the speed of the gradient method with small steps is that of a geometrical progression, $\|X_n - \theta\| = \mathcal{O}(q^n)$ with $0 < q < 1$. This is much more than we can hope for in stochastic approximation. But, of course, when we have this rate, it is proferable to have q small. A small q will result for suitable $a_n$ if $\lambda_1 \doteqdot \lambda_k$ (cf. (4)) and thus if $H(\theta)$ approximately equals I, the identity matrix. This in turn may be achieved by transforming the coordinate system and applying the gradient method then. Expressing back in original coordinates we obtain the recurrence formula (1.1.2) with $Y_n = B\,D(X_n)$ and the asymptotic results with $BH(\theta)$ substituted for $H(\theta)$. We would like to use $B = H(\theta)^{-1}$ but usually we only can compute $B = B_n = H(X_n)^{-1}$, at least for n large. Under some regularity conditions this so called Newton method preserves convergence, can eventually use $a_n = 1$ and gives $\|X_n - \theta\| = \mathcal{O}(q_n^n)$ with $q_n \to 0$. For a precise formulation and better estimates of error see, e.g., Tapia (1971). Presently, variations of the Newton Method are often used, which do not require inverting of the matrices $H(X_n)$. These are mostly related to the conjugate gradient methods for solving linear equations (Hestens (1951), Lanczos (1952), Stiefel (1952)); see Wolfe (OMIS symposium, 1971) for references.

(1.3) Stochastic approximations. Stochastic approximations resemble the gradient methods with small steps (rather than the Newton methods as some authors say when they look at the one-dimensional case only). [There have been modifications proposed which look like steepest descent methods (Fabian 1960) and a Newton-like stochastic approximation procedure is given here in Theorem 2.7.]

It is obvious that stochastic approximation procedures share all difficulties of corresponding numerical methods, among which are the following:

a) If condition (1.1.1) is omitted then $X_n$ may converge to a stationary point $x_0$ (i.e. $D(x_0) = 0$) different from $\theta$.

b) The procedure does not give any information about how close a given $X_n$ is to $\theta$.

c) For gradient methods the speed $\mathcal{O}(q^n)$ seems to hold too often with q near 1 and is then very slow. Slow speed occurs especially often if k, the dimension, is high, but also for k small ($= 2$) if f is obstinate enough. Newton-type methods usually do much better but may be impractical for k large.

Now stochastic approximations not only share the above difficulties but as far as the speed is concerned, are much worse. Fortunately experimenters, likely to face problems where stochastic approximation is applicable, are more modest than customers of numerical analysis; they usually also require much less as far as the ultimate approximation to $\theta$ is concerned.

An upper bound for the speed can be easily derived when accepting the plausible fact that instead of observing the function values, observing the values of the derivative is advantageous. If then $k = 1$, $f(x) = \frac{1}{2}(x - \theta)^2$, and $\xi_1, \xi_2, \ldots, \xi_n$ are unbiased estimates of $D(X_i)$ such that the errors $\eta_i = \xi_i - D(X_i)$ are independent normal $(0,\sigma^2)$ then $-\xi_i + X_i$ are independent normal $(\theta,\sigma^2)$ and the minimum square distance estimator available is

$$\xi = \frac{1}{n} \sum_{i=1}^{n} (-\xi_i + X_i)$$

with the property

$$E\,(\xi - \theta)^2 = \frac{\sigma^2}{n}\,.$$

Thus the best possible speed we can hope for is, in an obvious sense, $\mathcal{O}(n^{-\frac{1}{2}})$. We do not say, however, that this speed can be achieved.

(1.4) A choice of $Y_n$. As this question can well be illustrated on case $k = 1$, we shall restrict ourselves to this simple case. In case $k > 1$ the coordinates of D are estimated as here the scalar D is. The restriction to $k=1$ will be in effect throughout the Introduction.

An obvious method is to use differences $\Delta(x,c) = f(x+c) - f(x-c)$; if unbiased estimates of function values, with variances equal to $\sigma_0^2$, are available, we can obtain unbiased estimates of $(2c)^{-1}\Delta(x,c)$ with variances

$$\sigma^2 = \tfrac{1}{2}c^{-2}\sigma_0^2\ ^{*)}.$$

Now assume that in (1.1.2) every $Y_n$ has a conditional distribution, given $X_1, \ldots, X_n$, with mean and variance as given above with $x = X_n$, $c = c_n$. Add to our basic assumptions concerning f the following

$$\inf\{f(x) - f(\theta);\ \|x - \theta\| > \epsilon\} > 0 \text{ for every } \epsilon > 0 \qquad (1)$$

and assume that f has a continuous third derivative in a neighborhood of $\theta$. Then under some conditions on $a_n$, $c_n$, we obtain $X_n \to \theta$ (cf. Theorem 2.4).

(1.5) Choice of $a_n, c_n$. Let us limit our considerations to $a_n = an^{-\alpha}, c_n = cn^{-\gamma}$, $a, \alpha, c, \gamma$ positive constants. Consider $\eta$, the lowest upper bound of all such $\eta_0$ for which

$$n^{\eta_0}(X_n - \theta)$$

forms a mass preserving sequence, for fixed $a, c, \alpha, \gamma$ and all f, $Y_n$ satisfying our assumptions. Obviously, if $\eta_0 > \eta$ we

*With slight complication in formulation of results we may allow

$$\sigma^2 \leqq \tfrac{1}{2}c^{-2}\sigma_0^2.$$

cannot have the convergence of

$$n^{\eta_0}(X_n - \theta)$$

in distribution, nor in probability and so on. It turns out (Th. 2.5) that $\eta > \beta = 1/3$ unless $\alpha = 1$, $\gamma = 1/6$, $af''(\theta) > \beta$ when, if a mild Lindeberg type condition is required of the original estimates of function values, we obtain that $n^{\beta}(X_n - \theta)$ is asymptotically normal. (The addition of the Lindeberg condition when considering $\eta$ would not change the above statement about $\eta > \beta$ for all other choices of $a_n$, $c_n$.)

Thus the above values of $\alpha, \gamma$ may be considered to lead to optimal $a_n$, $c_n$ within the class considered. There are results on convergence and asymptotic normality for $a_n$, $c_n$ not necessarily of the form $an^{-\alpha}$, $cn^{-\gamma}$ (Sacks (1958), Schmetterer (1968)) but they do not lead to improvement in speed.

The speed $\mathcal{O}(n^{-1/3})$ is not very satisfactory. It can be improved if we restrict further the class of functions considered. Let s be an even number and suppose a continuous $f^{(s+1)}$ exists in a neighborhood of $\theta$ and $f^{(j)}(\theta) = 0$ for all odd j in $[3, s - 1]$. Then taking a different $\gamma$ one can obtain a speed of $\mathcal{O}(n^{-\beta})$ with $\beta = \frac{1}{2}(s/s + 1)$, thus close to the best value $\frac{1}{2}$, if s is large. The reason for this is, that near $\theta$, $(2c^{-1})\Delta(x,c)$ is a better estimate of $D(x)$ as the function is "approximately" even. However, the condition used is one rarely to be satisfied in applications.

(1.6) Another choice of $Y_n$: optimal $a_n$, $c_n$. Let us assume only the existence of a continuous

$$f^{(s+1)}$$

in a neighborhood of $\theta$ (for $k > 1$ existence of

$$\partial^{s+1} f/\delta(x^{(i)})^s$$

for every coordinate $x^{(i)}$). Then we still cannot obtain better speed than $\mathcal{O}(n^{-1/3})$ unless we change the method of estimating $D(x)$, which until now here, and also in the literature until 1967, was considered fixed.

Take $m = s/2$, choose $u_i$ such that (with Kronecker symbols $\delta_{ij}$) $0 < u_1 < \dots < u_m = 1$, find solutions $v_i$ to the equations

$$\sum_1^m v_i u_i^{2j-1} = \tfrac{1}{2}\delta_{ij} ,$$

and assume that the conditional, given $X_1, X_2, \dots, X_n$, mean and variance of $Y_n$ is

$$\sum_1^m v_i \Delta(x, c_n u_i) , \quad c_n^{-2}\sigma^2 \quad \text{with } \sigma^2 = 2\sum_1^m v_i^2 \sigma_0^2 . \tag{1}$$

Then all the statements following (1.4) hold again with the change that $\beta$ is now $\tfrac{1}{2}(s/s+1)$ (note that the previous situation is a special case of the present one with $s = 2$, $m = 1$, $u_1 = 1$) and the optimal choice of $\alpha$, $\gamma$ is

$$\alpha = 1, \qquad \gamma = (2s+2)^{-1}, \qquad af''(\theta) > \beta \tag{2}$$

(Fabian(1967)).

The reason for this improvement is that the choice of $v_i$ eliminates the effect of odd derivatives of f of order greater than 1 and less than s. Constructing any of the $Y_n$ with the required properties will require of course 2m (2mk if $k > 1$) observations instead of 2, but that is asymptotically offset by the increase in the rate of convergence.

We shall now add (2) to our assumptions.

(1.7) <u>The choice of designs and of</u> a, c. With the possibility of a more complicated choice of random variables $Y_n$ as specified by the design $(u_1, u_2, \dots, u_m)$ a number of questions arise: If we assume f has all derivatives, what value should one choose for m. Evidently the bigger m, the better asymptotic behaviour, but that is not true nonasymptotically, as the number of observations per step increases linearly with m. Even if m is chosen, what is the best design $(u_1, \dots, u_m)$?

The second question at least allows a formulation

within the asymptotic theory. Setting

$$N_0 = 2\,((s+1)!)^{-1}(\Sigma\, v_i u_i^{s+1}) f^{(s+1)}(\theta) \tag{1}$$

we obtain that

$$\Sigma\, v_i \Delta_i(x,c) = D(x) + c^s N(x,c) \tag{2}$$

with $N(x,c) \to N_0$ as $x \to \theta$, $c \to 0$ and this makes it possible to show that the mean and the variance of the asymptotic normal distribution of $n^\beta (X_n - \theta)$ are

$$\frac{-ac^s N_0}{af''(\theta) - \beta}, \quad \tfrac{1}{2}\,\frac{a^2 c^{-2} \sigma^2}{af''(\theta) - \beta}$$

and thus if $\xi$ stands for a random variable with this distribution,

$$E\,\xi^2 = \frac{a^2 c^{2s} N_0^2}{(af''(\theta) - \beta)^2} + \tfrac{1}{2}\,\frac{a^2 c^{-2} \sigma^2}{af''(\theta) - \beta} \tag{3}$$

(cf. Theorems 2.6 and 2.7 for $k > 1$). We shall consider one stochastic approximation procedure better than another if it has a larger $\beta$ or, with the same values of $\beta$, if it has a smaller $E\,\xi^2$. The latter is not an obvious or unique criterion. However it seems to be not unreasonable and under some additional assumptions it is possible to prove that

$$n^{2\beta} E(X_n - \theta)^2 \to E\xi^2.$$

Also not all the subsequent discussion actually depends on this convention.

It is a matter of elementary calculations (cf. Abdelhamid, 1971) that a and c, minimizing (3) are given by

$$a = (f''(\theta))^{-1}, \quad c = \left[\frac{\sigma^2 (1 - \beta)}{2sN_0^2}\right]^{\frac{1}{2s+2}} \tag{4}$$

resulting in

$$E\,\xi^2 = \left(\frac{2s}{1-\beta}\right)^{\frac{1}{s+1}} \frac{1}{4\beta(f''(\theta))^2} N_0^{\frac{2}{s+1}} \sigma^{4\beta} ; \qquad (5)$$

substituting from (1.6.1) and (1) we get a function of $\Sigma\, v_i^2$ and

$$\Sigma\, v_i u_i^{s+1}$$

and the optimal design should make its value minimum. This problem has not been solved yet.

Without having an optimal design it is of interest, to have some designs for which we can evaluate the $v_i$'s and, also, the important characteristics

$$\Sigma\, v_i^2 , \quad \Sigma\, v_i\, u_i^{s+1} .$$

Fortunately the second of these cannot be too large since elementary considerations show that

$$\Sigma\, v_i u_i^{s+1} = (-1)^{m-1}\, 2^{-1} \prod_{i=1}^{m} u_i^2$$

which is at most ½ in absolute value. We may thus rewrite (1) as

$$N_0 = (-1)^{m-1}((s+1)!)^{-1} f^{(s+1)}(\theta) \prod_{i=1}^{m} u_i^2 \qquad (6)$$

(cf. Remark 3.3, Fabian (1968a)).

It is then possible to minimize

$$\sum_{1}^{m} v_j^2$$

only, but before we talk about this problem let us make an observation which will be useful also later.

If somebody decides to take M independent observations

where one is required, and then to use the average, he will decrease $\sigma_0^2$ to $\sigma_0^2/M$ and similarly $\sigma^2$ to $\sigma^2/M$. At the stage n, after taking 2km Mn observations his $X_n - \theta$ will be approximately distributed as $n^{-\beta}\xi$ where $E\xi^2$ is given by (5) (if he is using optimal a, c) but with $\sigma$ replaced by $\sigma/\sqrt{M}$; alternatively as $n^{-\beta} M^{-\beta} \xi$ where $E\xi^2$ is exactly as in (5). Another person using the original observations (i.e. M = 1) is at stage Mn, after the same number of observations, and his $X_n - \theta$ will be approximately $(nM)^{-\beta}\xi$ with $E\xi^2$ as in (5). This shows that increasing the number of observations in the above manner leaves the asymptotic behavior unchanged. The same is true even for non-optimal a, c, if the value of c is adjusted properly.

With 2mM observations available for points $\pm u_i c_m$ we may make the variance even smaller by distributing these observations in an unequal way, using $M = 2mM\xi_i$ observations for $\Delta(x, c_n u_i)$ with the resulting conditional variance of $Y_n$ being

$$M^{-1}\, 2 \sum_1^m (v_i^2/\xi_i)\sigma_0^2.$$

The design $(u_1, \ldots, u_m)$ and the vector $(\xi_1, \ldots, \xi_m)$ minimizing $\Sigma\, v_i^2/\xi_i$ (under the obvious condition $\xi_i \geqq 0$, $\Sigma\, \xi_i = 1$) were obtained in Fabian (1968a), by using the results on optimum designs of experiments by Kiefer and Wolfowitz (1959) (a part of the proof can be replaced by a reference to results in Kiefer and Wolfowitz (1964), pp. 91-92). The optimal $u_j$ and $\xi_j$ are given by

$$u_j = \cos\,[(m - j)\pi/(2m - 1)] \tag{7}$$

$$\xi_j = [s(m - 1) + \tfrac{1}{2}]^{-1}\, u_j^{-2}(1 - \tfrac{1}{2}\delta_{mj}) \tag{8}$$

with the resulting

$$v_j = (s - 1)^{-1}\, (-1)^{j-1}\, u_j^{-2}(1 - \tfrac{1}{2}\delta_{mj}), \tag{9}$$

$$\Sigma\, v_j^2/\xi_j = \tfrac{1}{4}(s - 1)^2, \quad \pi\, u_j^4 = 2^{-4(m-1)} \tag{10}$$

where the last line gives quantities involved in the computation of $N_0^2$ and $\sigma^2$.

The numbers $2m\,M\xi_i$ may not be even integers but taking $M_i$ the least even integer greater or equal to $2m\,M\xi_i$ we only improve the variance with the number of observations increasing from 2mM to at most $2m(M + 1)$.

Using these designs we are able to find out how $E\xi^2$ in (3) depends on s, at least if we know $\sigma_0^2$ and $f^{(s+1)}(\theta)$. We may then decide possibly, what s to choose, to make $n^{-2\beta}E\xi^2$ minimal, for a fixed n, by a choice of s.

(1.8) Increasing the size of design. A more sensible approach seems to let s change through the approximation process. There seems to be one difficulty: we should know something about the sequence

$$\{f^{(2m+1)}(\theta)\}$$

because this sequence, or some bounds for it, will determine how suitable a sequence $s_n$ is. Fortunately there is a qualitative assumption which yields sufficient information within certain bounds: if f behaves as an analytic function in a neighborhood of θ then for a neighborhood $\mathcal{U}$ of θ and some positive K and r,

$$|f^{(s)}(x)| \leqq Ks\ r^s \quad \text{for all } x \text{ in } \mathcal{U}.$$

With this assumption it is then possible to use $s_n$ the smallest even integer greater or equal to $\log n \log_2^{-1} n$ and suitably choose $a_n$, $c_n$ to achieve a speed given by

$$E(X_n - \theta)^2 = (t_n^{-1} \log^3 t_n)$$

where $t_n$ is the total number of observations required to construct $X_1, X_2, \ldots, X_n$. It should be stressed that no knowlege of K, r or $f''(\theta)$ is required (see Fabian (1969); the result holds in the multidimensional case.)

On the other side no optimality was proved or even conjectured for the result or the choice of $s_n$, but the rate is close to the best possible.

(1.9) Procedure adapting a,c to their optimal values.

There are some other interesting questions and results if we return back to the procedure using a fixed design (possibly but not necessarily with $s = 2$). Namely we have found optimal values for the constants a and c but they depend on some, in general unknown, parameters of the function to which the procedure is applied. Can these be estimated during the approximation process? We obviously are accumulating a sequence of estimates of f(x) for many x's but the obvious attempt to use all the observations leads to difficulties as there is too much interdependence (for a study in this direction see Anbar (1971)).

But Venter (1967) (the result is strenghtened in Fabian (1968b)) used a fact we noticed above, that increasing number of observations and taking averages leaves the asymptotic properties unchanged. Venter's work concerns the Robbins-Monro case where at stage n one requires an estimate of $f(X_n)$, and the optimal a depends on $f'(\theta)$. Venter then showed that instead of estimating twice $f(X_n)$ one can estimate $f(X_n - d_n)$, $f(X_n + d_n)$, substitute the average for an estimate of $f(X_n)$ and use the difference, divided by $2d_n$, as an estimate of $f'(\theta)$. Keeping $d_n$ sufficiently small but not too small makes both estimates to be sufficiently good to achieve the asymptotic properties as if the original procedure were used with a optimal. This also disposes automatically of the difficulty of choosing a satisfying a condition analogous to (1.6.2) which is necessary to obtain the optimum rate of the convergence.

Another possibility to estimate the optimum a,c in our situation is to take additional observations used exclusively for this estimation but only at some stages so that the increase in the number of observations becomes negligible asymptotically. Theorem 2.7 gives a result which, specialized to $k = 1$, describes a procedure which itself estimates the best possible a. Without any difficulty we may similarly obtain a procedure which estimates both the optimal a and c.

(1.10) A Newton-like procedure. In the k-dimensional case Theorem 2.7 gives a Newton-like variant of the stochastic approximation method. The resulting method behaves asymptotically as the original stochastic approximation procedure applied after a change of coordinates in which

$H(\theta)$ becomes the unit matrix I, and using the optimal value for a. From the comparison with the gradient methods and the Newton method of numerical analysis it would seem that this modification may be very advantageous but no formal results of this type are offered at this time.

(1.11) Transformation of $Y_n$. Fabian (1960) noticed some difficulties with the original stochastic approximation method when $|f'|$ was large near $\theta$ and small away from $\theta$ so that the corrections $a_n Y_n$ tended to be too big when $X_n$ was near $\theta$ and too small otherwise. (Notice that this difficulty will not affect the asymptotic results when $X_n$ is near $\theta$.) He then proposed to replace $Y_n$ in (1.1.2) by $c_n^{-1}$ sign $(c_n Y_n)$ and proved the convergence to $\theta$ under some additional conditions. But the question remained how are the asymptotic properties affected or, in general, what is the optimal transformation h which, when used in

$$X_{n+1} = X_n - a_n c_n^{-1} h(c_n Y_n) \tag{1}$$

will not destroy the convergence to $\theta$ and make the speed maximal.

If the conditional distribution of

$$c_n Y_n - E_{X_1, \ldots, X_n} c_n Y_n$$

is independent of $X_1, \ldots, X_n$ and of n and given by a symmetric density g admitting a derivative, the optimal choice of h is $-\text{const}\; g'/g$ if some additional mild requirements are satisfied. This result was obtained, independently, by Abdelhamid (1971) and Anbar (1971); the first author considered both the Robbins-Monro and the Kiefer-Wolfowitz situation, the second the Robbins-Monro situation. The resulting asymptotic variance then achieves, in the Robbins-Monro situation, the bound given by the Cramér-Rao inequality for unbiased estimators. In this sense the modified stochastic approximation method is an asymptotically efficient estimator of $\theta$. An analogous but a little more complicated result holds for the modified Kiefer-Wolfowitz method.

(1.12) Limitation of asymptotic theory. The limits of

possible practical inference from asymptotical results should be apparent in general and per se. To this author they seem to be more menacing with the number of asymptotic results increasing. The improvement would be in obtaining results about the speed with which the behaviour of $n^{\beta}(X_n - \theta)$ approaches that of $\xi$, or, of course, results concerning fixed n. Such a result for the Robbins-Monro situation is contained in Dvoretzky (1956).

(1.13) Practical applications. I think they may be essentially of two kinds. Applications to finding optimal conditions by experimentation (chemical, biological research, etc.) where the number of steps will usually be very small or applications to models (e.g. realized on a computer) with random components, where the number of available observations may be very large.

Some friends of mine, working in chemical research, were rather enthusiastic about their experience with stochastic approximation procedures applied to problems like finding optimum amounts of different components in a catalyzer. The number of steps they have taken was usually very small but they often found conditions if not optimal then much better than the starting ones. But asymptotic theory was here hardly relevant because of small n and probably some satisfaction was derived from the pleasure of thinking no more to be obliged to use some more standard statistical procedures.

For application of the second kind see, e.g., Janáč (1971; see additional references there) who was using a sophisticated Monte-Carlo model and a stochastic approximation procedure to find optimum parameters of a suspension system of an automobile.

(1.14) Related results: Let us mention some of the related results which were not included in other reviews. Fabian (1960) considered the behaviour of $X_n$ when conditions (1.1.1) and (1.4.1) are not satisfied. One unpleasant thing which then can happen is that $X_n$ may stay in a region where f is constant. Fabian (1964) proposed a modified stochastic approximation which does not have this disadvantage if $k = 1$. Unfortunately this result cannot be simply extended to $k > 1$ because of the differences in the recurrence properties of the random walk for dimension 1 and higher. Dupač (1965,1966) has considered stochastic approximation in a presence of a trend and Fabian (1965) a stochastic approximation for minimization with convex constraints.

## 2. Theorems

(2.1) Preliminaries. Some of the results described in the Introduction will be reformulated in a greater generality. But those which were described with sufficient detail in the Introduction are not repeated here.

Let us summarize some notation. $\delta_{ij}$ denotes 0 or 1 according to whether $i \neq j$ or $i = j$, $k$ is a fixed positive integer, $e_i$ $(i = 1, \ldots, k)$ are the unit vectors in $R^k$, i.e., $e_i \in R^k$, $e_i^{(j)} = \delta_{ij}$. We shall talk about functions $f$ on $R^k$ and then $D_s^{i}$ will denote the vector of the s-th order derivatives of $f$ with respect to individual coordinates, i.e.

$$D_s^{(i)}(x) = \frac{\partial^s f(x)}{\partial (x^{(i)})^s},$$

$D = D_1$ and $H$ will be the Hessian, i.e. the symmetric matrix of the second order partial derivatives of $f$. The symbol $\Delta(x,c)$ will denote the vector defined by

$$\Delta^{(i)}(x,c) = f(x + ce_i) - f(x - ce_i).$$

All random variables are supposed to be defined on a probability space $(\Omega,\mathfrak{F},P)$, relations between random variables are meant with probability one. $E$ or $E_{\mathfrak{F}_0}$ denote expectation and conditional expectation, given $\mathfrak{F}_0$.

Next we formulate two assumptions, depending on $s$, which will always be an even positive integer; assumption (2.2) has two parts the first of which concerns $f$, while the second part concerns the stochastic procedure. Note the simple case when $s = 2$ which implies that (2.2.5) is

$$M_n(x) = c_n^{-1} \tfrac{1}{2}\Delta(x,c_n) \quad ;$$

also note that $A_n$ may be all equal to $I$, the identity

matrix, when (2.2.3) is satisfied whenever $\nu_n \leqq 1 \leqq \mu_n$.

(2.2) Assumption: f is a function on $R^k$ with an absolute minimum at a $\theta$ in $R^k$. The Hessian H exists and is bounded on $R^k$

$$\inf \{\|D(x)\|;\ x \notin S(\epsilon)\} > 0, \quad \inf \{f(x) - f(\theta); x \notin S(\epsilon)\} > 0 \tag{1}$$

for every $\epsilon > 0$ where $S(\epsilon)$ is the open $\epsilon$-sphere around $\theta$.

$X_1, X_2, \ldots, Y_1, Y_2 \ldots$ are random k-dimensional vectors, $\sigma^2$, $a_n$, $c_n$, $\nu_n$, $\mu_n$, are positive numbers, $A_n$ are non-singular symmetric (k x k) matrix valued measurable functions on $\Omega$, the eigenvalues of $A_n(\omega)$ lie in $[\nu_n, \mu_n]$ for all $\omega$,

$$X_{n+1} = X_n - a_n A_n Y_n. \tag{2}$$

With $\mathfrak{F}_n = \sigma(X_1, Y_1, \ldots, Y_{n-1}, A_1, \ldots, A_n)$ the smallest $\sigma$-algebra generated by the indicated transformations and with

$$Z_n = Y_n - E_{\mathfrak{F}_n} Y_n \tag{3}$$

we have

$$E_{\mathfrak{F}_n} \|Z_n\|^2 \leqq c_n^{-2} \sigma^2 \tag{4}$$

and

$$E_{\mathfrak{F}_n} Y_n = M_n(X_n)$$

with

$$M_n(x) = c_n^{-1} \sum_1^m v_i \, \Delta(x, c_n u_i). \tag{5}$$

Here $m = s/2$, $0 < u_1 < \ldots < u_m = 1$,

$$\sum_{i=1}^m v_i u_i^{2j-1} = \tfrac{1}{2}\delta_{1j} \qquad \text{for} \quad j = 1, 2, \ldots, m. \tag{6}$$

(2.3) Assumption: $D_{s+1}$ exists and is continuous in a neighborhood of $\theta$, $H(\theta)$ is positive definite with the smallest eigenvalue $\lambda$, H is continuous at $\theta$.

There is a matrix $\Sigma$ such that, for $V_n = c_n Z_n$,

$$\lim E_{\mathfrak{F}_n} V_n V_n' \to \Sigma \quad \text{on the set } \{X_n \to \theta\} \tag{1}$$

and, with $\chi$ the indicator function,

$$\lim n^{-1} \sum_{j=1}^{n} E\chi\{\| V_n \|^2 > rj\} \| V_n \|^2 \to 0 \text{ for every } r > 0. \tag{2}$$

The numbers $a_n$, $c_n$ are of the form

$$a_n = an^{-\alpha}, \quad c_n = cn^{-\gamma} \tag{3}$$

with $a$, $c$, $\alpha$, $\gamma$ positive.

(2.4) Theorem: If Assumption (2.2) holds with a positive integer s and if

$$c_n \to 0, \quad (c_n + a_n\mu_n)\mu_n\nu_n^{-1} \to 0, \qquad \Sigma a_n\nu_n = +\infty, \tag{1}$$

$$\Sigma a_n\mu_n c_n < +\infty, \qquad \Sigma a_n^2 \mu_n^2 c_n^{-2} < +\infty$$

then

$$X_n \to \theta.$$

(2.5) Theorem: Suppose $a_n$, $c_n$ are as in (2.3.3), s is a positive even integer, $\beta = (s\gamma) \wedge (\frac{1}{2}\alpha - \gamma)$ and $\lambda$ is a positive number.

Then, if $\eta > \beta$ or if ($\alpha = 1$ and $\lambda a \leq \eta = \beta$), there are f, $X_0$, $Y_n$ satisfying Assumptions (2.2) and (2.3), with the given $\lambda$ and with $A_n = I$ such that

$$\lim\sup P\{n^{\eta}\|X_n - \theta\| > K\} = +\infty$$

for every $K \in R$.

In the next theorem take $A_n = A = I$ to obtain a self contained result. The more general form with arbitrary $A_n$ will be used in Theorem (2.7). Notice that the choice $\alpha = 1$, $\gamma = (2s + 2)^{-1}$ leads to a rate which is optimal by the preceding result.

(2.6) Theorem: Suppose assumptions (2.2) and (2.3) hold with a positive even integer s, with $A_n$ converging to a positive definite matrix A and with

$$\alpha = 1, \quad \gamma = (2s + 2)^{-1}, \quad a\Lambda_0 > \beta, \quad \nu_n = (\log n)^{-1}, \mu_n = n^{\mu} \tag{1}$$

where $\beta = \frac{1}{2}s(s+1)^{-1}$, $\Lambda_0$ is the smallest eigenvalue of $AH(\theta)$ and $\mu$ is a positive number, $\mu < \gamma$.

Let P be an orthogonal matrix such that $\Lambda = P'AH(\theta)P$ is diagonal,

$$N_0 = (-1)^{m-1}((s+1)!)^{-1}\left(\prod_{i=1}^{m} u_i^2\right) D_{s+1}(\theta), \tag{2}$$

$$M^{(ij)} = a^2c^{-2} \frac{(P'A\Sigma AP)^{(ij)}}{a\Lambda^{(ii)} + a\Lambda^{(jj)} - 2\beta}. \tag{3}$$

Then $X_n \to \theta$ and the asymptotic distribution of $n^{\beta}(X_n - \theta)$ is normal with mean

$$-ac^{s}(aAH(\theta) - \beta I)^{-1}AN_0 \tag{4}$$

and covariance matrix

$$PMP'. \tag{5}$$

(2.7) Theorem: Suppose Assumptions (2.2) and (2.3) hold with a positive even integer s but with $\mathfrak{F}_n$ as given below in (2), with the requirements concerning $A_n$, except for (2.2.2), omitted. Let

$$\alpha = 1, \ \gamma = (2s + 2)^{-1}, \ a > \beta = \tfrac{1}{2}s(s+1)^{-1}, \\ \nu_n = (\log n)^{-1}, \ \mu_n = n^{\mu}, \quad \mu \in (0,\gamma). \tag{1}$$

Suppose $\{m_\ell\}$ is an increasing sequence of positive integers, $\{U_\ell\}$ is a sequence of random symmetric $k \times k$ matrices and

$$\mathfrak{F}_n = \sigma(\{X_1, Y_1, \ldots, Y_{n-1}\} \cup \{U_\ell; m_\ell < n\}). \tag{2}$$

Suppose

$$E_{\mathfrak{F}_{m_\ell}} U_\ell = (2d_\ell)^{-2}\Delta_2(X_{m_\ell}, d_\ell), \quad E\|U_\ell - E_{\mathfrak{F}_{m_\ell}} U_\ell\|^2 \leqq d_\ell^{-4}K \tag{3}$$

with K a constant,

$$d_\ell = d\ell^{-\delta}, \qquad d > 0 , \qquad 0 < \delta < \tfrac{1}{4} \tag{4}$$

and $\Delta_2(x,\eta)$ the second order difference,

$$\Delta_2^{(ij)}(x,\eta) = \Delta^{(j)}(x + \eta e_i,\eta) - \Delta^{(j)}(x - \eta e_i,\eta). \tag{5}$$

Denote by $\bar{U}_n$ the average of all $U_\ell$ with $m_\ell < n$ and set

$$A_n = \begin{cases} \bar{U}_n^{-1} & \text{if } \bar{U}_n^{-1} \text{ exists and is positive definite with all eigenvalues in } [\nu_n, \mu_n] \\ I & \text{otherwise.} \end{cases} \tag{6}$$

Then $X_n \to \theta$, $A_n \to H^{-1}(\theta)$, and the asymptotic distribution of

$$n^\beta(X_n - \theta)$$

is normal with mean and covariance

$$-\frac{ac^s}{a-\beta} H^{-1}(\theta)N_0, \quad \tfrac{1}{2}\frac{a^2c^{-2}}{a-\beta} H^{-1}(\theta)\,\Sigma\, H^{-1}(\theta) \tag{7}$$

where $N_0$ is given by (2.6.2).

## 3. Proofs and Comments on Methods of Proofs

The following is a slight generalization of a lemma by Chung (1954) who used it to derive asymptotic moments for the Robbins-Monro method. The same lemma was then used by other authors to derive asymptotic moments for other procedures (Derman (1956), Burkholder (1956), Dupač (1957)) and later (Fabian 1968b) to derive asymptotic properties of the characteristic function and to obtain so a general result about asymptotic normality. Some authors prefer to use other means (Sacks (1958), Schmetterer(1961 and 1969)). The present proof is essentially the same as in Fabian (1967):

(3.1) <u>Lemma (Chung)</u>: Let $\alpha, \beta$, B, $b_n$, $A_n$ be real numbers, $0 < \alpha \leq 1$, $\beta \geqq 0$, $B > 0$. Denote $\beta_+ = 0$ if $\alpha < 1$ and $\beta_+ = \beta$ if $\alpha = 1$ and assume that $c = \liminf (A_n - \beta_+)$, $C = \limsup (A_n - \beta_+)$ are finite.

If, eventually,

$$b_{n+1} \leqq b_n(1 - A_n n^{-\alpha}) + B_n^{-\alpha-\beta} \tag{1}$$

and if $c > 0$ then

$$\limsup n^{\beta} b_n \leqq B/c. \tag{2}$$

If (1) holds with the opposite inequality $\geqq$ and $b_n \geqq 0$ then

$$\liminf n^{\beta} b_n \begin{cases} \geqq B/C & \text{if } C > 0 \\ = +\infty & \text{if } C \leqq 0. \end{cases} \tag{3}$$

<u>Proof</u>: Writing $\xi_n = n^{\beta} b_n$ it is a matter of elementary computation* to derive

$$\xi_{n+1} = \xi_n(1 - \gamma_n n^{-\alpha}) + B_n n^{-\alpha} + Q_n \tag{4}$$

where for every $\epsilon$ positive eventually $c - \epsilon < \gamma_n < C + \epsilon$, $B - \epsilon < B_n < B + \epsilon$; $Q_n \leqq 0$ if (1) holds, $Q_n \geqq 0$ if (1) holds with the reversed inequality. We may assume the above inequalitites hold for all n and that

$$\gamma_n n^{-\alpha} < 1 \text{ and } \epsilon < \tfrac{1}{2}B.$$

Let us treat only the second case, as the first case is similar and simpler*: thus assume $Q_n \geqq 0$. Take a $C_1$ positive and larger than $C + \epsilon$ so that $\gamma_n C_1^{-1} < 1$.

Then

*If details are desired, see Fabian (1967), p. 196, where the proof of the second part was skipped and the assumption $b_n \geqq 0$ was inadvertently omitted.

$$\xi_{n+1} \geqq \xi_n + n^{-\alpha}\epsilon \quad \text{if} \quad \xi_n \leqq C_1^{-1}(B - 2\epsilon) \tag{5}$$

is obviously true if $\gamma_n \leqq 0$ since $\xi_n \geqq 0$. To show it for $\gamma_n \geqq 0$ observe that

$$\gamma_n\xi_n \leqq \gamma_n C_1^{-1}(B - 2\epsilon) \leqq B - 2\epsilon \quad \text{if} \quad \xi_n \leqq C_1^{-1}(B - 2\epsilon)$$

and then (5) follows from (4). On the other hand

$$\xi_{n+1} \geqq C_1^{-1}(B - 2\epsilon) \text{ if } \xi_n \geqq C_1^{-1}(B - 2\epsilon) \tag{6}$$

since this follows from (5) if $\xi_n = C_1^{-1}(B - 2\epsilon)$ and, for other $\xi_n$, from the fact that the right-hand side of (5) is increasing in $\xi_n$.

Since $\Sigma n^{-\alpha} = +\infty$, (5) and (6) imply $\liminf \xi_n \geqq C_1^{-1}(B - 2\epsilon)$; this easily implies (3). □

The following is a simple result, stated and used by Blum (1954). For a slightly different formulation see Schmetterer (1969).

(3.2) <u>Lemma</u>: Let $\xi_1, \xi_2, \ldots$ be a sequence of non-negative random variables, $\mathfrak{F}_n = \sigma\{\xi_1, \ldots, \xi_n\}$,

$$\theta_n = E_{\mathfrak{F}_{n-1}}(\xi_n - \xi_{n-1}) \quad , \quad n = 1,2, \ldots , \quad \xi_0 = 0 \tag{1}$$

and with $\theta_n^+$ denoting the positive part of $\theta_n$,

$$\sum_{n=1}^{\infty} E\,\theta_n^+ < +\infty. \tag{2}$$

Then the sequence $\{\xi_n\}$ converges to an integrable random variable.

<u>Proof</u>: From (1),

$$E_{\mathfrak{F}_{n-1}}[\theta_n^+ - (\xi_n - \xi_{n-1})] \geqq 0,$$

this formally rewritten gives

$$E_{\mathfrak{F}_{n-1}}\left[\sum_{j=1}^{n}\theta_j^+ - \xi_n - \left(\sum_{j=1}^{n-1}\theta_j^+ - \xi_{n-1}\right)\right] \geqq 0$$

so that, setting

$$\zeta_n = \sum_{j=1}^{n}\theta_j^+ - \xi_n,$$

$$\zeta_{n-1} \leqq E_{\mathfrak{F}_{n-1}}[\zeta_n]. \tag{3}$$

Noticing that $\theta_1 = \xi_1$, using the non-negativity of $\xi_n$ and a simple inductive argument one obtains

$$E\xi_n \leqq \sum_{j=1}^{n} E\,\theta_j^+ \tag{4}$$

which with (2) implies

$$E|\zeta_n| \leqq 2 \sum_{j=1}^{\infty} E\,\theta_j^+ < +\infty. \tag{5}$$

Thus $\{\zeta_n\}$ is a submartingale and (5) implies (cf.,e.g., Breiman (1968), Theorem 5.14) that $\{\zeta_n\}$ converges to an integrable random variable $\zeta$. Then

$$\xi_n \to \sum_{j=1}^{\infty}\theta_j^+ - \zeta. \; \square$$

The next proof is almost identical to that used by Blum (1954) to prove the convergence of the first version of the multi-dimensional stochastic approximation method for finding a minimum. The lemma is sufficiently general so that it can also be used to prove the convergence of the RM and KW one-dimensional methods (see Fabian (1960)).

(3.3) Lemma: Suppose f is a from below bounded function on $R^k$ with a bounded Hessian H, $X_n$, $U_n$ are k-dimensional random vectors, $\mathcal{X}_n = \sigma(X_1, \ldots, X_n)$,

$$X_{n+1} = X_n - U_n \tag{1}$$

and, with $N_n = E_{\mathcal{X}} U_n$, $\alpha_n$, $\beta_n$, $\gamma_n$, $\epsilon_n$ nonnegative numbers, $\alpha_n > 0$ and $B_n$ nonnegative random variables

$$N_n'D(X_n) \geqq \alpha_n B_n^2 - \beta_n - \gamma_n B_n \quad , \tag{2}$$

$$E_{\mathcal{X}_n} \|U_n\|^2 \leqq \beta_n + \gamma_n B_n + \epsilon_n B_n^2 \tag{3}$$

and

$$\Sigma \alpha_n = +\infty, \ \Sigma \beta_n < +\infty, \ \Sigma \frac{\gamma_n^2}{\alpha_n} < +\infty, \ \frac{\gamma_n}{\alpha_n} \to 0, \ \frac{\epsilon_n}{\alpha_n} \to 0. \tag{4}$$

Then a subsequence of $\{B_n\}$ converges to 0 and the sequence $\{f(X_n)\}$ converges to a random variable.

Proof: By a Taylor expansion, since H is bounded

$$f(X_{n+1}) \leqq f(X_n) - U_n'D(X_n) + C\|U_n\|^2$$

with a constant C. Then

$$E_{\mathcal{X}_n} f(X_{n+1}) - f(X_n) \leqq -(\alpha_n - C\epsilon_n)B_n^2 + (C+1)\gamma_n B_n + (C+1)\beta_n.$$

Writing $\epsilon_n' = C\epsilon_n$, $\gamma_n' = \frac{1}{2}(C+1)\gamma_n$, $\beta_n' = (C+1)\beta_n$, verifying that these also satisfy (4), and then omitting the primes,

$$E_{\mathcal{X}_n} f(X_{n+1}) - f(X_n) \leqq -(\alpha_n - \epsilon_n)B_n^2 + 2\gamma_n B_n + \beta_n$$

$$= -(\alpha_n - \epsilon_n)\left[B_n - \frac{\gamma_n}{\alpha_n - \epsilon_n}\right]^2 + \frac{\gamma_n^2}{\alpha_n - \epsilon_n} + \beta_n,$$

and, eventually,

$$E_{\mathcal{X}_n} f(X_{n+1}) - f(X_n) \leqq -\tfrac{1}{2}\alpha_n \left[B_n - 2\frac{\gamma_n}{\alpha_n}\right]^2 + \delta_n \tag{5}$$

with $\delta_n \geqq 0$ and $\Sigma\delta_n < +\infty$.

Thus in particular

$$E_{f(X_1), \ldots, f(X_n)}(f(X_{n+1}) - f(X_n))^+ \leqq \delta_n$$

and $\{f(X_n)\}$ is a convergent sequence by the preceding lemma since f - K is non-negative if K is the lower bound for f.

Further, if $P\{B_n > \epsilon\} > \epsilon$ eventually for a positive $\epsilon$ then from (5), for some $n_0$, $2\gamma_n/\alpha_n < \frac{1}{2}\epsilon$ for $n \geqq n_0$ and

$$E\, f(X_{N+1}) - E\, f(X_{n_0}) = \sum_{n=n_0}^{N} E\,[f(X_{n+1}) - f(X_n)]$$

$$\leqq - \frac{\epsilon^3}{8} \sum_{n=n_0}^{N} \alpha_n + \sum_{n=n_0}^{N} \delta_n \rightarrow -\infty \ .$$

This is impossible as f is bounded from below. Thus a subsequence of $\{B_n\}$ converges to 0 in probability and another one with probability one. □

(3.4) Proof of Theorem (2.4): Using the boundedness of H we obtain from (2.2.5) with a constant C that

$$M_n(x) = c_n^{-1}[c_n \sum_{i=1}^{m} v_i u_i 2D(x) + c_n^2 R_n(x)] = D(x) + c_n R_n(x)$$

where $\|R_n(x)\| \leqq C$ for a constant C. Thus

$$D'(x) A_n M_n(x) \geqq D'(x) A_n D(x) - C\, c_n \|D(x)\|\, \|A_n\|$$

$$\geqq \nu_n \|D(x)\|^2 - C\, c_n \|D(x)\| \mu_n \ .$$

Also

$$\|A_n M_n(x)\|^2 \leqq 2\mu_n^2 \|D(x)\|^2 + 2C^2 c_n^2 \mu_n^2$$

$$E_{\mathfrak{F}_n} \|A_n Z_n\|^2 \leqq \mu_n^2 c_n^{-2} \sigma^2 \ .$$

Using notation of the previous lemma with $U_n = a_n A_n Y_n$ and with $B_n = \|D(X_n)\|$ we have

$$N_n' D(X_n) \geqq a_n \nu_n B_n^2 - C a_n c_n \mu_n B_n$$

$$E_{\chi_n} \|U_n\|^2 \leqq a_n^2 \mu_n^2 c_n^{-2} \sigma^2 + 2C^2 a_n^2 c_n^2 \mu_n^2 + 2a_n^2 \mu_n^2 B_n^2 .$$

Thus (3.3.2) and (3.3.3) are satisfied with

$$\alpha_n = a_n \nu_n, \ \gamma_n = a_n c_n \mu_n C, \ \beta_n = a_n^2 \mu_n^2 c_n^{-2} \sigma^2 + 2C^2 a_n^2 c_n^2 \mu_n^2 ,$$

$$\epsilon_n = 2a_n^2 \mu_n^2$$

and (3.3.4) follows from (3.4.1).

It follows that $B_{n_i} \to 0$ for some $\{n_i\}$ and that $f(X_n)$ is convergent. It follows from (2.2.1) that $X_n \to \theta$. □

(3.5) Proof of Theorem (2.5): Suppose $k = 1$ first; without loss of generality assume $\lambda = 1$. Take

$$f(x) = \tfrac{1}{2}x^2 - b\, x^{s+1}$$

so that, by (2.2.5) and (2.2.6),

$$M_n(x) = x - Nc_n^s, \quad N = 2\left(\sum_1^m v_i u_i^{s+1}\right) b.$$

Treat the case $\beta = s\gamma$. Choose $Z_n = 0$, $X_0 = 0$, to obtain

$$X_{n+1} = X_n(1 - an^{-\alpha}) + ac^s n^{-\alpha-\beta} N.$$

Increase the value of N, if neccssary, to guarantee that all $X_n$ are positive until $an^{-\alpha} < 1$. Then all $X_n$ are positive and Lemma (3.1) implies

$$\liminf n^{\beta} X_n \geqq N/(a - \beta_+)^+ .$$

This is the desired result if $\eta > \beta$ or $\alpha = 1$, $\eta = \beta$, $a\lambda \leqq \beta$.

Treat the case $\beta = \tfrac{1}{2}\alpha - \gamma$. Choose $b = 0$, $Z_n = c_n^{-2}\xi_n$ with $\xi_1, \xi_2, \ldots$ independent normal $(0,\sigma^2)$ random variables, take $X_0 = 0$. Since $X_{n+1} = X_n(1 - an^{-\alpha}) + c_n^{-2}\xi_n$, the $X_n$'s are normal, with

$$EX_{n+1}^2 = (1 - 2A_n n^{-\alpha})EX_n^2 + a^2c^{-2}n^{-2\alpha+2\gamma}\sigma^2, \quad A_n \to a$$

and Lemma (3.1) gives

$$\liminf n^{2\beta}EX_n^2 \geq \frac{a^2c^{-2}\sigma^2}{2(a-\beta_+)}^+$$

and we obtain the desired result again, both if $\eta > \beta$ or $\alpha = 1$, $\beta = \eta$, $a\lambda < \beta$.

For $k > 1$, use the construction $f(x) = \sum_{i=1}^{k} f_i(x^{(i)})$.□

The results concerning the asymptotic normality will be based on the following theorem given in Fabian (1968b). The proof is readily accessible and will not be reproduced here. The method of proof consists (i) of an argument due to Hodges and Lehmann (1956) and a repeated use of Lemma (3.1) to simplify the original recurrence formula and (ii) of an application of Lemma (3.1) to $\varphi_n - \Psi_n$, where $\varphi_n$ is the characteristic function of the random variable $n^\beta(X_n - \theta)$ and $\Psi_n$ is a function defined by a recursion formula similar to that for $\varphi_n$, but dependent only on $\Sigma$ and $\Gamma$. The result is $\varphi_n - \Psi_n \to 0$ and an invariance type argument shows that $\varphi_n$ converges to the appropriate normal characteristic function.

The following additional notation is used: $R^{k\times k}$ is the space of all $k \times k$ matrices and $\underline{R}^k$, $\underline{R}^{k\times k}$ denote families of all measurable transformations on $(\Omega,\mathfrak{F})$ into $R^k$, $R^{k\times k}$, respectively.

(3.6) Theorem: Suppose $k$ is a positive integer, $\{\mathfrak{F}_n\}$ a nondecreasing sequence of $\sigma$-fields, $\mathfrak{F}_n \subset \mathfrak{F}$; suppose $U_n$, $V_n$, $T_n \in \underline{R}^k$, $T \in R^k$, $\Gamma_n$, $\Phi_n \in \underline{R}^{k\times k}$, $\Sigma$, $\Gamma$, $\Phi$, $P \in R^{k\times k}$, $\Gamma$ is positive definite, $P$ is orthogonal and $P'\Gamma P = \Lambda$ diagonal. Suppose $\Gamma_n$, $\Phi_{n-1}$, $V_{n-1}$ are $\mathfrak{F}_n$-measurable, $C, \alpha$, $\beta \in R$ and

$$\Gamma_n \to \Gamma, \quad \Phi_n \to \Phi, \quad T_n \to T \quad \text{or } E\|T_n - T\| \to 0, \tag{1}$$

$$E_{\mathfrak{F}_n} V_n = 0, \quad C > \|E_{\mathfrak{F}_n} V_n V_n' - \Sigma\| \to 0, \tag{2}$$

and, with $\sigma^2_{j,r} = E_{\chi}\{\|V_j\|^2 \geqq rj^{\alpha}\} \|V_j\|^2$, let either

$$\lim_{j \to \infty} \sigma^2_{j,r} = 0 \quad \text{for every } r > 0, \tag{3}$$

or

$$\alpha = 1, \; \lim_{n \to \infty} n^{-1} \sum_{j=1}^{n} \sigma^2_{j,r} = 0 \quad \text{for every } r > 0. \tag{4}$$

Suppose that, with $\lambda = \min_i \Lambda^{(ii)}$, $\beta_+ = \beta$ if $\alpha = 1$, $\beta_+ = 0$ if $\alpha \neq 1$,

$$0 < \alpha \leqq 1, \quad 0 \leqq \beta, \quad \beta_+ < \lambda \tag{5}$$

and

$$U_{n+1} = (I - n^{-\alpha}\Gamma_n)U_n + n^{-\alpha/2-\beta}\Phi_n V_n + n^{-\alpha-\beta} T_n. \tag{6}$$

Then the asymptotic distribution of $n^{\beta}U_n$ is normal with mean $(\Gamma - \beta_+ I)^{-1} T$ and covariance matrix PMP' where

$$M^{(ij)} = (P'\Phi\Sigma\Phi'P)^{(ij)} (\Lambda^{(ii)} + \Lambda^{(jj)} - 2\beta_+)^{-1}. \tag{7}$$

(3.7) <u>Proof of Theorem (2.6)</u>: From Theorem (2.4), its (2.4.1) implied by (2.6.1), we obtain $X_n \to \theta$. Define the $\mathfrak{F}_n$-measurable random vector $N_n$ by

$$M_n(X_n) = D(X_n) + c_n^{\,s} N_n. \tag{1}$$

For x sufficiently near to $\theta$ and $c_n$ small expand

$$\sum_{i=1}^{m} v_i \Delta(x, u_i c_n)$$

up to the s+1 derivative and use (2.5.5) to obtain

$$N_n \to N_0 = 2((s+1)!)^{-1}) \left( \sum_{i=1}^{m} v_i u_i^{s+1} \right) D_{s+1}(\theta). \tag{2}$$

This is the same value as given in (2.6.2) (cf. Fabian (1968a), relation (3.3.1)).

Next, interpreting 0/0 as 0, and setting

$$e_n = \|X_n - \theta\|^{-1}[D(X_n) - H(\theta)(X_n - \theta)]$$

we obtain $e_n \to 0$ and

$$D(X_n) = H(\theta)(X_n - \theta) + e_n\|X_n - \theta\|$$

and this can be written as

$$D(X_n) = H_n(X_n - \theta) \tag{3}$$

with $H_n$ in $\underline{R}^{k\times k}$, $\mathfrak{F}_n$-measurable, and $H_n \to H(\theta)$.

(The measurability may be obtained by letting $H_n = H(\theta) + E_n$ and

$$E_n^{(ij)} = \delta_{jj_0}(X_n^{(j_0)} - \theta)^{-1}\|X_n - \theta\|e_n^{(i)}$$

with $j_0$ maximizing $|X_n^{(j_0)} - \theta|$.) Then

$$X_{n+1} - \theta = (I - n^{-1}aA_nH_n)(X_n - \theta) - n^{-1+\gamma}ac^{-1}A_nV_n$$

$$- n^{-1-\gamma s}\, ac^sA_nN_n = (I - n^{-1}\Gamma_n)(X_n - \theta) + n^{-\frac{1}{2}-\beta}\,\Phi_nV_n$$

$$+ n^{-1-\beta}T_n \quad \text{with } \Gamma_n \to \Gamma = aAH(\theta),\ \Phi_n \to -ac^{-1}A,$$

$$T_n \to T = -ac^sAN_0 \text{ and } \Gamma_n,\ \Phi_{n-1},\ V_{n-1} \text{ are}$$

measurable with respect to $\mathfrak{F}_n$.

The result follows from the preceding theorem. □

(3.8) <u>Proof of Theorem (2.7)</u>: Denote the $\mathfrak{F}_n$ from the Theorem by $\mathcal{S}_n$ and set

$$\mathfrak{F}_n = \sigma(X_1, Y_1, \ldots Y_{n-1}, A_1, \ldots, A_n)$$

as in Assumption (2.2). From the measurability condition on $U_\ell$, we obtain $\mathfrak{F}_n \subset \mathcal{B}_n$ and all the relations involving conditional expectations given $\mathcal{B}_n$ hold with $\mathfrak{F}_n$ as well. For example the most difficult transition is in (2.3.1) where we obtain the desired result by taking

$$E_{\mathfrak{F}_n}$$

of both sides and using the dominated convergence theorem as

$$E_{\mathcal{B}_n} \| V_n \|^2 \leqq \sigma^2 .$$

Since the requirements on $A_n$ follow immediately from the definition (2.7.6.) of $A_n$, Assumptions (2.2) and (2.3) hold. Since (2.4.1) follows from (2.7.1), Theorem 2.4 yields $X_n \to \theta$

If we then prove $A_n \to H(\theta)$ all conditions of Theorem (2.6) will be satisfied with $\Lambda = I$, $\Lambda_0 = 1$ and the result will follow from this theorem.

Denote $W_\ell = U_\ell - d_n^{-2} \Delta_2(X_{m_\ell}, d_n)$. Then $\{W_\ell\}$ is an orthogonal sequence

$$\Sigma (\log \ell)^2 \ell^{-2} E\| W_\ell \|^2 < +\infty$$

by (2.7.4). Thus by Theorem 4.5.2, Doob (1953), applied coordinatewise, we obtain

$$\ell^{-1} \sum_{j=1}^{\ell} W_j \to 0, \quad \bar{U}_n \to \lim d_\ell^{-2} \Delta_2(X_{m_\ell}, d_\ell) = H(\theta).$$

Thus the eigenvalues of $\bar{U}_n$ will converge to those of $H(\theta)$, so that eventually they will lie in a finite interval $[a,b]$, not containing 0. Then $\bar{U}_n^{-1}$ exists and has eigenvalues in $[b^{-1}, a^{-1}]$ and, eventually

$$[b^{-1}, a^{-1}] \subset [\nu_n, \mu_n] \quad \text{and} \quad \bar{U}_n^{-1} = A_n \to H^{-1}(\theta)$$

which completes the proof. □

## References


1. Abdelhamid, S. (1971): Transformation of observations in stochastic approximation, Thesis, Michigan State University, East Lansing, Michigan.
2. Anbar, D. (1971): On optimal estimation using stochastic approximation procedures, Thesis, University of California, Berkeley, California.
3. Blum, J. R. (1954): Multidimensional stochastic approximation methods, Ann. Math. Statist. 25, 737-744.
4. Breiman, L. (1968): Probability, Addison-Wesley, Reading, Massachusetts.
5. Burkholder, D. L. (1956): On a certain class of stochastic approximation procedures, Ann. Math. Statist. 27, 1044-1059.
6. Chung, K. L. (1954): On a stochastic approximation method, Ann. Math. Statist. 25, 463-483.
7. Crockett, J. B. and H. Chernoff (1955): Gradient methods of maximization, Pacif. J. Math. 5, 33-50.
8. Derman, C. (1956): An application of Chung's lemma to the Kiefer-Wolfowitz stochastic approximation procedure, Ann. Math. Statist. 27, 532-536.
9. Doob, J. L. (1953): Stochastic Processes, Wiley, New York.
10. Dupač, V. (1957): O Kiefer-Wolfowitzově aproximační metodě. Časopis Pěst. Mat. 82, 47-75; (also: Selected Translations in Mathematical Statistics and Probability 4, 43-69.).
11. Dupač, V. (1965): A dynamic stochastic approximation method, Ann. Math. Statist. 36, 1695-1702.
12. Dupač, V. (1966): Stochastic approximations in the presence of trend, Czech. Math. J. 16 (91), 454-461.
13. Dvoretzky, A. (1956): On stochastic approximation, Proc. Third Berkeley Symp. Math. Statist. Prob. 1, 39-56. University of California Press.
14. Fabian, V. (1960): Stochastic approximation methods, Czech, Math. J. 10, 123-159.
15. Fabian, V. (1964): A new one-dimensional stochastic approximation method for finding a local minimum of a function, Trans. Third Prague Conf. Inf. Theor., Statist. Dec. Functions, Random Processes. Czechoslovak Academy of Sciences, Prague, 85-105.
16. Fabian, V. (1965): Stochastic approximation of constrained minima, Trans. Fourth Prague Conf. Information

Th., Decision Functions, Random Processes, 277-290. Academia, Prague, 1967.
17. Fabian, V. (1967): Stochastic approximation of minima with improved asymptotic speed, Ann. Math. Statist. 38, 191-200.
18. Fabian, V. (1968a): On the choice of design in stochastic approximation methods, Ann. Math. Statist. 39, 457-465.
19. Fabian, V. (1968b): On asymptotic normality in stochastic approximation, Ann. Math. Statist. 39, 1327-1332.
20. Fabian, V. (1969): Stochastic approximation for smooth functions, Ann. Math. Statist. 40, 299-302.
21. Goldstein, A. A. (1962): Cauchy's method of minimization, Numer. Math. 4, 146-150.
22. Hestens, M. R. (1951): Iterative methods for solving linear equations, NAML Report 52-9, Nat. Bureau of Standards, Los Angeles, 19 pp.
23. Hodges, J. L., Jr. and E. L. Lehmann (1956): Two approximations to the Robbins-Monro process, Proc. Third Berkeley Symp. Math. Stat. Prob. 1, 95-104. Berkeley and Los Angeles, University of California Press.
24. Janáč, K. (1971): Adaptive stochastic approximations, Simulation 16, 2, February, 51-58.
25. Kiefer, J. and J. Wolfowitz (1952): Stochastic estimation of the maximum of a regression function, Ann. Math. Statist. 23, 462-466.
26. Kiefer, J. And J. Wolfowitz (1959): Optimum designs in regression problems, Ann. Math. Statist. 30, 271-294.
27. Kiefer, J. and J. Wolfowitz (1964): Optimum extrapolation and interpolation designs, I, Annals of the Inst. of Statistical Mathematics 16, 79-108.
28. Lanczos, C. (1952): Solution of systems of linear equations by minimized iterations, J. Res. Nat. Bureau of Standards 49, 33-53.
29. Robbins, H. and S. Monro (1951): A stochastic approximation method, Ann. Math. Statist. 22, 400-407.
30. Sacks, J. (1958): Asymptotic distribution of stochastic approximation procedures, Ann. Math. Statist. 29, 373-405.
31. Schmetterer, L. (1953): Bemerkungen zum Verfahren der stochastischen Iteration, Osterreich. Ing. Archiv. 7,

111-117.

32. Schmetterer, L. (1954a): Sur l'approximation stochastique, Bull. Inst. Internat. Statist. 24, 203-206.
33. Schmetterer, L (1954b): Zum Sequentialverfahren von Robbins und Monro, Monatsh. Math. 58, 33-37.
34. Schmetterer, L. (1961): Stochastic approximation, Proc. Fourth Berkeley Symp. Math. Statist. Prob. 1, 587-609, University of California Press.
35. Schmetterer, L. (1968): Multidimensional stochastic approximation, Multivariate Analysis, II, Proc. 2nd Int. Symp., Dayton, Ohio, 443-460, Academic Press.
36. Stiefel, E. (1952): Über einige Methoden der Relaxationsrechnung, Zeit. Angew. Math. Phys. 3, 1-33.
37. Tapia, R. A. (1971): The Kantorovich theorem for Newton's method, Amer. Math. Monthly 78, 4, 389-392.
38. Venter, J. H. (1967): An extension of the Robbins-Monro procedure, Ann. Math. Statist. 38, 181-190.

# ALLOCATION OF OBSERVATIONS IN RANKING AND SELECTION WITH UNEQUAL VARIANCES

Edward J. Dudewicz[1] and Siddhartha R. Dalal

The University of Rochester

Abstract

Suppose given k populations $\pi_1, \ldots, \pi_k$, where observations from $\pi_i$ are normally distributed $N(\mu_i, \sigma_i^2)$ $(1 \leq i \leq k)$ with $\mu_1, \ldots, \mu_k$ unknown, and a goal (e.g. of selecting that population which has the largest mean $\mu_{[k]} = \max(\mu_1, \ldots, \mu_k)$). Previous work on this problem, with various assumptions about $\sigma_1^2, \ldots, \sigma_k^2$, is reviewed (Section 1). Some results on optimum design when $\sigma_1^2, \ldots, \sigma_k^2$ are known are given (Section 2), along with asymptotic considerations (Section 3). We consider generalized Stein-type procedures and develop a procedure which solves the general ranking problem with $\sigma_1^2, \ldots, \sigma_k^2$ unknown (Section 4); this problem was mentioned in an abstract of Bechhofer, Dunnett and Sobel (1953). Other procedures and some applications, e.g., to simulation, are studied (Section 5). Modified procedures for other goals (e.g. selection of the t best populations and selection of a subset of $\{\pi_1, \ldots, \pi_k\}$) are considered (Section 6). Variations of the procedure of Section 4 for other problems (e.g. multiple comparisons with a control, partitioning of populations, estimation of ranked parameters, and the Behrens-Fisher problem) are noted briefly (Section 7); separate papers on these aspects are in preparation. Tables needed to apply the procedure of Section 4 are given (Section 8).

Let $\pi_1, \ldots, \pi_k$ be k given populations, where observations from $\pi_i$ are normally distributed

[1]This author's research was supported by ONR contract N00014-68A-0091.

$N(\mu_i, \sigma_i^2)$ $(1 \leq i \leq k)$ with $\mu_1, \ldots, \mu_k$ unknown. Suppose one has a goal (as discussed by e.g. Bechhofer (1954), p. 19). We will assume without loss that the goal is to select that population which has the largest mean $\mu_{[k]}$, where

$$\mu_{[1]} \leq \cdots \leq \mu_{[k]}$$

denote the ordered means; in Section 6 we note how our work generalizes in a straightforward manner to any of the more complex goals with regard to

$$\mu_{[1]} \quad \cdots \quad \mu_{[k]}$$

which have been considered in the literature of ranking and selection.

Bechhofer (1954) gave a solution to this problem for the case $\sigma_1^2 = \cdots = \sigma_k^2 = \sigma^2$ (say) with $\sigma^2$ known. For the case of

$$\sigma_1^2, \ldots, \sigma_k^2$$

unequal but known he noted a method of solution (which we will call the "usual" $k \geq 2$ sample size allocation). He also noted that the "usual" $k \geq 2$ sample size allocation differs from the "optimal" $k \geq 2$ sample size allocation (and specified the latter for the case $k = 2$). The "optimal" allocation is not known for the case $k > 2$, and no one has yet compared the "optimal" and "usual" allocations for the case $k = 2$; this we do in Section 2, for two reasons. First, to give insight on how far from optimal the "usual" $k \geq 2$ allocation is in the case $k = 2$ (where the "optimal" allocation is also known); this furnishes a guide as to how valuable it would be to know the "optimal" allocation for case $k > 2$. Second, to call attention to the "optimal" $k = 2$ solution, since (because this problem is virtually untouched since Bechhofer's 1954 paper proposed the "usual" $k \geq 2$ allocation) there is some danger experimenters with $k = 2$ may use the "usual" allocation instead of the "optimal" allocation. (Stein (1948) did give a sequential procedure for the case $\sigma_1^2, \ldots, \sigma_k^2$ known but unequal; it has apparently never been studied in detail, although on p. 242 Bechhofer, Kiefer and Sobel (1968) note that it is currently being investigated further.)

Zinger and St. Pierre (1958) and Zinger (1961) also considered the case $\sigma_1^2, \ldots, \sigma_k^2$ unequal but known. They considered procedures which sample in the same proportions (but not the same numbers) as the "usual" allocation, which need not reach any decision (but may, if desired, be forced to reach a decision), and which utilize certain special prior information. The latter considerations (contained in the 1958 paper for three populations) lead to an improvement over the "usual" allocation in certain cases.

Various authors considered other aspects of the case $\sigma_1^2 = \ldots = \sigma_k^2 = \sigma^2$ known, and a large body of literature on this problem exists. We shall make no further explicit references to that work at this time, but these aspects (such as subset selection, prior distributions on $\mu_1, \ldots, \mu_k$, and a decision theoretic approach for specific loss functions) are areas of interest for the procedure of Section 4 below also.

Bechhofer, Dunnett and Sobel (1954) gave a two-sample Stein-type solution for the case $\sigma_1^2 = \cdots = \sigma_2^2 = \sigma^2$ with $\sigma^2$ unknown (actually they assumed

$$(\sigma_1^2/a_1) = \cdots = (\sigma_k^2/a_k) = \sigma^2$$

with $a_1, \ldots a_k$ known, which is slightly more general), which led them to introduce the multivariate-t distribution. This paper (which did not consider the case of $\sigma_1^2, \ldots, \sigma_k^2$ unequal and unknown) was abstracted in Bechhofer, Dunnett and Sobel (1953). In that abstract (and in Bechhofer (1954), pp. 24-25) the authors noted that they were working on the case $\sigma_1^2, \ldots, \sigma_k^2$ unequal and unknown via a generalized Stein two-sample procedure, and referred to a need for tables of a (k - 1)-variate analogue of the distribution of the difference between two independent Student's-t statistics. No further details were given, and no work has appeared (e.g., see Bechhofer, Kiefer and Sobel (1968), which references most ranking and selection work up to 1968). In our work we found several pitfalls arose which could have led to abandonment of this line of inquiry. We have also found that Maurice (1958) has (see Remark (4.12)) considered the case $k = 2$.

Tables and approximations for use with the procedure

of Bechhofer, Dunnett and Sobel (1954) have been given by Dunnett and Sobel (1954), Dunnett and Sobel (1955), and Krishnaiah and Armitage (1966). These applied mainly to the case where one's goal was to select the population with the largest mean $\mu_{[k]}$. Freeman, Kuzmack and Maurice (1967) gave additional tables for the goal of a complete ranking of the populations, for $k = 3,4,5$ populations, and noted an interesting application to ranking of competing consumer or producer goods. The Freeman-Kuzmack-Maurice paper also considered the approximate choice of initial sample size.

Finally, Dudewicz (1971) gave three proofs of the fact, long a part of the folklore, that a single-sample procedure cannot satisfy the usual requirements in such a problem with $\sigma_1^2, \ldots, \sigma_k^2$ unknown. This proof is generalized in Section 4.

# SEQUENCES OF MINIMAL FRACTIONS OF $2^n$ DESIGNS OF RESOLUTION V

Peter W. M. John

University of Texas

Abstract

A minimal resolution V fraction of a $2^n$ factorial design is a fraction with exactly

$$1 + n + \binom{n}{2} = (n^2 + n + 2)/2 = N$$

points from which we can estimate the mean, all $n$ main effects and all

$$\binom{n}{2}$$

two factor interactions when the higher order interactions are suppressed from the model. Two obvious examples are the complete $2^2$ factorial and the half replicate of the $2^5$ factorial defined by $I = \pm$ ABCDE.

John and Joula (1970) have shown that any resolution V design for n factors with N points can be augmented to give a resolution V design for $n + 1$ factors in $N + n + 1$ points. It follows that the design $\mathcal{D}_1$ which consists of 1, all one letter words, and all two letter words is a minimal design for $n$ factors. For $n = 3$ we have (1), a, b, c, ab, ac, bc. Unfortunately, this series is of only academic interest because

$$8V(\hat{A}) = (n^2 - 5n + 8)\sigma^2 ,$$

so that $V(\hat{A})$ increases as $n^2$.

A preferable minimal resolution design for $n$ factors in $\mathcal{D}_2$ which consists of (1), all two letter words and all (n-1) tuples. For designs of series $\mathcal{D}_2$ we have

$$V(\hat{A}) = V(\hat{A}B) = (n^4-10n^3+39n^2-66n+44)\sigma^2/16(n-1)^2(n-3)^2.$$

An important example of the series is the $2^{n-1}$ defined by I = -ABCDE for which

$$V(\hat{A}) = \sigma^2/16.$$

For $6 \leq n \leq 10$, $V(\hat{A})$ is approximately 0.05. For larger n the variance increases to the asymptotic value

$$\sigma^2/16.$$

Some other minimal designs are also discussed including the following design for n = 6:

$\mathcal{D}_b$: (1), abcdef together with all twenty triples. For this design we have

$$V(\hat{A}) = V(\hat{A}B) \cong 0.05\sigma^2.$$

OPTIMUM INTERVAL ESTIMATION FOR THE LARGEST SCALE PARAMETER

K. M. Lal Saxena
Yung Liang Tong

University of Nebraska

Abstract

Let $X_1, \dots, X_k$ be k independent random variables $(k \geq 1)$ with distributions $F(x,\theta_i)$, $i = 1, \dots, k$, respectively. Let

$$\theta^* = \max_{1 \leq i \leq k} \theta_i \qquad X_i^* = \max_{1 \leq i \leq k} X_i$$

and $h_1$, $h_2$ be suitably chosen continuous increasing functions with respective inverses $g_2$ and $g_1$. Then an interval estimate for $\theta^*$ is

$$I = (h_1(X^*), h_2(X^*)).$$

It is proved that if

(a) $(\partial/\partial\theta)\, F(x,\theta) \leq 0$ for all x,

(b) $\{(\partial/\partial\theta)\, F(x,\theta)\}/\{(\partial/\partial\theta)\, F(x',\theta)\}$ is an increasing function of $\theta$ for every $x > x'$, then

$$\inf_{\theta \in \Omega} P(\theta^* \in I) = \min(H_1, H_k),$$

where

$$H_r = F^r(g_2(\theta^*),\theta^*) - F^r(g_1(\theta^*),\theta^*),\ r = 1, \dots, k,$$

and $\Omega$, is the set of all possible k-triples $(\theta_1, \dots, \theta_k)$

with $\theta^*$ fixed. Analogous result is obtained for

$$\theta^* = \min_{1 \leq i \leq k} \theta_i.$$

An optimization criterion is used when $\theta$ is a scale parameter, and tables are constructed. The case when $\theta$ is a location parameter (Dudewicz and Tong (1971), "Optimal confidence intervals for the largest location parameter," *Proceedings of the Symposium on Statistical Decision Theory and Related Topics*, Academic Press) is a special case of the above result.

# c-SAMPLE TESTS OF HOMOGENEITY AGAINST ORDERED ALTERNATIVES

Z. Govindarajulu
H. Smith Haller, Jr.

University of Kentucky
B. F. Goodrich Co.

## Abstract

Let population $\pi_i$ have an unknown continuous cumulative distribution function $F_i(x)$, $(i = 1,2, \ldots ,c)$. The problem considered is to test the null hypothesis, $H_0$, that the $F_i$'s are equal against the alternative: $F_1(x) \geq F_2(x) \geq \cdots \geq F_c(x)$ (or $F_1(x) \leq F_2(x) \leq \cdots \leq F_c(x)$) for all $x$ with at least one inequality strict. Nonparametric tests for this problem were proposed by Jonckheere (Biometrika 41 (1945) 135-145), Chacko (Ann. Math. Statist. 34 (1963) 945-956), and Puri (Comm. Pure Appl. Math. 18 (1965) 51-63). These tests are for location shift alternatives and are based on the asymptotic distribution of the statistics. For tests of $H_0$ against two classes of ordered alternatives of the Lehmann type, we obtain a locally most powerful rank order test and the test statistic is found to be a weighted sum of c-sample rank order statistics of the type introduced by Savage (Ann. Math. Statist. 27 (1956) 590-615). We obtain a general result pertaining to the optimal weights to be assigned in a weighted sum of c-sample linear rank statistics such that the Pitman efficacy of the test procedure is maximized. Applying this result we propose a class of test statistics for location shift and change of scale alternatives which are weighted sums of rank order statistics of the type introduced by Chernoff-Savage (Ann. Math. Statist. 29 (1958) 972-994), Bhapkar (Ann. Math. Statist. 32 (1961) 1108-1117), and Deshpande (J. Indian Statist. Assoc. 3 (1965) 20-29). These Chernoff-Savage statistics are easier to compute than and are asymptotically equivalent to those proposed by Puri (1965) for equally spaced location shift alternatives.

# PARTICIPANTS

*Abdenhamid, Sami,* Michigan State University, East Lansing, Michigan

*Agnew, Robert A,* Air Force Institute of Technology, Wright-Patterson Air Force Base, Dayton, Ohio

*Ahlswede, Rudolf,* The Ohio State University, Columbus, Ohio

*Anderson, Peter,* Division of Statistics, The Ohio State University, Columbus, Ohio

*Anderson, R. L.,* Statistics Department, University of Kentucky, Lexington, Kentucky

*Armitage, John V.,* U. S. Air Force Headquarters, Washington, D. C.

*Arvesen, James,* Purdue University, Lafayette, Indiana

*Aucoin, Clayton,* Clemson University, Clemson, South Carolina

*Bai, Do Sun,* The Ohio State University, Columbus, Ohio

*Barlow, R. E.,* Department of Statistics and Operations Research Center, University of California, Berkeley, California

*Berk, Kenneth,* Department of Mathematics, Illinois State University, Normal, Illinois

*Bilikam, John,* Wright-Patterson Air Force Base, Dayton, Ohio

*Broom, Albert,* Wright-Patterson Air Force Base, Dayton, Ohio

*Bushnell, Robert,* The Ohio State University, Columbus, Ohio

*Cain, Joseph P.,* Air Force Institute of Technology, Wright-Patterson Air Force Base, Dayton, Ohio

*Chandrangsu, Srisook,* The Ohio State University, Columbus, Ohio

*Chaney, R. W.,* Department of Mathematics, Western Washington State College, Bellingham, Washington

*Charnes, A.,* Center of Cybernetic Studies, Graduate School of Business, University of Texas, Austin, Texas

*Chen, You C.,* Children's Hospital of Michigan, Detroit, Michigan

*Cheng, Smiley,* Department of Mathematics, University of Western Ontario, London, Ontario, Canada

*Chernoff, Herman,* Department of Statistics, Stanford University, Stanford, California

*Chitgopekar, Sharad,* University of Wisconsin, Madison, Wisconsin

*Chu, Poo-Sen,* Indianapolis, Indiana

*Conley, Charles,* Michigan Department of Social Sciences, Lansing, Michigan

*Cue, Pablo,* Ross Laboratories, The Ohio State University, Columbus, Ohio

*Dahlgren, Dirk,* Sandia Corporation Division, Livermore, California

*Dalal, Siddhartha R.,* Department of Statistics, University of Rochester, Rochester, New York

*Davis, John,* The Ohio State University, Columbus, Ohio

*Deverman, J. N.,* Sandia Laboratories, Albuquerque, New Mexico

*Deyonker, Leo,* General Motors Research Laboratories, General Motors Technical Center, Warren, Michigan

*Dudewicz, Edward J.,* Department of Statistics, University of Rochester, Rochester, New York

## PARTICIPANTS

*Eitner, Emerson,* Chemical Abstract Service, Columbus, Ohio

*Fabian, Vaclav,* Department of Statistics, Michigan State University, Lansing, Michigan

*Francis, Richard,* Department of Industrial and Systems Engineering, University of Florida, Gainesville, Florida

*Freisner, Gregory,* Chemical Abstracts Service, Columbus, Ohio

*Gandhi, Devinder,* Department of Statistics and Operations Research, University of Pennsylvania, Philadelphia, Pennsylvania

*Gastwirth, Joseph,* Harvard University, Cambridge, Massachusetts

*Ghurye, S. G.,* Stanford University, Stanford, California

*Golhar, Madhukar,* The Ohio State University, Columbus, Ohio

*Govindarajulu, Z.,* Department of Statistics, University of Kentucky, Lexington, Kentucky

*Gupta, Radhey,* The Ohio State University, Columbus, Ohio

*Gupta, Shanti,* Department of Statistics, Purdue University, Lafayette, Indiana

*Hall, W. J.,* University of Rochester, Rochester, New York

*Harris, Bernard,* Mathematics Research Center, University of Wisconsin, Madison, Wisconsin

*Harter, H. Leon,* Applied Mathematics Laboratories, Wright-Patterson Air Force Base, Dayton, Ohio

*Hartley, H. O.,* Institute of Statistics, Texas A & M University,College Station, Texas

*Harville, David,* Wright-Patterson Air Force Base, Dayton, Ohio

*Hensel, Jean,* The Ohio State University, Columbus, Ohio

*Hern, Thomas,* Department of Mathematics, Bowling Green State University, Bowling Green, Ohio

*Jackson, Clement,* University of Wisconsin, Madison, Wisconsin

*John, Peter W. M.,* Department of Statistics, University of Kentucky, Lexington, Kentucky

*Kallianpur, G.,* Department of Mathematics, University of Minnesota, Minneapolis, Minnesota

*Katz, Leo,* Michigan State University, East Lansing, Michigan

*Kemperman, J. H. B.,* Department of Mathematics, University of Rochester, Rochester, New York

*Klimko, Eugene,* Purdue University, Lafayette, Indiana

*Klimko, Lawrence,* Pennsylvania State University, University Park, Pennsylvania

*Klinger, Allen,* Department of Computer Science, University of California, Los Angeles, California

*Krieger, Wolfgang,* The Ohio State University, Columbus, Ohio

*Krishnaiah, P. R.,* Aerospace Research Laboratories, Wright-Patterson Air Force Base, Dayton, Ohio

*Kulldorff, Gunnar,* Department of Statistics, Purdue University, Lafayette, Indiana, and University of Umea, Sweden

*Kumar, Arunod,* Michigan State University, Lansing, Michigan

*Kurtz, Freda,* Wright-Patterson Air Force Base, Dayton, Ohio

*Larson, Dennis,* Chemical Abstracts Service, Columbus, Ohio

*Lawrence, Willard,* Marquette University, Milwaukee, Wisconsin

*Lee, Jae C.,* The Ohio State University, Columbus, Ohio

*Li, Yuku,* The Ohio State University, Columbus, Ohio

*Lientz, Bennet P.,* Department of Industrial Engineering, University of Southern California, Los Angeles, California

*Lowrimore, Gene,* Environmental Protection Agency, Raleigh, North Carolina

*Lutz, L. H.,* Ross Laboratories, Columbus, Ohio

*Meeks, Howard,* Iowa State University, Ames, Iowa

*Minucci, Lambert,* PPG Industries, Allison Park, Pennsylvania

*Mond, Bertram,* La Trobe University, Melbourne, Australia

*Munro, Sutton,* Industrial Engineering Department, Lehigh University, Bethlehem, Pennsylvania

*Murty, P. W. Ramachandra,* Department of Mathematics, University of Alberta, Edmonton, Alberta, Canada

*Nanda, P.,* Syracuse University, Syracuse, New York

*Neuhardt, John,* The Ohio State University, Columbus, Ohio

*Neyman, Jerzy,* Department of Statistics, University of California, Berkeley, California

*Nickens, Paul,* Joint Systems Integrated Planning Staff, Fairfax, Virginia

*Nylander, John,* General Mills, Minneapolis, Minnesota

*Olkin, Ingram,* Stanford University, Stanford, California

*Oprian, Charles,* Western Illinois University, Macomb, Illinois

*Papaioannou, Takis,* Statistics Department, University of Georgia, Athens, Georgia

*Papangelou, F.,* Department of Mathematics, The Ohio State University, Columbus, Ohio

*Paroff, John,* Babcock & Wilcox, Alliance, Ohio

*Phadia, Eswar,* The Ohio State University, Columbus, Ohio

*Pledger, Gordon,* Department of Statistics, Florida State University, Tallahassee, Florida

*Polk, Dwight W.,* Wright-Patterson Air Force Base, Dayton, Ohio

*Poola, Jack,* PPG Industries, Inc., Pittsburgh, Pennsylvania

*Pound, Cynthia,* General Motors Proving Grounds, Milford, Michigan

*Presser, Marshall,* Department of Statistics and Operations Research, University of Pennsylvania, Philadelphia, Pennsylvania

*Rao, B. R.,* Department of Biostatistics, University of Pittsburgh, Pittsburgh, Pennsylvania

*Reklaitis, G. V.,* School of Chemical Engineering, Purdue University, Lafayette, Indiana

*Richmond, Elmer,* Wright-Patterson Air Force Base, Dayton, Ohio

*Rubin, Herman,* Department of Statistics, Purdue University, Lafayette, Indiana

*Rustagi, Jagdish S.,* The Ohio State University, Columbus, Ohio

*St. John, Ralph C.,* University of Wisconsin, Madison, Wisconsin

*Sage, Andrew P.,* Information and Control Center, Southern Methodist University, Dallas, Texas

*Saltzer, Charles,* The Ohio State University, Columbus, Ohio

*Sammons, William H.,* CTU, SCS, UDA, Hyattsville, Maryland

*Samuels, S. M.,* Department of Statistics, Purdue University, Lafayette, Indiana

*Sanderson, D. F.,* Department of Mathematics, Western Washington State College, Bellingham, Washington

*Saxena, K. M. Lal,* Department of Mathematics, University of Nebraska, Lincoln, Nebraska

*Schoenfeldt, Charles,* Carnegie-Mellon University, Pittsburgh, Pennsylvania

*Sclove, Stanley,* Statistics Department, Carnegie-Mellon University, Pittsburgh, Pennsylvania

*Selig, Seymour,* Office of Naval Research, Arlington, Virginia

*Shaw, Douglas E.,* Department of Statistics, University of Wisconsin, Madison, Wisconsin

*Siegmund, D.,* Department of Mathematical Statistics, Columbia University, New York, New York

*Singh, Jagbir,* The Ohio State University, Columbus, Ohio

*Smith, Dennis,* HRB-Singer, Inc., State College, Pennsylvania

*Studden, W. J.,* Department of Statistics, Purdue University, Lafayette, Indiana

*Sullivan, James,* Department of Quantitive Analysis and Control, Bowling Green State University, Bowling Green, Ohio

*Thayer, Phillip,* U. S. Air Force Weapons Laboratory, (ADDS) Kirtland AFB, New Mexico

*Thompson, W. A., Jr.,* Statistics Department, University of Missouri, Columbia, Missouri

*Tong, Yung Liang,* Department of Mathematics, University of Nebraska, Lincoln, Nebraska

*Trott, William R.,* AFOSR, Arlington, Virginia

*Troy, Daniel,* Purdue University, Lafayette, Indiana

*Ubhaya, V.,* Department of Industrial Engineering, University of California, Berkeley, California

*Waikar, Vasant,* Wright-Patterson Air Force Base, Dayton, Ohio

*Walker, Leroy,* Mathematical Sciences and Computer Building, Brigham Young University, Provo, Utah

*Wang, Yao Hung,* The Ohio State University, Columbus, Ohio

*Weiss, L.,* Department of Industrial Engineering and Operations Research, Cornell University, Ithaca, New York

*Whitney, Ransom,* Division of Statistics, The Ohio State University, Columbus, Ohio

*Wilke, Thomas,* The Ohio State University, Columbus, Ohio

*Wolaver, Lynn,* Air Force Institute of Technology, Fairborn, Ohio

*Wolfe, Philip,* Thomas J. Watson Research Center, IBM, Yorktown Heights, New York

*Wolfowitz, J.,* Department of Mathematics, University of Illinois, Urbana, Illinois

*Wong, H. H.,* Ohio Northern University, Ada, Ohio

*Yen, Vincent,* The Ohio State University, Columbus, Ohio